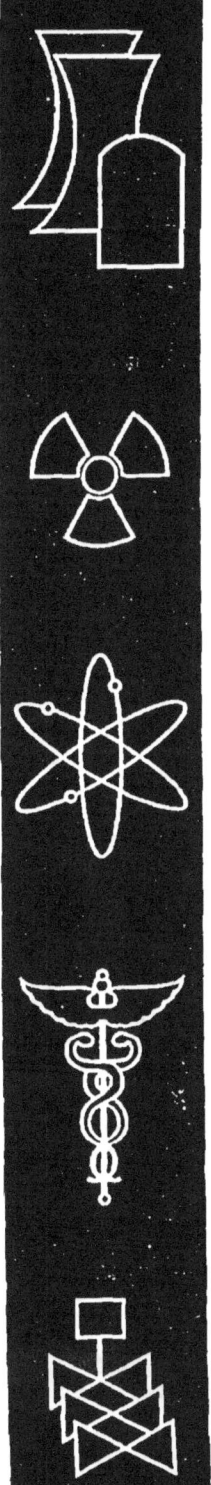

Safety Evaluation Report
for the National Enrichment Facility in Lea County, New Mexico

I0482621

Docket No. 70-3103

Louisiana Energy Services

U.S. Nuclear Regulatory Commission
Office of Nuclear Material Safety and Safeguards
Washington, DC 20555-0001

AVAILABILITY OF REFERENCE MATERIALS
IN NRC PUBLICATIONS

Safety Evaluation Report
for the National Enrichment Facility
in Lea County, New Mexico

Docket No. 70-3103

Louisiana Energy Services

Manuscript Completed: June 2005
Date Published: June 2005

**Division of Fuel Cycle Safety and Safeguards
Office of Nuclear Material Safety and Safeguards
U.S. Nuclear Regulatory Commission
Washington, DC 20555-0001**

ABSTRACT

The report documents the U.S. Nuclear Regulatory Commission (NRC) staff review and safety and safeguards evaluation of the Louisiana Energy Services' (LES) (the applicant) application for a license to construct a gas centrifuge uranium enrichment facility and possess and use special nuclear material (SNM), source material, and byproduct material in a gas centrifuge uranium enrichment facility. LES proposes that the gas centrifuge uranium enrichment facility be located in Lea County, New Mexico, near the city of Eunice, New Mexico. The facility will possess natural, depleted, and enriched uranium, and will enrich uranium up to a maximum of 5 weight percent uranium-235.

The objective of this review is to evaluate the potential adverse impacts of operation of the facility on worker and public health and safety under both normal operating and accident conditions. The review also considers physical protection of SNM and classified matter, material control and accounting of SNM, and the management organization, administrative programs, and financial qualifications provided to ensure safe design and operation of the facility.

The NRC staff concludes, in this safety evaluation report, that the applicant's descriptions, specifications, and analyses provide an adequate basis for safety and safeguards of facility operations and that operation of the facility does not pose an undue risk to worker and public health and safety.

TABLE OF CONTENTS

TABLES

EXECUTIVE SUMMARY

On December 12, 2003, Louisiana Energy Services (LES) (the applicant) submitted, to the U.S. Nuclear Regulatory Commission (NRC), an application requesting a license, under 10 CFR Parts 30, 40, and 70, to possess and use byproduct, source, and special nuclear material (SNM) in a gas centrifuge uranium enrichment facility. LES proposes that the facility be located in Lea County, New Mexico, and have a nominal capacity of 3 million separative work units (SWUs). (A SWU is a unit of enrichment that measures the effort required to separate isotopes of uranium.) The facility will possess natural, depleted, and enriched uranium, and will enrich uranium up to a maximum of 5 percent uranium-235. The applicant also requested a facility clearance for classified information, under 10 CFR Part 95.

The NRC staff conducted its safety review in accordance with NUREG-1520, "Standard Review Plan for the Review of a License Application for a Fuel Cycle Facility." The staff's safeguards review involved reviews of the applicant's Fundamental Nuclear Materials Control Plan (FNMCP); the Physical Security Plan, which includes transportation security; and a "Standard Practice Procedures Plan for the Protection of Classified Matter." The staff also reviewed the applicant's Quality Assurance Program Description and Emergency Plan. Where the applicant's design or procedures should be supplemented, the NRC staff has identified license conditions to provide assurance of safe operation.

The applicant also submitted an Environmental Report, which was used to prepare, in a separate document, an Environmental Impact Statement for the facility.

A summary of NRC's review and findings in each of the review areas is provided below:

General Information

The applicant provided an adequate description of the facility and processes so that the staff has an overall understanding of the relationships of the facility features as well as the function of each feature. Financial qualifications were properly explained and outlined in the application. The description of the site included important information about regional hydrology, geology, meteorology, the nearby population, and potential effects of natural phenomena at the facility.

Organization and Administration

The applicant adequately described the responsibilities and associated resources for the design, construction, and operation of the facility and its plans for managing the project. The plans and commitments described in the application provide reasonable assurance that an acceptable organization, administrative policies, and sufficient competent resources have been established or committed for the design, construction, and safe operation of the facility.

"Integrated Safety Analysis" (ISA) and ISA Summary

The applicant provided sufficient information about the site, facility processes, hazards, and types of accident sequences. The information provided addressed each credible event, the potential radiological and chemical consequences of the event, and the likelihood of the event.

For nuclear criticality safety safe-by-design components, the applicant identified the hazards and demonstrated that the failure of those components would be highly unlikely. No mitigated event consequence exceeds the performance requirements of 10 CFR 70.61. The applicant also provided adequate information about items relied on for safety (IROFS). License conditions have been added to the license to ensure that IROFS boundaries will be defined using the applicant's IROFS boundary definition procedure and that the applicant will submit license amendment requests if digital instrumentation and controls are used in IROFS.

Radiation Protection

The applicant provided sufficient information to evaluate the Radiation Protection Program. The application adequately describes: (a) the qualification requirements; (b) written radiation protection procedures; (c) the radiation work permit (RWP) program; (d) the program for ensuring that worker and public doses are as low as is reasonably achievable (ALARA); and (e) necessary training for all personnel who have access to radiologically restricted areas. The radiation survey and monitoring program is adequate to protect workers and members of the public who may be potentially exposed to radiation.

Nuclear Criticality Safety

The applicant provided adequate information to evaluate the Nuclear Criticality Safety (NCS) program. The applicant committed to having an adequate group of qualified staff to develop, implement, and maintain the NCS program in accordance with the facility organization and administration and management measures. The program meets the regulatory requirements.

Chemical Process Safety

The applicant adequately described and assessed accident consequences that could result from the handling, storage, or processing of licensed materials and that could have potentially significant chemical consequences and effects. The applicant performed hazard analyses that identified and evaluated those chemical process hazards and potential accidents and established safety controls that meet the regulatory requirements.

Fire Safety

The applicant committed to reasonable engineered and administrative controls to minimize the risk of fires and explosions. The IROFS and defense-in-depth protection discussed in the applicant's ISA Summary, along with safety basis assumptions and the planned programmatic commitments in the license application, meet safety requirements and provide reasonable assurance that the facility is protected against fire hazards.

Emergency Management

The applicant provided an adequate Emergency Plan, for the facility, that meets the regulatory requirements. The applicant commits to maintaining and executing an Emergency Plan for responding to the radiological and chemical hazards resulting from potential release of radioactive or chemically hazardous materials incident to the processing of licensed material. The requirements of the Emergency Plan are implemented through approved written procedures.

Environmental Protection

The applicant committed to adequate environmental protection measures, including,"1) environmental and effluent monitoring; and (2) effluent controls to maintain public doses ALARA as part of the radiation protection program. The applicant's proposed controls are adequate to protect the environment and the health and safety of the public and comply with the regulatory requirements.

Decommissioning

The applicant provided a conceptual decommissioning plan, for the facility, that addresses: (a) contamination control; (b) control of worker exposures and waste volumes; (c) waste disposal; (d) the final radiation survey; (e) control of SNM; (f) control of classified matter; and (g) record-keeping for decommissioning.

The applicant provided a decommissioning funding plan, for the facility, that demonstrates that adequate funding will be available for decommissioning and that decommissioning will not pose a threat to public health and safety or the environment. The applicant also submitted an exemption request to allow for incremental funding for depleted uranium disposition based on depleted uranium tails generation rates. The decommissioning funding plan and the incremental approach for funding depleted uranium disposition costs will provide adequate assurance for decommissioning funding because sufficient funding will be available to decommission the facility and disposition the inventory of depleted uranium on-site at any point in time. The applicant also provided proposed language for a surety bond, with a standby trust agreement. The surety bond and standby trust agreement will be executed before the applicant takes possession of licensed material. The applicant will update the site-specific cost estimate at least every 3 years, to reflect inflation and changes in site inventories and conditions, that could affect the cost of decommissioning. A license condition has been added to the license to ensure that the applicant takes possession of no licensed material until the surety bond and standby trust agreement are executed and are acceptable to NRC. The decommissioning funding plan is acceptable because it provides sufficient funding to ensure decommissioning and decontamination of the facility can be accomplished even if the licensee is unable to meet its financial obligations.

Management Measures

The applicant provided information about management measures that will be applied to the project. The information describes: (a) the overall configuration management program and policy; (b) the maintenance program; (c) training; and (d) the process for the development, approval, and implementation of procedures. The applicant explained the audits and assessments program as well as incident investigations and records management system. The applicant committed to establishing and documenting surveillances, tests, and inspections to provide reasonable assurance of satisfactory performance of the IROFS. The proposed management measures are acceptable and meet the regulatory requirements in 10 CFR 70.62(d).

Materials Control and Accountability

The applicant provided information describing the Fundamental Nuclear Material Control Plan (FNMCP) for the project. The FNMCP describes the programs to be used to control and account for SNM in the facility. The program meets the applicable regulatory requirements in Part 74.

Physical Protection

The applicant provided information on the policies, methods, and procedures to be implemented to protect SNM of low strategic significance used and possessed at the facility. This information is acceptable and meets the requirements in Part 73.

The applicant also provided information on the protection of classified matter, including security controls and procedures, to ensure that classified matter is used, processed, stored, reproduced, transmitted, transported, and destroyed. This program is acceptable and in accordance with the regulatory requirements in Part 95 for a facility clearance.

Transportation Security

The applicant provided information in the Physical Security Plan on the policies, methods, and procedures to be implemented to protect SNM of low strategic significance in transit to and from the facility. This information is acceptable and meets the requirements in Part 73.

LIST OF ACRONYMS AND ABBREVIATIONS

ACI	American Concrete Institute
AEGL	Acute Exposure Guideline Level
AHU	Air handling unit
AISC	American Institute of Steel Construction
ALARA	As low as is reasonably achievable
Al_2O_3	Aluminum oxide
ANSI	American National Standards Institute
ARF	Airborne release fraction
ASCE	American Society of Civil Engineers
ASM	Additional Safety Measures
ASME	American Society of Mechanical Engineers
BDC	Baseline design criteria
BNFL-EL	BNFL Enrichment Limited
CAA	Controlled access area
CAP	Corrective Action Program
CAS	Central alarm station
CCTV	Closed circuit television
CEC	Claiborne Enrichment Center
CEDE	Committed Effective Dose Equivalent
CFR	Code of Federal Regulations
cm	centimeter
CM	Configuration management
CPD	Core Plant Design
CRDB	Cylinder Receipt and Dispatch Building
CS	Chemical safety
CUB	Central Utility Building
DOE	U.S. Department of Energy
DOEQAP	U.S. Department of Energy Quality Assurance Program
DR	Damage ratio
EECP	Entry/exit control point
EO	Emergency Organization
EOC	Emergency Operations Center
EP	Emergency Plan
EPA	U.S. Environmental Protection Agency
EPIP	Emergency Plan Implementing Procedures
ER	Environmental Report
ERO	Emergency response organization

FAA	Federal Aviation Administration
FNMCP	Fundamental Nuclear Material Control Plan
FOCI	Foreign ownership, control, or influence
ft	feet
ft/s	feet per second
g	acceleration of gravity
GEVS	Gaseous Effluent Ventilation System
gpm	gallons per minute
ha	hectare
HAZOP	hazard and operability
HEPA	High efficiency particulate air
HF	Hydrogen fluoride
HPS	Health Physics Society
HS&E	Health, safety, and environment
HVAC	Heating, ventilating, and air conditioning
ICBO	International Conference of Building Officials
IEC	International Electrotechnical Commission
in	inch
IRB	Industrial revenue bond
IROFS	Items relied on for safety
ISA	Integrated safety analysis
ISO	International Organization for Standardization
kg	kilogram
km	kilometer
kPa	kiloPascals
kPa/s	kiloPascals per second
L	liter
lb	pound
LEL	Lower explosive limit
LES	Louisiana Energy Services
LLW	Low-level radioactive waste
LPF	Leak path factor
Lpm	liters per minute
LTTS	Low temperature takeoff station
m	meter
m^3	cubic meter
MAPEP	Mixed Analyte Performance Evaluation Program
MAR	Material at risk
MC&A	Material control and accounting
mg	milligram
mi	mile
mm	millimeter
MOU	Memorandum of Understanding

m/s	meter per second
MT	metric ton
NAC	National Advisory Committee
Na_2CO_3	Sodium carbonate
NaF	Sodium fluoride
NCS	Nuclear criticality safety
NELAC	National Environmental Laboratory Accreditation Conference
NFPA	National Fire Protection Association
NIOSH	National Institute for Occupational Safety and Health
NMAC	New Mexico Administrative Code
NOAA	National Oceanic and Atmospheric Administration
NRC	Nuclear Regulatory Commission
OSHA	U.S. Occupational Safety and Health Administration
PFPE	Perfluorinated polyether
PM	Preventive maintenance
psf	pounds per square foot
psf/s	pounds per square foot per second
psi	pounds per square inch
psia	pounds per square inch absolute
QA	Quality assurance
QAPD	Quality Assurance Program Description
RAI	Request for additional information
RASCAL	Radiological Assessment System for Consequence Analysis
rem	Roentgen equivalent man
REMP	Radiological Environmental Monitoring Program
RF	Respirable fraction
RP	Radiation protection
RWP	Radiation Work Permit
SAR	Safety Analysis Report
SER	Safety Evaluation Report
SM	Source Material
SNM	Special Nuclear Material
SNM-LSS	Special Nuclear Material - Low Strategic Significance
SOP	Standard Operating Procedure
SPPP	Standard Practice Procedures Plan
SRC	Safety Review Committee
SSC	Structures, systems, and components
Sv	Sievert
SWU	Separative Work Unit
T	Total Likelihood Index
TEEB	Treated Effluent Evaporative Basin
TID	Tamper indicating device

TLD	Thermoluminescent dosimeter
TSB	Technical Services Building
TWA	Time-weighted average
UBC	Uranium byproduct cylinder
UCN	Ultra-Centrifuge Nederland NV
μg	microgram
UF_4	Uranium tetrafluoride
UF_6	Uranium hexafluoride
UO_2F_2	Uranyl fluoride
UPS	Uninterruptible power supply
USEC	U.S. Enrichment Corporation
USGS	U.S. Geological Survey
wt	weight

1.0 GENERAL INFORMATION

1.1 FACILITY AND PROCESS DESCRIPTION

The purpose of the U.S. Nuclear Regulatory Commission's (NRC's) review of the proposed Louisiana Energy Services (LES) facility and process description is to determine whether the application includes an overview of the facility layout and a summary description of the proposed processes. A more detailed description of the facility and processes is contained in the "Integrated Safety Analysis (ISA) Summary" (LES, 2005b).

1.1.1 REGULATORY REQUIREMENTS

The regulations in 10 CFR 30.33, 10 CFR 40.32, and 10 CFR 70.22 require each application for a license to include information on the proposed activity and the equipment and facilities that will be used by the applicant to protect health and minimize danger to life and property. In addition, the regulations in 10 CFR 70.65 require each application to include a general description of the facility, with emphasis on those areas that could affect safety, including identification of the controlled area boundaries.

1.1.2 REGULATORY ACCEPTANCE CRITERIA

The acceptance criteria applicable to NRC's review of the facility and process description section of the application are contained in Section 1.1.4.3 of the "Standard Review Plan for the Review of a License Application for a Fuel Cycle Facility," NUREG-1520 (NRC, 2002).

1.1.3 STAFF REVIEW AND ANALYSIS

In Section 1.1 of the Safety Analysis Report (SAR) (LES, 2005a), the applicant provides a summary description of the proposed gas centrifuge uranium enrichment plant and processes. This description includes discussion of the major chemical and mechanical processes to be used in the facility. The facility is proposing to use a gas centrifuge enrichment process to enrich uranium from its natural isotopic concentration of about 0.7 percent uranium-235 (U-235) to 5 percent U-235. The proposed plant will have a nominal enrichment capacity of 3 million Separative Work Units (SWUs). (A SWU is a measure of the effort required to perform isotopic separation.) The process uses uranium in the chemical form of uranium hexafluoride (UF_6). Gaseous UF_6 enters a high-speed rotor at subatmospheric conditions where centrifugal forces press the heavier isotope of uranium, uranium-238 (U-238), to the outer wall of the rotor. The lighter isotope, U-235, remains closer to the center, away from the rotor wall. Internal scoops are used to collect the heavier and lighter fractions and circulate them to other centrifuges piped in a cascade arrangement.

The proposed plant will be constructed to have three Separations Building Modules, each having two Cascade Halls, with each Cascade Hall having eight cascades. Each Separations Building consists of a UF_6 Feed System, Cascade Systems, a Product Take-off System, and a Tails Take-off System. The plant also has a Product Liquid Sampling System and a Product Blending System.

Natural uranium feed is shipped to the plant primarily by truck in cylindrical steel containers having a capacity of up to 12.7 metric tonnes (MT) (14 tons) of UF_6. Under ambient conditions, the UF_6 is a solid. Feed containers are vented to remove air and hydrogen fluoride (HF) gases and then heated to sublime the solid UF_6 to a gas. The feed system is designed to preclude the UF_6 from becoming liquid. The light gases and gaseous UF_6 pass through the Feed Purification Subsystem to remove the light gases that are directed through the Gaseous Effluent Ventilation System (GEVS) to ensure that HF and UF_6 are removed and not released to the atmosphere. After the venting is complete, the UF_6 feed from the Solid Feed Stations is directed to a cascade for enrichment.

After enrichment in gas centrifuge machines, both depleted and enriched products are withdrawn from the cascade and desublimed at subatmospheric pressure in the Tails Take-off System and the Product Take-off System, respectively. Tails and Product Take-off Systems are designed to preclude UF_6 from becoming a liquid.

Sampling to verify the assay level is performed in the Product Liquid Sampling Autoclave. In the autoclave, UF_6 is heated to a liquid; the cylinder is tilted so that UF_6 can flow into sample manifold and sample bottles; and the cylinder is returned to its original horizontal position. This is the only system in the plant where UF_6 is in a liquid form.

To produce enriched uranium meeting customer-assay specifications, the Product Blending System is used to mix enriched uranium at two different enrichment levels to one meeting the customer specifications. This system can also be used to transfer product between cylinders.

Facility information contained in the ISA Summary (LES, 2005b) is provided in the application in layout drawings of the plant buildings and the location of plant systems within the buildings. Geographical features and transportation routes are also provided on these drawings.

The proposed facility is expected to possess natural, enriched, and depleted uranium. It is expected to handle, on an annual basis, approximately 690 nominal 12.5-MT (14-ton) or 9.5-MT (10.5-ton) natural uranium feed cylinders; 350 nominal 2-MT (2.5-ton) enriched-uranium product cylinders; and 625 nominal 12.5-MT (14-ton) depleted uranium tails cylinders.

Gaseous airborne effluents will be released from the proposed facility. The applicant estimates that less than 10 grams (0.35 ounces (oz)) of uranium and less than 1 kilogram (kg) (2.2 pounds (lbs)) of HF will be released annually in 2.47×10^9 cubic meters of air discharge. These effluents are significantly below 10 CFR Part 20 and U.S. Environmental Protection Agency National Emission Standards for Hazardous Air Pollutants (NESHAPs) airborne release limits.

Liquid discharges include contaminated process effluents, cooling tower blowdown, and stormwater discharges. Liquid effluents will be significantly below Part 20 liquid effluent requirements.

Wastes expected to be generated include non-hazardous industrial, Class A radioactive, hazardous, and mixed wastes. Construction wastes will also be generated in construction of the plant. Radioactive wastes will be disposed of at properly licensed low-level radioactive waste disposal facilities. Hazardous chemical wastes will be properly treated and disposed of at permitted treatment and disposal facilities. Mixed low-level radioactive and chemically hazardous wastes will be treated and disposed of at facilities having the proper licenses and permits for these wastes. Depleted uranium tails will be stored on-site on the Uranium Byproduct Cylinder (UBC) pad until they are transferred to another licensee for commercial use or they are designated for disposal as waste. If designated as waste, the applicant is proposing to use either a commercial disposition path or the U.S. Department of Energy (DOE) disposition path set out in the USEC Privatization Act of 1996. The applicant has committed to not store depleted uranium tails for longer than the 30-year life of the plant.

1.1.4 EVALUATION FINDINGS

The staff has reviewed the proposed general facility and process descriptions according to Section 1.1 of the Standard Review Plan. The applicant has adequately described: (1) the facility and processes so that the staff has an overall understanding of the relationships of the facility features; and (2) the function of each feature. The staff concludes that the applicant has met the requirements and acceptance criteria applicable to this section.

1.2 INSTITUTIONAL INFORMATION

The purpose of NRC's review of institutional information is to establish whether the license application includes adequate information identifying the applicant, the applicant's characteristics, and the proposed activity.

1.2.1 REGULATORY REQUIREMENTS

The regulations in 10 CFR 30.32 and 10 CFR 40.31 require each application for a license to include: (a) information on the identity of the applicant; (b) name, chemical and physical form, and maximum amount that will be possessed; and (c) purpose for which the licensed material will be used. The regulations in 10 CFR 70.22 require each application for a license to include: (a) information on the corporation applying for a license; (b) the location of the principal office; (c) the names and citizenship of the principal officers; (d) information concerning ownership and control; (e) the proposed site activities; (f) financial qualifications; and (g) the name, amount, and specifications of the licensed material to be used. The regulations in 10 CFR Part 95 contain provisions for obtaining a facility security clearance. The regulations in 10 CFR 140.13b require applicants for uranium enrichment facilities to provide and maintain liability insurance.

1.2.2 REGULATORY ACCEPTANCE CRITERIA

The acceptance criteria applicable to NRC's review of the facility and process description section of the application are contained in Section 1.2.4.3 of the "Standard Review Plan for the Review of a License Application for a Fuel Cycle Facility," NUREG-1520 (NRC, 2002).

1.2.3 STAFF REVIEW AND ANALYSIS

1.2.3.1 Corporate Identity

In Section 1.2.1 of the SAR (LES, 2005a), the applicant provides information on the corporate organization. LES is a Limited Partnership chartered in Delaware. If was formed solely to provide uranium enrichment services to the commercial nuclear power sector. LES has a 100-percent-owned subsidiary, NEF Series 2004, LLC, a limited liability company, organized under the laws of the State of Delaware, formed to purchase Industrial Revenue Bonds issued by Lea County. The General Partners are as follows:

1.   Urenco Investments, Inc., a Delaware corporation and wholly owned subsidiary of Urenco Limited, a corporation formed under laws of the United Kingdom and owned equally by BNFL Enrichment Limited (BNFL-EL); Ultra-Centrifuge Nederland NV (UCN); and Uranit GmbH (Uranit) companies, formed under English, Dutch, and German law, respectively. BNFL-EL is wholly owned by British Nuclear Fuels plc, which is wholly owned by the Government of the United Kingdom. UCN is 99 percent owned by the Government of the Netherlands and 1 percent owned by the Royal Dutch Shell Group, DSM, Koninklijke Philips Electronics N.V., and Stork N.V. Uranit is owned equally by Eon Kernkraft GmbH and RWE Power AG, which are corporations formed under laws of the Federal Republic of Germany.

2.   Westinghouse Enrichment Company, LLC, a Delaware limited liability company and wholly owned subsidiary of Westinghouse Electric Company LLC, also a Delaware limited liability company, whose ultimate parent, through two intermediary Delaware corporations and one corporation formed under the laws of the United Kingdom, is British Nuclear Fuels plc, which is wholly owned by the Government of the United Kingdom.

The Limited Partners are as follows:

1.   Urenco Deeinemingen B.V., a Netherlands corporation and wholly owned subsidiary of Urenco Nederlands B.V.

2.   Westinghouse Enrichment Corporation, LLC, a Delaware limited liability company and wholly owned by Westinghouse Electric Company, LLC.

3.   Entergy Louisiana, Inc., a Louisiana corporation and wholly owned subsidiary of Entergy Corporation, a publicly owned Delaware corporation and a public utility holding company.

4.   Claiborne Energy Services, Inc., a Louisiana corporation and wholly owned subsidiary of Duke Energy Corporation, a publicly owned North Carolina corporation.

5.   Cenesco Company, LLC, a Delaware limited liability company and wholly owned subsidiary of Exelon Generation Company, LLC, a Pennsylvania limited liability company.

6.   Penesco Company, LLC, a Delaware limited liability company and wholly owned subsidiary of Exelon Generation Company, LLC, a Pennsylvania limited liability company.

Urenco owns 70.5 percent of the partnership while Westinghouse Electric Company owns 19.5 percent. Entergy, Duke Energy Corporation, and Exelon Generation Company own the remaining 10 percent in equal shares.

No other companies will be present or operating on the uranium enrichment plant property other than where the applicant has contracted such services. The principal location for business is Albuquerque, New Mexico.

The applicant provided the name of the President of LES, who is a citizen of the United States.

1.2.3.2   Foreign Ownership, Control, or Influence

With respect to a foreign ownership, control, or influence (FOCI) determination for LES' National Enrichment Facility, the NRC staff received a letter from the Department of Energy (DOE) dated March 31, 2005 (DOE, 2005), which states in part that "...any additional FOCI mitigation measures placed on LES would provide no additional benefit to the National Security of the U.S." The letter further recommends that the NRC waive the requirement for FOCI mitigation associated with the granting of a nuclear facility licence to LES. The NRC accepts this finding by DOE based on an Interagency Agreement between NRC and DOE dated May 6, 2002. (DOE, 2002).

1.2.3.3   Financial Qualifications

1.2.3.3.1   Project Costs

The applicant estimates the total construction cost of the facility to be approximately $1.2 billion, in 2002 dollars, which excludes escalation, interest during construction, tails disposition, decommissioning, and any replacement equipment required during the operating life of the facility. The facility SAR (LES, 2005a) and supporting supplements addressing NRC's April 19, 2004, Request for Additional Information (RAI), provide detailed bases that supported the $1.2 billion estimate. The supporting supplements included detailed proprietary construction cost estimates for the facility.

As part of the financial review, before starting the detailed review of the cost estimate, the staff conferred with the technical reviewers assigned to evaluate the support systems/structures necessary to support the safe operation of the facility to confirm that the necessary systems had been identified in the SAR (LES, 2005a). The staff also conducted a detailed review of the SAR, Section 1.1.1 (LES, 2005a), which provided a detailed description of each supporting structure/system, and then compared the support systems for each building with the systems identified in the cost estimate, to confirm that the cost estimate and the facility description were consistent. The cost estimate is based on a reasonable estimate of the cost of the supporting systems and structures, as well as confirming that all the major equipment necessary to support safe operation were included.

The applicant identified the principal buildings necessary to support the operation of the facility. The buildings are: (a) Separation Building/3-modules; (b) Technical Services Building (TSB); (c) Centrifuge Assembly Building; (d) site infrastructure; (e) central utilities; (f) Cylinder Receipt and Dispatch Building (CRDB); and (g) blending and liquid sampling area. The applicant also identified each of the principal components/systems necessary to support its operation and further divided the cost estimate by components/systems and buildings necessary to support the operation of the facility. For each of the major structures, the applicant typically divided the systems into the following: 1) engineering and project management support; 2) centrifuge mounting and equipment; 3) UF_6 systems; 4) control and instrumentation; 5) auxiliary systems;

6) electrical systems; 7) site, building, and landscape; 8) miscellaneous costs including start-up cost; and 9) a contingency of approximately of 10 percent. The staff reviewed each of the detailed costs for the major facilities and its supporting components, and based on its review, the staff has concluded the costs for each of the major structures is reasonable.

The validity of the estimated cost and its supporting assumptions were also key factors in determining if the applicant is financially qualified to construct the facility. The applicant stated that it plans to meet contingencies for cost overruns and construction revenue shortfalls in several ways. Unforeseen construction contingencies will be minimized by the use of a turnkey contractor for the engineering, procurement, and construction of the facility. For cost overruns not covered under the turnkey provisions of the contract, the applicant will seek additional partner equity contributions. However, if cost overruns are much higher than could be anticipated, the applicant would cancel the project and leave an allowance for site stabilization. The facility will begin operation in a phased approach, with each separation module capable of production of one-third of the plant's capacity. This will allow the facility to generate income while coming up to full production.

The staff considers that the construction cost estimate, as presented in the application, and supporting supplements in response to the staff's RAIs, is reasonable and, therefore, the staff concluded that the $1.2 billion estimate is a reasonable estimate to construct the facility.

1.2.3.3.2   Financial Qualifications

The applicant made commitments that construction of the facility will not begin before funding is fully committed. Of this funding (equity and debt), the applicant will have in place, before construction, a minimum of equity contributions of 30 percent of the project's estimated costs of $1.2 billion from the parents and affiliates of the partners, and firm commitments ensuring funds for the remaining project costs. The applicant plans to fund the construction phase of the project with a mix of approximately 50 percent debt and 50 percent equity contributions by the two major partners. The applicant's reliance on approximately 50 percent equity is viewed as a positive endorsement because, by contrast, some analogous construction projects rely on 100 percent financing, which often proves to be difficult to secure from financial institutions.

The applicant has no reported income statements. However, the partners have assets to support their respective equity ownership portions of LES. Urenco Investments, Inc., is a wholly owned subsidiary of Urenco Limited, which in turn is owned in equal shares by BNFL EL; UCN; and Uranit - companies that are formed under English, Dutch and German law, respectively. BNFL EL is wholly owned by British Nuclear Fuels plc, which is wholly owned by the Government of the United Kingdom. UCN is 99 percent owned by the Government of the Netherlands, with the remaining 1 percent owned collectively by private consortiums. Uranit is equally owned by Eon Kernkraft GmbH and RWE Power AG, which are both German companies.

For the year ending December 31, 2003, Urenco Group had total assets of €1.49 billion, with cash assets of €14.4 million. Urenco Group's net income in 2003 was €107.9 million. Urenco Limited is the holding company for the Urenco Group.

Cenesco Company and Penesco Company are both wholly owned subsidiaries of Exelon Generation Company. For the year ending December 31, 2003, Exelon Generation Company had total assets of $14.76 billion, with cash or near-cash assets of $233 million. The company sustained a net loss of $133 million in 2003. Net losses in 2003 can be attributed primarily to operating expenses, in particular the costs of purchased fuel, purchased power, impairment of long-lived assets, and other operating and maintenance expenses. Furthermore, for the 9 months ending September 30, 2004, the company had a positive net income of $599 million.

Duke Energy Corporation, a publicly held North Carolina corporation, is the owner of Claiborne Energy Services, Inc., which is also a 3.33 percent owner of LES. For the year ending December 31, 2003, Duke Energy Corporation had total assets of $56.2 billion, with cash or near-cash assets of $1.16 billion. Duke Energy Corporation sustained a net loss of $1.3 billion in 2003. Net losses in 2003 can be attributed primarily to operating expenses, in particular the costs of purchased natural gas and petroleum products. Furthermore, for the 9 months ending September 30, 2004, the corporation had a positive net income of $1.13 billion.

Entergy Corporation, a public utility holding company, is the owner of Entergy Louisiana Inc., which is a 3.33 percent owner of LES. For the year ending December 31, 2003, Entergy Corporation had total assets of $28.55 billion, with cash or near-cash assets of $692 million. Entergy Corporation's net income in 2003 was $950 million.

Westinghouse Enrichment Company, LLC, is a wholly owned subsidiary of Westinghouse Electric Company, LLC, whose ultimate parent, through two intermediary Delaware corporations and one corporation under United Kingdom laws, is British Nuclear Fuels plc. British Nuclear Fuels plc is wholly owned by the Government of the United Kingdom.

For the year ending March 31, 2004, Westinghouse Electric Company had total assets of £1.02 billion. The company's pre-tax net income was £17 million for that financial year. Westinghouse Electric Company is a subsidiary of British Nuclear Fuels, which had total assets of £23.94 billion for the year ending March 31, 2004. British Nuclear Fuels sustained a net loss of £194 million after taxes for that financial year (pre-tax losses were £299 million).

The remaining 50 percent of the estimated $1.2 billion construction costs will be financed through financial institutions and bond holders.

Lea County will serve as the lessor-owner of the facility during the 30-year term of Industrial Revenue Bonds (IRB) issuance by the State of New Mexico. In this capacity, Lea County will hold the legal title to the uranium enrichment facility, including all related buildings, storage, infrastructure, and equipment, and will hold legal title or a possessory interest in the site on which the facility is located during the term of the IRB. This financial structure will allow the applicant to take advantage of certain tax abatements, tax avoidance, and make other payments in lieu of taxes available under New Mexico law. The IRB is not a vehicle for financing the plant.

Lea County will have no authority to operate the facility as a business or otherwise use or acquire the facility for any purpose, except in its limited role as lessor. During the term of the lease, the applicant is solely responsible, on behalf of, and as agent for, the County, for acquiring, constructing, and installing the equipment into the facility. At the conclusion of the 30-year lease, which corresponds to the 30-year term of the IRB, the applicant will purchase the land and facility from Lea County for the sum of $1.00.

On December 3, 2003, the applicant announced that the first round of contracts with several U.S. nuclear power plants, including Exelon, were signed. These contracts represent at least 70 percent of the facility's first 10 years of production. As the project construction progresses, LES will make a decision to continue, based on a comparison of future incremental construction and operations and maintenance costs to the expected revenues generated from enrichment services sales.

The NRC staff finds that LES and its partner-owners appear to be financially qualified to build and operate the proposed facility, in accordance with 10 CFR 70.23(a)(5). The applicant identified sources of debt and equity for construction, and has reasonable assurance of securing additional financial resources, if needed.

1.2.3.3.3   Liability Insurance

Under 10 CFR 140.13b, a uranium enrichment facility is required to carry liability insurance to cover public claims arising from any occurrence, within the U.S. that causes, within or outside the U.S., bodily injury, sickness, disease, death, loss of, or damage to, property, or loss of use of property arising from the radioactive, toxic, explosive, or other hazardous properties of chemicals containing licensed material. The applicant is proposing to have and maintain up to $300 million to satisfy the 10 CFR 140.13b requirement. The applicant has already obtained a nuclear energy liability policy with a limit of $1 million as a standby policy until the facility is ready to begin operations. At that time, the applicant will increase the amount to approximately $300 million.

Because full liability insurance coverage will not be provided until prior to receipt of licensed material, NRC staff is imposing the following license condition:

> "The licensee shall provide proof of full liability insurance as required under 10 CFR 140.13b, at least 30 days prior to the planned date for obtaining licensed material. If the licensee is proposing to provide less than $300 million of liability insurance coverage, the licensee shall provide, to the NRC for review and approval, an evaluation supporting liability insurance coverage in amounts less than $300 million at least 120 days prior to the planned date for obtaining licensed material."

1.2.3.4   Type, Quantity, and Form of Licensed Material

Table 1.2-1 of the SAR (LES, 2005a) lists the type, quantity, and form of the licensed material proposed for possession. The applicant proposes to use and possess the amounts of special nuclear material, source material, and byproduct material given in Table 1.2-1. The quantities of Tc-99 and transuranics from residual contamination as a consequence of the historical feed of recycled uranium at other facilities are expected to have no significant radiological impact.

1.2.3.5   Authorized Uses

The application is for the issuance of licenses under 10 CFR Parts 30, 40, and 70. The applicant is proposing to use special nuclear material and source material in the enrichment of uranium. The uranium enrichment services would be sold to clients for the production of low-enriched uranium that would be ultimately used in the manufacture of fuel for commercial nuclear power plants. Byproduct material would be used in instrument-calibration sources and may be present as contamination as a consequence of the historical feed of recycled uranium at other enrichment facilities. Feed cylinders that have been previously used to transport or store recycled uranium must be decontaminated before being allowed on the facility site. In addition, natural UF_6 supplied to the facility will meet American Society for Testing and Materials (ASTM) ASTM C787, "Standard Specification for Uranium Hexafluoride for Enrichment" (ASTM, 2003), and periodic audits of suppliers will be performed to ensure that these conditions are met. The applicant intends to identify specific byproduct calibration sources in future license amendment requests. The applicant proposed a 30-year license term. The applicant also requested approval of a classified-matter facility clearance, under 10 CFR Part 95.

Table 1.2-1
Proposed Possession Limits

Source or Special Nuclear Material	Physical and Chemical Form	Maximum Amount to be Possessed at Any One Time
Uranium (natural and depleted) and daughter products	Physical: Solid, Liquid, and Gas Chemical: UF_6, UF_4, UO_2F_2, oxides and other compounds	136,120,000 kg (300,093,231 lbs)
Uranium enriched in isotope U-235 up to 5 percent by weight and uranium daughter products	Physical: Solid, Liquid, and Gas Chemical: UF_6, UF_4, UO_2F_2, oxides and other compounds	545,000 kg (1,201,519 lbs)
Tc-99, transuranic isotopes and other contamination	Any	Amount that exists as contamination as a consequence of the historical feed of recycled uranium at other facilities

Note: Tc-99 - Technetium-99
UF_4 - Uranium Fluoride
UO_2F_2 - Uranyl Fluoride

1.2.3.6   Special Exemptions or Special Authorizations

In Section 1.2.5 of the SAR (LES, 2005a), the applicant addressed an exemption request to 10 CFR 40.36 and 10 CFR 70.25 to provide incremental funding for decommissioning to reflect its phased approach for enrichment capacity at the facility and its expected depleted uranium tails generation rate. As discussed in Section 10.2.2 of the SAR (LES, 2005a), the applicant stated that it would initially provide funding for the projected cost of facility decontamination and decommissioning, assuming operation at full capacity, and disposition of the tails generated during the first three years of operation. Thereafter, the applicant will provide NRC with revised funding instruments for depleted uranium disposition on an annual forward-looking incremental basis. In the event that the applicant does not employ all projected modules as expected, updates required under 10 CFR 40.36 and 10 CFR 70.25 could reflect a corresponding reduction in the anticipated facility decommissioning costs based on the actual number of modules used. NRC staff will review revisions to the cost estimate and the financial instrument, which are presented in Section 10.2.2 of the SAR (LES, 2005a), before the applicant takes possession of licensed material. NRC staff will also review all subsequent revisions to the cost estimate and financial instruments.

Under 10 CFR 40.14 and 10 CFR 70.17, the Commission may grant exemptions from the requirements of the regulations as it determines are authorized by law and will not endanger life or property or the common defense and security and are otherwise in the public interest. NRC staff evaluated the exemption request and determined that such exemption is not prohibited by law. Staff also determined that, because the incremental funding approach proposed by the applicant will provide funding for the all applicant's decommissioning obligations at any point time, the approach will not endanger life or property or the common defense and security. Because the incremental funding approach will reduce the applicant's expenses from having to fund a 30-year decommissioning obligation when, in actuality, the decommissioning obligations prior to the end of the 30-year operating period are less, the staff has determined that the proposed approach will be in the public interest by reducing unnecessary regulatory costs. Therefore, the staff grants the requested exemption as provided in Section 1.2.5 of the SAR. A license condition will be included in the license that will address the applicant's commitments for updating the decommissioning funding plan over time. This license condition is discussed further in Section 10.3.1.10 of this SER.

1.2.3.7   Security of Classified Matter

The purpose of this review is to verify that the applicant provided sufficient information to conclude that there is an adequate Standard Practice Procedures Plan (SPPP) for the protection of classified matter at the proposed facility to be located in Lea County, New Mexico, and a facility clearance can be issued.

1.2.3.7.1   Regulatory Requirements

10 CFR 70.22(m) provides the regulatory requirements for the SPPP that describes the facility's proposed security procedures and controls, as set forth in 10 CFR 95.15(b).

The applicable portion of 10 CFR 70.22(m) identifies that the requirements to protect against unauthorized viewing of classified enrichment equipment and unauthorized disclosure of classified matter are contained in 10 CFR Parts 25 and 95.

1.2.3.7.2   Regulatory Acceptance Criteria

The LES SPPP was reviewed for compliance with the requirements of 10 CFR Parts 25 and 95, by using "Standard Practice Procedures Plan Standard Format and Content for the Protection of Classified Matter for NRC Licensee, Certificate Holder and Others Regulated by the Commission" (NRC, 1999).

1.2.3.7.3   Staff Review and Analysis

The staff reviewed and evaluated information provided by LES in the facility's proposed security procedures and controls to ensure that classified matter is used, processed, stored, reproduced, transmitted, transported, and destroyed in accordance with the requirements of 10 CFR Parts 25 and 95.

NRC staff reviewed the LES SPPP and found it satisfied the requirements of 10 CFR Parts 25 and 95. The applicant has made commitments that meet the requirements of 10 CFR Parts 25 and 95 by providing an acceptable SPPP that establishes controls to ensure that classified matter is used, processed, stored, reproduced, transmitted, transported, and destroyed only under conditions that will provide adequate protection and prevent access by unauthorized persons. By meeting these requirements, the applicant complies with the requirements of 10 CFR 70.22(m). On the basis of these findings, the staff concludes that the SPPP is acceptable for implementation.

1.2.4 EVALUATION FINDINGS

The staff reviewed the institutional information for the proposed LES uranium enrichment facility, according to Section 1.2 of the Standard Review Plan. The applicant has adequately described and documented the corporate identity, structure, and financial information, and is in compliance with those parts of 10 CFR 30.32, 10 CFR 40.31, 10 CFR 70.22, and 10 CFR 70.65 related to institutional information.

The staff reviewed the information provided by the applicant on liability insurance. This information meets the requirements of 10 CFR 140.13b. Because full liability insurance coverage will not be provided until prior to receipt of licensed material, NRC staff is imposing the following license condition:

> "The licensee shall provide proof of full liability insurance as required under 10 CFR 140.13b, at least 30 days prior to the planned date for obtaining licensed material. If the licensee is proposing to provide less than $300 million of liability insurance coverage, the licensee shall provide, to the NRC for review and approval, an evaluation supporting liability insurance coverage in amounts less than $300 million at least 120 days prior to the planned date for obtaining licensed material."

In addition, in accordance with 10 CFR 30.32, 10 CFR 40.31, and 10 CFR 70.22(a)(2) and (4), the applicant has adequately described the types, forms, and quantities and proposed purpose and authorized uses of licensed materials to be permitted at the facility. The applicant provided information on an exemption request related to decommissioning funding that meets the requirements of 10 CFR 40.14 and 10 CFR 70.17. The applicant has also adequately described information related to FOCI and its plans to secure classified matter for a facility clearance under 10 CFR Parts 25 and 95. The staff concludes that the applicant has met the requirements and acceptance criteria applicable to this section.

1.3 SITE DESCRIPTION

The purpose of a site description review is to determine whether the information provided by an applicant adequately describes the geographic, demographic, meteorological, geologic, hydrologic, and seismologic characteristics of the site and the surrounding area. The site description is a summary of the information that the applicant used in preparing the environmental report, emergency plan, and integrated safety analysis summary.

1.3.1 REGULATORY REQUIREMENTS

The regulations in 10 CFR 30.33, 10 CFR 40.32, 10 CFR 70.22, and 10 CFR 70.65(b)(1) require each application to include a general description of the site, with emphasis on those factors that could affect safety (i.e., nearby facilities, meteorology, and seismology).

1.3.2 REGULATORY ACCEPTANCE CRITERIA

The acceptance criteria applicable to the NRC review of the site description section of the application are contained in Section 1.3.4.3 of NUREG–1520 (NRC, 2002).

1.3.3 STAFF REVIEW AND ANALYSIS

1.3.3.1 Site Geography

1.3.3.1.1 Location

The proposed site is in Southeastern New Mexico in Lea County, approximately 1.6 km (1 mi) west of the New Mexico-Texas border on the north side of New Mexico Highway 234. Andrews County, Texas, lies across the border from the site. The site is about 8 km (5 mi) east of Eunice, New Mexico, and 32 km (20 mi) south of Hobbs, New Mexico. The site is 220 ha (543 acres) in size and is located within County Section 32, Township 21 South, Range 38 East. The site is owned by Lea County.

The proposed site is relatively flat with elevations between 1033 and 1045 m (3,390 to 3,430 ft) above sea level. The site slopes to the southwest, is undeveloped, and is used for domestic livestock grazing.

1.3.3.1.2 Nearby Highways

Information concerning public roads is provided in Section 1.3.2.4 of the SAR (LES, 2005a) and Section 3.2.1.2 of the ISA Summary (LES, 2005b). The New Mexico State Highway 234 passes along the southern boundary of the proposed facility.

Based on review of the information provided on nearby highways, staff concludes that the data used in the analysis are accurate and are from acceptable sources.

1.3.3.1.3 Nearby Gas Pipelines

Information concerning gas pipelines passing through or located near the proposed facility site is provided in Sections 1.3.2.4 of the SAR (LES, 2005a) and 3.2.2.4 of the ISA Summary (LES, 2005b).

Natural Gas Pipeline

The applicant identified an underground natural gas pipeline located along the south property line running parallel to New Mexico State Highway 234. A parallel gas pipeline is also identified, but is not in use.

Carbon Dioxide Pipeline

An underground carbon dioxide pipeline currently runs across the property. The ISA Summary (LES, 2005b) indicated this pipeline will be relocated along the western and southern boundary of Section 32 so the pipeline will be positioned at least 396.2 m (1300 ft) from the facility restricted area and is approximately 945m (3100 ft) [estimated from Figure 3.2-3 of the ISA Summary (LES, 2005b)] from a Separations Building Module, a safety-significant structure that houses two cascade halls. The applicant concluded that, at this distance from the proposed facility, the pipeline was not a safety concern. Staff agrees with the applicant's assessment that the carbon dioxide pipeline is not a safety concern to the Separations Building Module.

Onsite Natural Gas Pipeline

The proposed facility will include an on-site natural gas pipeline (Harper, 2003a). This pipeline will be used to provide natural gas for heating the boiler in the Central Utility Building (CUB).

Summary

Staff reviewed information provided in the SAR (LES, 2005a) and ISA Summary (LES, 2005b) on nearby highways, natural gas pipelines, and carbon dioxide pipelines and finds the data used to be accurate and from reliable sources.

1.3.3.1.4   Nearby Air Transportation

Information relating to local air transportation is provided in Sections 1.3.2.4 of the SAR (LES, 2005a), and Section 3.2.1.2.4 of the ISA Summary (LES, 2005b). The information included the number of operations and holding patterns of six local airports. These airports include:

- Lea County Regional Airport—40 km (25 statute miles) northwest of the proposed facility site;

- Eunice Airport—24 km (15 statute miles) west of the proposed facility site;

- Lea County/Jal Airport—40 km (25 statute miles) south-southwest of the proposed facility site;

- Andrews County Airport—48 km (30 statute miles) east of the proposed facility;

- Gaines County Airport—48 km (30 statute miles) northeast of the proposed facility; and

- Seminole Spraying Services (private)—48 km (30 statute miles) northeast of the proposed facility.

The information about the number of operations and holding patterns of each airport was obtained from the Federal Aviation Administration (FAA) (Yeung, 2003). Military flights are operated from the Lea County Regional Airport. The number of military operations is included in the number of operations for the Lea County Regional Airport.

Based on information from the FAA (Yeung, 2003), the applicant concluded the holding patterns for four of the six airports were, in general, more than the 3.2-km (2-statute miles) proximity criterion (third criterion) provided in Section 3.5.1.6 of NUREG–0800 (NRC, 1981) for airway distance from the site of interest. The applicant pointed out no specific holding patterns existed for Eunice or Lea County/Jal Airport (LES, 2004a; LES, 2005a). For the Eunice Airport, the annual operations are small, approximately 480 flights per year. This number is substantially smaller than the threshold limit (225,000 annual operations) provided in NUREG–0800 (NRC, 1981). For the Lea County/Jal Airport, the applicant indicated the airport is more than 32 km (20 mi) away from the proposed facility, and the landing procedure usually will not be initiated until an aircraft is within 32 km (20 mi) of the airport (LES, 2004a; LES, 2005a). Therefore, even if an aircraft is placed in a holding pattern, it will not bring the aircraft near the proposed facility.

The applicant identified a low-level Federal airway passing within 9 km (~6 statute miles) northeast of the proposed facility. This airway was analyzed in the 'ISA Summary" and shown to pose no hazard to the proposed facility.

Three military training routes were identified in the region. The closest route to the proposed facility was approximately 26 km (16 statute miles) southwest of the site.

Based on the review of aircraft-crash risk assessment, staff concluded that the aircraft transportation information used in the analysis is accurate and was obtained from a reliable source.

1.3.3.2   Demographics

Information about demographics is provided in Section 1.3.2 of the SAR (LES, 2005a).

1.3.3.2.1   Local Population and Land Use

The proposed site is in Lea County, New Mexico, and is about 1.6 km (1 mi) from the New Mexico-Texas border. Andrews County, Texas abuts the border on the Texas side. Together, the counties have a combined population of 68,515, based on the 2000 census. In 1990, the combined population was 70,130. This decrease is counter to the trends within New Mexico and Texas, which had state-wide population increases of 20.1 percent and 22.8 percent, respectively, over that 10-year period. The population decreases in Lea and Andrews Counties are caused by decrease in petroleum industry jobs since the mid-1980s. It is expected that population growth in these two counties in the next 30 years will be at a lower rate than the overall rates in New Mexico and Texas.

Lea County covers 11,378 km^2 (4393 mi^2), which is about three times the size of the State of Rhode Island. Andrews County covers 3,895 km^2 (1504 mi^2).

Major population centers near the proposed site include the following:

Eunice, New Mexico, about 8 km (5 mi) west of the site;
Hobbs, New Mexico, about 32 km (20 mi) north of the site;
Jal, New Mexico, about 37 km (23 mi) south of the site;
Lovington, New Mexico, about 64 km (39 mi) north-northwest of the site;
Andrews, Texas, about 51 km (32 mi) east of the site;
Seminole, Texas (in Gaines County) about 51 km (32 mi) east-northeast of the site; and
Denver City, Texas (in Gaines County) 65 km (40 mi) north-northeast of the site.

Outside of these population centers, population density is very low. The nearest residences are located about 4.3 km (2.6 mi) west of the proposed facility site.

Within 8 km (5 mi) of the site, land is primarily open land used for cattle grazing. Oil and gas potentials are absent within this range, although operations are widespread beyond this area. Nearby industrial activities include a quarry and a "produced-water" reclamation company. Lea County operates a county landfill on the south side of New Mexico Highway 234, and about 1.6 km (1 mi) east of the proposed site, Waste Control Specialists operates a hazardous chemical waste disposal facility and has licenses for the treatment and storage of low-level radioactive and mixed wastes. A natural gas processing plant is located about 6 km (4 mi) from the site.

1.3.3.2.2   Local Public Services

Fire fighting services are provided locally by Eunice Fire and Rescue, which is located 8 km (5 mi) from the proposed site. It is staffed by a full-time fire chief and 34 volunteer firefighters. Equipment includes three pumpers, one tanker, and three grass trucks. Eunice Fire and Rescue also has agreements for mutual assistance with all Lea County fire departments.

Police and law enforcement services are provided by the Eunice Police Department, which has five full-time officers. The Lea County Sheriff's Department also has a substation in Eunice. Agreements between Lea and Andrews Counties provide mutual support when needed. The New Mexico State Police can also provide support.

Educational institutions in Eunice include an elementary school, a middle school, a high school, and a private K-12 school. The nearest other schools are in Hobbs, New Mexico, 32 km (20 mi) north of Eunice. The nearest schools in Andrews County, Texas, are in Andrews, Texas, about 51 km (32 mi) from the proposed site.

There are two hospitals in Lea County – one is located in Hobbs, New Mexico, 32 km (20 mi) north of the proposed site, and the other in Lovington, New Mexico, 64 km (39 mi) north-northwest of the site. The hospital in Hobbs is a 250-bed facility capable of handling acute and stable chronic care patients. The hospital in Lovington is a full-service, 27-bed facility. The Eunice clinic is the nearest medical center to the proposed facility site. The nearest nursing home facilities are in Hobbs, New Mexico.

There are no recreation facilities near the site. The Eunice Golf Course is located approximately 15 km (9.2 mi) west of the site. A historical marker and picnic area are located about 3.2 km (2 mi) west of the proposed site at the intersection of New Mexico Highways 234 and 18.

1.3.3.2.3   Water Use

Southeast New Mexico has a semi-arid climate with an average annual precipitation of 33 to 38 cm (13 to 15 in.). The proposed site has no surface water and/or drainage features. Essentially all precipitation either infiltrates the soil or is evapotranspirated. There are no significant bodies of water or navigable waterways in the vicinity of the proposed site. There is also no agricultural activity in the site vicinity although there are various crops grown in Lea and Andrews Counties. Cattle grazing does occur at the proposed site and in the nearby vicinity. Dairy farming is important in Lea County, although none takes place near the site and/or in Andrews County.

Known sources of water near the site include: a man-made pond at the quarry, adjacent to the proposed site, stocked with fish for private use; Baker Spring, which is an intermittent surface-water feature, located about 1.6 km (1 mi) northeast of the proposed site; and several cattle watering holes, where groundwater is pumped by windmills and stored in above-ground tanks.

1.3.3.2.4   Summary

The staff reviewed the site demographic information presented by the applicant and finds that the applicant has adequately described and summarized general site demographical information related to local population, identification of population centers, schools, commercial facilities, land use, and water use. Population information is provided based on the latest census information.

1.3.3.3   Meteorology

1.3.3.3.1   Tornado Hazard and Tornado-Generated Missiles

Information about the tornadoes and design-basis tornado at the proposed facility is provided in Sections 1.3.3.3 of the SAR (LES, 2005a), and Section 3.2.3.4.1 of the ISA Summary (LES, 2005b).

There is an average of nine tornadoes a year in New Mexico, and the occurrence of tornadoes in the vicinity of the proposed facility is rare. Tornadoes are classified using the Fujita Tornado Damage Scale (F-scale) with classifications ranging from F0-F5 (NOAA, 2005). Eighty-seven tornadoes of low magnitude (F0 to F2) were reported in Lea County, New Mexico, between January 1, 1950, and December 31, 2004. Only one additional tornado was reported as F3 on May 17, 1954. Two tornadoes, one in 1998 and the second in 1999, had a magnitude of F0 and were located near Eunice. All the reported tornadoes were associated with very light damage (NCDC, 2005).

The tornado-generated missiles the applicant considered for the proposed facility included three classes of missiles. These missiles were: (i) a 6.8-kg (15-lb), 10.2- x 30.5-cm (2- x 4-in.) timber plank; (ii) 34-kg (75-lb), 7.6-cm (3-in.)-diameter steel pipe; and (iii) 1361-kg (3000-lb) automobile. The associated vertical and horizontal impact velocities for each missiles also were provided in the ISA Summary (LES, 2005b). According to the applicant (Harper, 2003b), the tornado-generated missiles were determined based on DOE–STD–1020–2002 (DOE, 2002).

Based on the review of the information concerning tornados and tornado-generated missiles, NRC concludes: (i) the information is accurate and is from reliable sources; and (ii) the design-bases tornado-generated missiles are acceptable because they were determined based on an appropriate DOE standard. The use of a DOE standard is an acceptable approach to NRC staff.

1.3.3.3.2   High Winds and Hurricanes

Information about high winds at the proposed facility is provided in Sections 1.3.3.1 of the SAR (LES, 2005a); Section 3.2.3.4.2 of the ISA Summary (LES, 2005b), and Section 3.6.1.4 of the Environmental Report (ER) (LES, 2005c).

According to the SAR (LES, 2005a), no meteorological data were available for the proposed facility site. Although the measured wind data at Midland–Odessa, Texas, and Roswell, New Mexico, were discussed in the SAR (LES, 2005a) and ER (LES, 2005c), the Midland–Odessa annual extreme wind data were used exclusively to estimate the high-wind hazard at the proposed facility site (LES, 2005b; Harper, 2003b). The annual extreme data used range from 1973 through 1999. The wind speeds were 3-second gust speeds measured at 10 m (32.8 ft) above ground. The Midland–Odessa weather station is located at the regional airport approximately 103 km (64 mi) east-southeast of the proposed site, whereas the Roswell station is approximately 161 km (100 mi) northwest of the proposed site. The climate data for both locations were collected by the National Oceanic and Atmospheric Administration (NOAA) (LES, 2005c).

The largest wind speed for the annual extreme straight-line winds from 1973 through 1999 at Midland–Odessa was 140 km/h (87 mph) and the smallest annual extreme straight-line wind speed was 84 km/h (52 mph) (Harper, 2003b). The mean and standard deviation wind speeds were 111.5 and 16.6 km/h (69.3 and 10.3 mph).

The high-wind hazard assessment was performed by fitting the annual extreme wind data using the Fisher–Tippett Type I distribution model. The applicant chose the speed of a wind with an annual probability of 1.0×10^{-5} for the design-basis straight-line wind speed for the proposed facility. This design-basis straight-line wind speed was 252 km/h (157 mph) (LES, 2005b).

Because the proposed facility is not located near the coastal area [805 km (500 mi) from the coast], hurricanes affecting the coastal area will have no effect on the performance of the proposed facility. Consequently, consideration of hurricane hazards on the design of the proposed facility is not needed.

Based on the review of the information concerning high winds, the staff concludes that high-wind hazards and the associated design-basis straight-line winds have been addressed acceptably because the data used for assessment were from a recognized source and the method used for assessing high-wind hazards is a commonly used and accepted method.

1.3.3.3.3   Temperature Extremes

Information about the temperature at the proposed facility site is provided in Section 3.6.1.2 of the ER (LES, 2005c).

The regional temperatures in Hobbs, New Mexico [32 km (20 mi) north of the proposed facility site]; Midland–Odessa, Texas; and Roswell, New Mexico, are discussed in the ER. The discussions are based on 30-year records (from 1971 through 2000). As indicated previously, NOAA collected the climate data for Midland–Odessa and Roswell. However, the Western Regional Climate Center collected the climate data for Hobbs (LES, 2005c).

The highest recorded monthly mean maximum temperature was 38.9 °C (102.1 °F), and the lowest recorded monthly mean minimum temperature was -5.1 °C (22.8 °F) for Hobbs, New Mexico. No such data were presented in the ER for Midland–Odessa or Roswell. The highest daily maximum and lowest daily minimum temperatures were 46.7 °C (116.0 °F) and -23.9 °C (-11.0 °F) for Midland–Odessa, and 45.6 °C (114.0 °F) and -22.8 °C (-9.0 °F) for Roswell. No such data were presented for Hobbs. As indicated, the highest daily maximum and the lowest daily minimum temperatures for Midland–Odessa and Roswell were similar.

The staff reviewed the temperature information and find the information acceptable because recognized data sources were used and the temperature extremes are properly determined.

1.3.3.3.4   Extreme Precipitation

Section 1.3.3.2 of the SAR (LES, 2005a); and Sections 3.2.3.2 and 3.2.3.4.4 of the ISA Summary (LES, 2005b) discuss the rainfall precipitation at the proposed facility site. The precipitation data for Hobbs, Midland–Odessa, and Roswell were listed in Tables 3.2-14 through 3.2-16 of the ISA Summary (LES, 2005b). These data were collected from the Western Regional Climate Center and NOAA and are based on data from 1971 through 2000 (LES, 2005c). The maximum monthly totals were 35.13 cm (13.83 in.) for Hobbs; 24.6 cm (9.7 in.) for Midland–Odessa; and 17.5 cm (6.88 in.) for Roswell. The minimum monthly totals were zero for all locations. The highest 24-hour precipitation was 15.2 cm (5.99 in.) for Midland–Odessa and 12.5 cm (4.91 in.) for Roswell.

According to the SAR (LES, 2005a), the local intense probable maximum precipitation was estimated from NOAA data (NOAA, 1982). The local intense probable maximum precipitation was approximately 43.9 cm (17.3 in.) in 1 hour, over a 2.6-km^2 (1-mi^2) area.

The staff reviewed the information concerning regional precipitation and local intense probable maximum precipitation presented in the SAR (LES, 2005a) and the ISA Summary (LES, 2005b) and find the information acceptable because recognized data sources, such as NOAA, were used.

1.3.3.3.5   Snow

Section 1.3.3.2 of the SAR (LES, 2005a), and Section 3.2.3.3 of the ISA Summary (LES, 2005b) discuss the regional snowfall. NOAA collected the snowfall data. The maximum monthly snowfall/ice pellets were 24.9 cm (9.8 in.) for Midland–Odessa and 53.3 cm (21.0 in.) for Roswell. The maximum snowfall/ice pellets during a 24-hour period were 12.47 cm (4.91 in.) for Midland–Odessa and 41.91 cm (16.5 in.) for Roswell. No snowfall information was available for Hobbs, New Mexico.

The staff reviewed the information concerning snow precipitation presented in the SAR (LES, 2005a) and the ISA Summary (LES, 2005b) and find the information acceptable because recognized data sources, such as NOAA, were used.

1.3.3.3.6   Lightning and Thunderstorms

Section 1.3.3.3 of the SAR (LES, 2005a), and Section 3.2.3.4.5 of the ISA Summary (LES, 2005b) describe the potential of thunderstorms and lightning strikes at the proposed facility site. The applicant indicated thunderstorms occur every month and are most common in spring and summer at the proposed facility site.

The applicant estimated the lightning strike frequency at the proposed facility site to be 1.36 flashes per year. The applicant also stated in the ISA Summary (LES, 2005b) that the proposed facility will be designed for lightning protection.

The staff reviewed the information about lightning and find the lightning strike frequency determined for the site is acceptable and appropriate. Staff further concludes the design approach proposed by the applicant to protect the proposed facility from lightning effects is acceptable.

1.3.3.3.7   Sandstorms

Section 1.3.3.3 of the SAR (LES, 2005a) describes the potential of sandstorms at the proposed facility site.

According to the SAR (LES, 2005a), blowing sand and dust may occur occasionally. Large dust storms with the potential of covering a large region are rare. Staff reviewed the information about sandstorms presented in the SAR (LES, 2005a) and find the information sufficient and acceptable.

1.3.3.4   Geology

1.3.3.4.1   Seismic Hazard

Seismic hazards are discussed in Section 1.3.5 of the SAR (LES, 2005a); and Sections 3.2.5 and 3.2.6 of the ISA Summary (LES, 2005b).

The following areas concerning the seismic hazard applicable to the safety analysis and design of the proposed facility were reviewed:

- Seismic source characterization;
- Ground motion attenuation;
- Seismic hazard calculation;
- Development of site-specific spectra; and
- Surface faulting.

1.3.3.4.1.1   Seismic Source Characterization

Geological and Tectonic Settings

Section 1.3.5 of the SAR (LES, 2005a) and Section 3.2.5 of the ISA Summary (LES, 2005b) provide a description of the regional and local geological and tectonic settings. The proposed facility site is located within the Central Basin Platform area. The Central Basin Platform Area is situated between the Midland and Delaware Basins, all of which are part of the Permian Basin, a 250-million-year-old structure. The Permian Basin is a downward flexure of a large thickness of originally flat-lying bedded, sedimentary rock. The base of the Permian Basin sediments extends to approximately 1525 m (5000 ft) beneath the proposed facility site. The top of the Permian section is approximately 434 m (1425 ft) below ground surface. These sediments are overlain by sedimentary strata of the Triassic Age Dockum Group. The upper formation of the Dockum Group is the Chinle Formation, locally overlain by either the Tertiary Ogallala, Gatuña, or Antlers Formations, or Quaternary alluvium. At the proposed facility site, geotechnical borings identified up to 0.6 m (2 ft) of loose eolian sand underlain by dense to very dense, fine- to medium-grained sand and silty sand of the Gatuña Formation. The sands of the Gatuña Formation are locally cemented with caliche. Beneath the Gatuña Formation, the Chinle claystone, a hard and highly plastic clay, was encountered in geotechnical borings at depths from 10.7 to 12.2 m (35 to 40 ft).

As noted in the ISA Summary (LES, 2005b), the Southeast New Mexico–West Texas area is presently structurally stable. The Laramide Orogeny (late Cretaceous to Early Tertiary time) uplifted the region to its present elevation, and there has been no substantial tectonic activity since this early Tertiary deformation. The Permian Basin has subsided slightly since the Laramide Orogeny. However, this subsidence is believed to be a result of dissolution of the Permian evaporite layers by groundwater or possibly compaction from oil and gas extraction. As stated in the ISA Summary (LES, 2005b), no active faults have been identified at the site. Faulting consists of geologically older subsurface faults in the Permian Basin subregion related to the development of the Permian Basin and the Laramide Orogeny. The nearest evidence of Quaternary faulting is 161 km (100 mi) west of the site, in the Basin and Range tectonic province.

Historical Seismicity

Section 1.3.5.2 of the SAR (LES, 2005a) and Section 3.2.6.1 of the ISA Summary (LES, 2005b) summarizes the historical seismicity at the proposed facility site. As stated in the ISA Summary (LES, 2005b), the assessment of historical seismicity included earthquakes in the region of interest known from felt or damage records and from more recent instrumental records (since the early 1960s). The largest earthquake known to occur within 322 km (200 mi) of the site was the August 16, 1931, earthquake near Valentine, Texas. This earthquake had an estimated M_L (Local Magnitude) of 6.0 to 6.4 and produced a maximum epicentral intensity of VIII on the Modified Mercalli intensity scale. This earthquake occurred approximately 237 km (147 mi) from the proposed site location. Within 80 km (50 mi) of the site, the largest historical earthquake was a M_L 5.0 event in 1992, approximately 16 km (10 mi) southwest of the site. Other significant events between 322 km (200 mi) and 80 km (50 mi) of the proposed facility site ranged in M_L from 4.0 to 5.7.

1-20

Earthquakes in the region of the proposed facility site include isolated and small clusters of low-to-moderate-magnitude events toward the Rio Grande Valley of New Mexico and in Texas, southeast of the proposed site. According to the ISA Summary (LES, 2005b), no earthquakes in the site region are known to be correlated to specific faults. An earthquake catalog based on the historic seismicity in the region [322-km (200-mi) radius] was presented in the ISA Summary (LES, 2005b). This catalog was composed of data from: the Advanced National Seismic System (NCEDC, 2004); University of Texas Institute for Geophysics (UTIG, 2002); New Mexico Tech Historical Catalog (NMIMT, 2003); and New Mexico Tech Regional catalogs. The catalog identified a substantial cluster of seismic activity that has occurred on and near the Central Basin Platform since the mid-1960s. It was suggested by DOE (DOE, 2003) and noted in the ISA Summary (LES, 2005b) that Central Basin Platform earthquakes are not tectonic in origin but instead are related to water injection and withdrawal resulting from secondary recovery operations in oil fields in the Central Basin Platform area. The ISA Summary (LES, 2005b) noted, however, the January 2, 1992, event was attributed to a tectonic origin because of its determined focal depth of approximately 12 km (7 mi) and is not correlated with oil or gas drilling. At the proposed facility site, postulated earthquakes that could impact safe operation of the proposed facility are associated with zones of crustal weakness in the Central Basin Platform and the Basin and Range tectonic province.

The staff concludes that information concerning seismic source characterization presented in Section 1.3.5 of the SAR (LES, 2005a) and the ISA Summary (LES, 2005b) is acceptable. The information provides a complete summary of seismicity and potential fault and tectonic sources and thereby demonstrates compliance with regulatory requirements in 10 CFR 30.33, 10 CFR 40.32, 10 CFR 70.22, and 10 CFR 70.65(b)(1).

1.3.3.4.1.2   Ground Motion Attenuation

Details of ground motion attenuation functions used to compute the hazard are described in Section 3.2.6.4.1 of the ISA Summary (LES, 2005b). Several attenuation models were used in the ISA Summary. The Nuttli attenuation model developed by the U.S. Department of the Army, Waterways Experiment Station (USDA, 1973) was primarily selected because it was used in the DOE (DOE, 2003) seismic hazard assessment. The Toro, et al. (Toro, 1997) attenuation model also was used in the hazard calculations for comparison.

The attenuation models used in the ISA Summary were applicable to locations within the Central U.S.. The proposed facility site is located at 103° west longitude, slightly east of the 105° west longitude cutoff for Central and Eastern U.S. sites, as specified in Regulatory Guide 1.165 (NRC, 1997). In addition, Frankel, et al. (Frankel, 1996) specified attenuation zones for the U.S. in its hazard mapping project. The U.S. Geological Survey (USGS) boundary separating the Western U.S. and the Central and Eastern U.S. attenuation zones also is located at approximately 105° west longitude and slightly to the west of the proposed facility site. The proposed facility site is thus situated within the area in which both the Central and Eastern U.S. attenuation models are applicable.

1.3.3.4.1.3   Surface Faulting

There is no geologic, geophysical, or seismological evidence of active surface faulting in the vicinity of the proposed facility site. As stated in the ISA Summary (LES, 2005b), the nearest recent faulting is located more than 161 km (100 mi) west of the site. Therefore, surface faulting was not considered a credible disruptive event for the proposed facility.

Recently, a fault was discovered at the nearby Waste Control Specialists (WCS) site. However, subsequent fault investigations revealed that the faulting is inactive because no faults exist in formations younger than Triassic age (205 to 240 million years old) (LES, 2004a).

1.3.3.4.2   Slope Stability

Section 1.3.1.2 of the SAR (LES, 2005a), and Section 3.2.1.1 of the ISA Summary (LES, 2005b) describe the topography at the proposed facility site. The SAR (LES, 2005a) and ISA Summary (LES, 2005b) indicated the site topography is relatively flat, with a gradual elevation increase from southwest to northeast. The staff site visit on May 27–28, 2004 (NRC, 2004), confirmed the area at the proposed facility is relatively flat. Consequently, slope stability is not a safety concern for this proposed facility.

1.3.3.4.3   Liquefaction

Liquefaction potential of soils beneath the proposed facility is discussed in Section 3.2.7.1 of the ISA Summary (LES, 2005b). According to the ISA Summary (LES, 2005b), except for a top layer of loose sand [up to 0.6m (2ft)], the soils at the proposed facility site are dense to very dense and the groundwater level is at least 30 m (98 ft) below ground surface. In Section 3.3.2.1 of the ER, the applicant (LES, 2005c) indicates the groundwater table at the site is 65 to 68 m (214 to 222 ft) below ground surface. Consequently, the applicant concluded the potential for liquefaction was remote (LES, 2005b, c).

Geotechnical investigation indicated the soil beneath the proposed facility site is a layer of loose eolian sand underlain by the Gatuña Formation (dense to very dense sand and silty sand). Below the Gatuña Formation is the Chinle claystone, a very hard, highly plastic clay. The Chinle claystone was encountered at depths approximately 10.7 to 12.2 m (35 to 40 ft). For the top 7.6 m (25 ft) of soils, the blow-count values ranged from 20 to 76. Beneath the 7.6-m (25-ft) horizon, typical blow-count values were more than 60, with even larger blow-count values for the Chinle claystone.

The staff reviewed the geotechnical investigation information presented in the ISA Summary (LES, 2005b, c) and concurs with the applicant that the potential for liquefaction of soils at the site may not be a safety concern for the proposed facility. The applicant committed in Section 3.3.9 of the SAR (LES, 2005a) to perform additional geotechnical investigations at the site to confirm that liquefaction is not a safety concern for the proposed facility. Additional site testing will be evaluated in accordance with NRC Regulatory Guide 1.198, "Procedures and Criteria for Assessing Seismic Soil Liquefaction at Nuclear Power Plant Sites" (NRC, 2003).

1.3.3.4.4   Settlement

Settlement of foundations for the proposed facility is discussed in Section 3.2.7 of the ISA Summary (LES, 2005b). In its ISA Summary, the applicant stated that only five borings were drilled at the proposed facility site to determine the suitability of the site. The applicant recognized the geotechnical results obtained from the five borings were not sufficient for final design purposes. The applicant committed in Section 3.3.9 of the SAR (LES, 2005a) that the settlement and differential settlement for the design of the proposed facility will be determined based on the information that will be obtained from the additional geotechnical investigations. Allowable soil bearing pressures will be evaluated in accordance with Naval Facilities Engineering Command Design Manual NAVFAC DM-7.02, "Foundations and Earth Structures" (NAVFAC, 1996). Building settlement analyses will be performed in accordance with Naval Facilities Engineering Command Design Manual NAVFAC DM-7.01; "Soil Mechanics" (NAVFAC, 1986) and Winterkorn and Fang, "Foundation Engineering Handbook" (Winterkorn, 1975).

The staff reviewed the information presented concerning differential settlements and find the applicant's commitment to perform additional geotechnical investigations using acceptable geotechnical standards for final facility design to be acceptable.

1.3.3.5   Hydrology

Site surface water and groundwater hydrology is discussed in Sections 1.3.4 of the SAR (LES, 2005a) and 3.2.4 of the ISA Summary (LES, 2005b). The applicant obtained hydrological data principally from previous investigations conducted by WCS, which is located 1.6 km (1 mi) east of the proposed site. WCS operates a hazardous chemical treatment and disposal facility. The applicant performed a limited number of geotechnical studies that demonstrate that the WCS data are applicable to the proposed site.

The proposed site contains no surface water and/or surface water drainage features, with essentially all precipitation subject to either infiltration or evapotranspiration.

The applicant performed subsurface studies of the alluvial material that overlies the Chinle red bed clays. These alluvial deposits are 9 to 18 m (30 to 60 ft) thick. The Chinle formation consists of a low-permeability clay unit having a thickness of 323 to 333 m (1060 to 1092 ft) and is the upper formation within the Triassic Age Dockum Group. No perched water systems in the alluvial deposits were found, although one well produced water samples, because of a limited groundwater occurrence.

The low permeability Chinle formation essentially isolates the deep and shallow groundwater systems. Within the Chinle formation are two distinct groundwater systems, with no interconnections. The first is a siltstone or silty sandstone unit with some saturation at 65 to 68 m (214 to 222 ft) below the surface. This unit is a low-permeability formation that does not yield groundwater easily. The second unit is a saturated siltstone layer approximately 30.5 m (100 ft) thick, at an elevation of 183 m (600 ft) below the surface. The Santa Rosa formation lies below the Chinle Unit, but within the Dockum Group, at 340 m (1115 ft) below the surface. The Santa Rosa unit is the first occurrence of a well-defined aquifer system. However, this system is considered non-potable because of high concentrations of dissolved solids.

At the quarry site, north of the proposed site, there are shallow groundwater occurrences. These shallow perched systems, however, are intermittent and limited and caused by a layer caliche or caprock at the surface that in places is fractured and can lead to rapid infiltration of precipitation forming the perched water system. Caprock, however, is not present at the proposed site, and, therefore, it is not expected that significant perched water systems would be produced.

Baker Spring is located about 1.6 km (1 mi) northeast of the proposed site. However, this spring is intermittent and flows only after precipitation events.

Several localized shallow perched groundwater systems exist to the east of the proposed site and are used to supply water pumped by windmills to tanks for grazing livestock. These perched systems are located above the Chinle clays, but the volume of water produced is limited.

Because of the lack of sufficient surface and groundwater supplies, the applicant will not make withdrawals of groundwater at the site. Instead, the applicant is proposing to obtain water for plant use using Eunice and Hobbs municipal supplies. These water supplies are obtained from well fields near Hobbs, New Mexico. The applicant is also not proposing to inject water into groundwater systems at the site.

Since there are no surface water bodies in the immediate vicinity of the site, flooding is not a design objective. The only potential flooding at the plant would occur from intense local precipitation events. Flood protection is provided by establishing building floor levels above the calculated depth of ponded water caused by intense precipitation events (see Sections 1.3.3.3.4 and 1.3.3.3.5 of this SER).

The staff reviewed the applicant's hydrological data and finds that it provides sufficient information to assess site flooding hazards and ground- and surface water impacts, and is consistent with information in the ISA Summary.

1.3.4   EVALUATION FINDINGS

The staff has reviewed the site description for the proposed LES uranium enrichment facility according to Section 1.3 of the Standard Review Plan. The applicant has adequately described and summarized general information pertaining to: (1) the site geography, including its location relative to prominent natural and man-made features such as mountains, rivers, airports, population centers, schools, and commercial and manufacturing facilities; (2) population information on the basis of the most current available census data to show population distribution as a function of distance from the facility; (3) meteorology, hydrology, and geology for the site; and (4) applicable design basis events. The reviewer verified that the site description is consistent with the information used as a basis for the ER, emergency management plan, and ISA Summary; and that it demonstrates compliance with regulatory requirements in 10 CFR 30.33, 10 CFR 40.32, 10 CFR 70.22, and 10 CFR 70.65(b)(1).

1.4   REFERENCES

(ASTM, 2003) American Society for Testing and Materials (ASTM). ASTM C787, "Standard Specification for Uranium Hexafluoride for Enrichment," 2003.

(DOE, 2002) U.S. Department of Energy (DOE). Interagency Agreement Between the U.S. Department of Energy Oak Ridge Operations and the U.S. Nuclear Regulatory Commission Concerning Foreign Ownership, Control, and Influence Determinations Associated with Certain NRC Licensees. May 6, 2002.

(DOE, 2003) U.S. Department of Energy (DOE). DOE/WIPP–95–2065, Rev. 7, "Waste Isolation Pilot Plant Contact-Handled (CH) Waste Safety Analysis Report," 2003.

(DOE, 2004) U.S. Department of Energy (DOE). DOE–STD–1024, "Natural Phenomena Hazards Design and Evaluation Criteria for Department of Energy Facilities," 2004.

(DOE, 2005) U.S. Department of Energy (DOE). Letter to U.S. Nuclear Regulatory Commission on Louisiana Energy Services Foreign Ownership, Control, or Influence issues, March 31, 2005.

(Frankel, 1996) Frankel, A., et al. "National Seismic Hazards Maps, June 1995 Documentation." U.S. Geological Survey. 1996.

(Harper, 2003a) Harper, G.A. "Assessment of Other External Event Hazards at NEF for ISA and Design Basis." Document Identifier 51-2400552-00. Framatome ANP, Inc. 2003.

(Harper, 2003b) Harper, G.A. "Assessment of Tornado, Tornado Missiles and High Wind Loads at NEF for ISA and Design Basis—Attachment 2." Document Identifier 51-2400548-00. Framatome ANP, Inc. 2003.

(LES, 2004a) Louisiana Energy Services (LES), letter to U.S. Nuclear Regulatory Commission, "Clarifying Information Related to External Hazards Analysis," August 31, 2004.

(LES, 2004b) Louisiana Energy Services (LES), letter to U.S. Nuclear Regulatory Commission, "Clarifying Information Related to Probabilistic Seismic Hazard Analysis," November 22, 2004.

(LES, 2005a) Louisiana Energy Services (LES). "National Enrichment Facility Safety Analysis Report," Revision 6, 2005.

(LES, 2005b) Louisiana Energy Services (LES). "Integrated Safety Analysis Summary," Revision 4, 2005.

(LES, 2005c) Louisiana Energy Services (LES). "National Enrichment Facility Environmental Report," Revision 5, 2005.

(NCDC, 2005) National Climatic Data Center (NCDC). "Storm Events." <http://www4.ncdc.noaa.gov/cgi-win/wwcgi.dll?wwevent~storms> (Accessed 3/17/05).

(NAVFAC, 1986) U.S. Naval Facilities Engineering Command (NAVFAC). Design Manual NAVFAC DM-7.01, "Soil Mechanics," 1986.

(NAVFAC, 1996) U.S. Naval Facilities Engineering Command (NAVFAC). Design Manual NAVFAC DM-7.02, "Foundations and Earth Structures," 1996.

(NCEDC, 2004) Northern California Earthquake Data Center (NCEDC). "Composite Earthquake Catalog," 2004, <http://quake.geo.berkeley.edu/anss/>

(NMIMT, 2003) New Mexico Institute of Mining and Technology (NMIMT). "Earthquake Catalogs for New Mexico and Bordering Areas: 1869–1998," 2003.

(NOAA, 1982) National Oceanic and Atmospheric Administration (NOAA). "Application of Probable Maximum Precipitation Estimates—United States East of the 105th Meridian," Hydrometeorological Report No. 52, 1982.

(NOAA, 2005) National Oceanic and Atomospheric Administration. "Tornadoes." <http://www.noaa.gov/tornadoes.html> (Accessed 3/17/05).

(NRC, 1981) U.S. Nuclear Regulatory Commission (NRC). NUREG–0800, "Standard Review Plan for the Review of Safety Analysis Reports for Nuclear Power Plants," 1981.

(NRC, 1997) U.S. Nuclear Regulatory Commission (NRC). Regulatory Guide 1.165, "Identification and Characterization of Seismic Sources and Determination of Safe Shutdown Earthquake Ground Motion," 1997.

(NRC, 1999) U.S. Nuclear Regulatory Commission (NRC). "Standard Practice Procedures Plan Standard Format and Content for the Protection of Classified Matter for NRC Licensee, Certificate Holder and Others Regulated by the Commission," 1999.
(NRC, 2002) U.S. Nuclear Regulatory Commission (NRC). NUREG–1520, "Standard Review Plan for the Review of a License Application for a Fuel Cycle Facility," 2002.

(NRC, 2003) U.S. Nuclear Regulatory Commission (NRC). Regulatory Guide 1.198, "Procedures and Criteria for Assessing Seismic Soil Liquefaction at Nuclear Power Plant Sites," 2003.

(NRC, 2004) Graves, H.L., U.S. Nuclear Regulatory Commission (NRC). Summary of Meeting with Louisiana Energy Services on Structural, Seismic, High Winds, and Tornado Safety Analysis. June 29, 2004.

(Toro, 1997) Toro, G.R., N.A. Abrahamson, and J.H. Schneider, "Model of Strong Ground Motions from Earthquakes in Central and Eastern North America: Best Estimates and Uncertainties," *Seismological Research Letters*, Vol. 68, No. 1, pp. 41–5, 1997.

(USDA, 1973) U.S. Department of Army, Waterways Experiment Station (USDA). "Design Earthquake for the Central United States," Miscellaneous Paper S–731–1, 1973.

(UTIG, 2002) The University of Texas Institute for Geophysics. "Compendium of Texas Earthquakes," 2002.

(Winterkorn, 1975) Winterkorn, H.F., and Fang, H.Y., "Foundation Engineering Handbook," 1975.

(Yeung, 2003) Yeung, W.S., "Aircraft Hazard Risk Determination," Document Identifier 32-2400569-00, Framatome ANP, Inc., 2003.

2.0 ORGANIZATION AND ADMINISTRATION

The purpose of the review of the applicant's organization and administration is to ensure that the proposed management policies will provide reasonable assurance that the licensee plans, implements, and controls site activities in a manner that ensures the safety of workers, the public, and the environment. The review also ensures that the applicant has identified and provided adequate qualification descriptions for key management positions.

2.1 REGULATORY REQUIREMENTS

10 CFR 30.33, 10 CFR 40.32, 10 CFR 70.22, 10 CFR 70.23, and 10 CFR 70.62(d) require a management system and administrative procedures for the effective implementation of health, safety, and environment (HS&E) protection functions concerning the applicant's corporate organization, qualifications of the staff, and adequacy of the proposed equipment, facilities, and procedures to provide adequate safety for workers, the public, and the environment.

2.2 REGULATORY ACCEPTANCE CRITERIA

The acceptance criteria applicable to the U.S. Nuclear Regulatory Commission's (NRC's) review of the organization and administration section of the application are contained in Section 2.4.3 of the "Standard Review Plan for Fuel Cycle Facilities," NUREG-1520 (NRC, 2002).

2.3 STAFF REVIEW AND ANALYSIS

In Section 2.1 of the applicant's Safety Analysis Report (SAR) (LES, 2005), the applicant provides a functional description of specific organization groups responsible for managing the design, construction, and operation of the facility. Included in this section are the plans for the transition from the start-up phase to operations.

In Section 2.2 of the applicant's SAR (LES, 2005), the applicant describes the qualifications, responsibilities, and authorities for key supervisory and management personnel, along with a listing of the shift crew composition.

In Section 2.3 of the applicant's SAR (LES, 2005), the applicant includes administration procedures for effective implementation of HS&E functions, using written procedures and reporting of unsafe conditions or activities, along with written agreements with offsite emergency resources and a commitment to establish formal management measures to ensure availability of Items Relied on for Safety (IROFS).

Figure 2.1-1 of the applicant's SAR (LES, 2005) shows the LES organization during design and construction phase of the facility. Before beginning operations, this organization transitions to the one shown in Figure 2.1-2.

2.3.1 Organizational Responsibilities and Qualifications

The Chief Operating Officer is appointed by the President and is responsible for ensuring that the facility complies with all applicable regulatory requirements. The Chief Operating Officer directs these responsibilities through the Plant Manager.

The Plant Manager will be appointed by, and reports to, the LES Chief Operating Officer. The plant manager has direct responsibility for operation of the facility in a safe, reliable, and efficient manner. He/she is responsible for proper selection of staff for all key positions, including positions on the Safety Review Committee (SRC). The Plant Manager is responsible for the protection of the facility staff and the general public from radiation and chemical exposure or any other consequences of an accident at the facility and also bears the responsibility for compliance with the facility license. He/she or designee(s) has the authority to approve and issue procedures. The Plant Manager will have, as a minimum, a bachelor's degree (or equivalent) in an engineering or scientific field and 10 years of responsible nuclear experience.

The Quality Assurance (QA) Director is appointed by, and reports to, the President, and has overall responsibility for development, management, and implementation of the LES QA Program. He/she will have, as a minimum, a bachelors degree (or equivalent) in an engineering or scientific field, and at least 6 years of responsible nuclear experience in the implementation of a QA program. The QA Director will have at least 4 years experience in a QA organization at a nuclear facility.

The QA Manager reports to the Plant Manager and is responsible for establishing and maintaining the QA Program for the facility. The facility line managers and their staff who are responsible for performing quality-affecting work are responsible for ensuring implementation of and compliance with the QA Program. The QA Manager position is independent from other management positions at the facility, to ensure that the QA Manager has access to the Plant Manager for matters affecting quality. In addition, the QA Manager has the authority and responsibility to contact the LES President through the QA Director, with any QA concerns. The QA Manager will have, as a minimum, a bachelor's degree (or equivalent), in an engineering or scientific field, and at least 5 years of responsible nuclear experience in the implementation of a QA program. The QA Manager shall have at least 2 years experience in a QA organization at a nuclear facility.

The HS&E Manager reports to the Plant Manager and has the responsibility for assuring safety at the facility through activities including maintaining compliance with safeguards, appropriate rules, regulations, codes – and has the responsibility for implementation and control of the Fundamental Nuclear Material Control Plan (FNMCP). This includes HS&E activities associated with nuclear criticality safety, radiation protection, chemical safety, environmental protection, emergency preparedness, and industrial safety. The HS&E Manager works with the other facility managers to ensure consistent interpretations of HS&E requirements, performs independent reviews, and supports facility and operations change control reviews. This position is independent from other management positions at the facility to ensure objective HS&E audit, review, and control activities. The HS&E Manager has the authority to shut down operations if they appear to be unsafe, and must consult with the Plant Manager with respect to restart of shutdown operations after the deficiency, or unsatisfactory condition, has been resolved. Changes to the facility or to activities of personnel that require prior NRC approval are reviewed and approved by the HS&E Manager or designee. The HS&E Manager will have, as a minimum, a bachelor's degree (or equivalent) in an engineering or scientific field and at least 5 years of responsible nuclear experience in HS&E or related disciplines. The HS&E Manager will also have at least 1 year of direct experience in the administration of nuclear criticality safety evaluations and analyses.

The Operations Manager reports to the Plant Manager and has the responsibility of directing the day-to-day operation of the facility. This includes such activities as ensuring the correct and safe operation of uranium hexafluoride (UF_6) processes, proper handling of UF_6, and the identification and mitigation of any off-normal operating conditions. In case of the absence of the Plant Manager, the Operations Manager may assume the responsibilities and authorities of the Plant Manager. The Operations Manager will have, as a minimum, a bachelor's degree (or equivalent) in an engineering or scientific field and 4 years of responsible nuclear experience.

The Uranium Management Manager reports to the Plant Manager and has the responsibility for UF_6 cylinder management (including transportation licensing) and directing the scheduling of enrichment operations to ensure smooth production. This includes activities such as ensuring that proper feed material and maintenance equipment are available for the facility. In case of the absence of the Plant Manager, the Uranium Management Manager may assume the responsibilities and authorities of the Plant Manager. The Uranium Management Manager will have, as a minimum, a bachelor's degree (or equivalent) in an engineering or scientific field and 4 years of responsible nuclear experience.

The Technical Services Manager reports to the Plant Manager and has the responsibility of providing technical support to the facility. This includes technical support for facility modifications (including administration of the configuration management system); engineering support for operations and maintenance; performance; operation of the chemistry laboratory; maintenance activities; and computer support. In case of the absence of the Plant Manager, the Technical Services Manager may assume the responsibilities and authorities of the Plant Manager. The Technical Services Manager will have, as a minimum, a bachelor's degree (or equivalent) in an engineering or scientific field and 4 years of responsible nuclear experience.

The Human Resource Manager reports to the Plant Manager and has the responsibility for community relations; ensuring adequate staffing; ensuring training is provided for facility employees; providing administrative support services to the facility, including document control; and for the physical security of the facility. The Human Resource Manager will have as a minimum, a bachelor's degree in Personnel Management, Business Administration, or related field, and 3 years of appropriate, responsible experience in implementing and supervising human resource responsibilities at an industrial facility.

The QA Inspectors report to the QA Manager (via a designated supervisory position, if applicable) and have the responsibility for performing inspections related to the implementation of the LES QA Program. QA Inspectors performing QA Level 1 activities will be certified in accordance with American Society of Mechanical Engineers (ASME) NQA-1 (ASME, 1994) and ASME NQA-1a (ASME, 1995) inspector qualification requirements.

The QA Auditors report to the QA Manager (via a designated supervisory position, if applicable) and have the responsibility for performing audits related to the implementation of the LES QA Program. QA Auditors require certification under the LES QA Program. This certification includes training on the LES QA Program, audit fundamentals, objectives and techniques for performing audits, and on-the-job training.

The QA Technical Support personnel report to the QA Manager (via a designated supervisory position, if applicable) and have the responsibility for providing technical support related to the implementation of the LES QA Program. QA Technical Support staff receive QA Indoctrination Training and training in the specific QA procedures needed to perform their jobs.

The Emergency Preparedness Manager reports to the HS&E Manager and has the responsibility for ensuring that the facility remains prepared to react and respond to any emergency situation that may arise. This includes emergency preparedness training of facility personnel; facility support personnel; the training of, and coordination with, offsite emergency response organizations; and conducting periodic drills to ensure facility-personnel and offsite-response-organization-personnel training is maintained up to date. The Emergency Preparedness Manager will have a minimum of 5 years of experience in the implementation and supervision of emergency plans and procedures at a nuclear facility. No credit for academic training may be taken toward fulfilling this experience requirement.

The Licensing Manager reports to the HS&E Manager and has the responsibility for coordinating facility activities to ensure that compliance is maintained with applicable NRC requirements. The Licensing Manager is also responsible for ensuring abnormal events are reported to NRC in accordance with NRC regulations. The Licensing Manager will have a minimum of 5 years of appropriate, responsible experience in implementing and supervising a nuclear licensing program.

The Environmental Compliance Manager reports to the HS&E Manager and has the responsibility for coordinating facility activities to ensure all local, State, and Federal environmental regulations are met. This includes submission of periodic reports to appropriate regulating organizations of effluents from the facility. The Environmental Compliance Manager will have a minimum of 5 years of appropriate, responsible experience in implementing and supervising a nuclear environmental compliance program.

The Radiation Protection Manager reports to the HS&E Manager and has the responsibility for implementing the Radiation Protection program. These duties include: the training of personnel in use of equipment; control of radiation exposure of personnel; continuous determination of the radiological status of the facility; and conducting the radiological environmental monitoring program. During emergency conditions, the Radiation Protection Manager's duties may also include:

- Providing Emergency Operations Center personnel information and recommendations concerning chemical and radiation levels at the facility;

- Gathering and compiling onsite and offsite radiological and chemical monitoring data;

- Making recommendations concerning actions at the facility and offsite deemed necessary for limiting exposures to facility personnel and members of the general public; and

- Taking prime responsibility for decontamination activities.

2-4

In matters involving radiological protection, the Radiation Protection Manager has direct access to the Plant Manager. The Radiation Protection Manager will have, as a minimum, a bachelor's degree (or equivalent), in an engineering or scientific field, and 3 years of responsible nuclear experience associated with implementation of a Radiation Protection program. At least 2 years of experience will be at a facility that processes uranium, including uranium in soluble form.

The Industrial Safety Manager reports to the HS&E Manager and has the responsibility for the implementation of facility industrial safety programs and procedures. This will include programs and procedures for training individuals in safety and maintaining the performance of the facility fire protection systems. The Industrial Safety Manager will have, as a minimum, a bachelor's degree (or equivalent) in either an engineering or a scientific field and 3 years of appropriate, responsible nuclear experience associated with implementation of a facility safety program.

Criticality Safety Engineers report to the HS&E Manager (via a designated supervisory position, if applicable) and are responsible for the preparation or review of nuclear criticality safety evaluations and analyses, and conducting and reporting periodic nuclear criticality safety assessments. Nuclear criticality safety evaluations and analyses require independent reviews by a Criticality Safety Engineer. Criticality Safety Engineers shall have a minimum of 2 years experience in the implementation of a criticality safety program. These individuals will hold a bachelor's degree in an engineering or scientific field and have successfully completed a training program, applicable to the scope of operations, in the physics of criticality and in associated safety practices.

Should a change to the facility require a nuclear criticality safety evaluation, an individual who, as a minimum, possesses the equivalent qualifications of the Nuclear Criticality Engineer will perform the evaluation or analysis. In addition, this individual will have at least 2 years of experience performing criticality safety analyses and implementing criticality safety programs. An independent review of the evaluation or analysis will be performed by a qualified Criticality Safety Engineer.

The Chemical Safety Engineer reports to the HS&E Manager (via a designated supervisory position, if applicable) and is responsible for the preparation or review of chemical safety programs and procedures for the facility. The Chemical Safety Engineer will have a minimum of 2 years experience in the preparation or review of chemical safety programs and procedures. This individual will hold a bachelor's degree (or equivalent) in an engineering or scientific field and have successfully completed a training program, applicable to the scope of operations, in chemistry and in associated safety practices.

The Shift Managers report to the Operations Manager and have the responsibility for ensuring safe operation of enrichment equipment and support equipment. Each Shift Manager directs assigned personnel, to provide enrichment services in a safe, efficient manner. Shift Managers will have a minimum of 5 years of appropriate, responsible experience in implementing and supervising a nuclear operations program.

The Production Scheduling Manager reports to the Uranium Management Manager and has the responsibility for developing and maintaining production schedules for enrichment services. This individual will have a minimum of 3 years of appropriate, responsible experience in implementing and supervising a continuous production scheduling program.

The Cylinder Management Manager reports to the Uranium Management Manager and has the responsibility for ensuring that cylinders of UF_6 are received and routed correctly at the facility, and is also responsible for all transportation licensing. This individual will have a minimum of 3 years of appropriate, responsible experience in implementing and supervising a continuous production scheduling program.

The Warehouse and Materials Manager reports to the Uranium Management Manager and has the responsibility for ensuring spare parts and other materials needed for operation of the facility are ordered, received, inspected, and stored properly. This individual will have a minimum of 3 years of appropriate, responsible experience in implementing and supervising a purchasing and inventory program.

The Safeguards Manager reports to the HS&E Manager and has the responsibility for ensuring the proper implementation of the FNMCP. This position is separate from, and independent, of the Operations, Technical Services, and Human Resources departments, to ensure a definite division between the safeguards group and the other departments. In matters involving safeguards, the Safeguards Manager has direct access to the Plant Manager. The Safeguards Manager will have as a minimum, a bachelor's degree in an engineering or scientific field, and 5 years of experience in the management of a safeguards program for Special Nuclear Material, including responsibilities for material control and accounting. No credit for academic training may be taken toward fulfilling this experience requirement.

The Chemistry Manager reports to the Technical Services Manager and has the responsibility for the implementation of chemistry analysis programs and procedures for the facility. This includes effluent sample collection, chemical analysis of effluents, comparison of effluent analysis results to limits, and reporting of chemical analysis of effluents to appropriate regulatory agencies. The Chemistry Manager will have, as a minimum, a bachelor's degree (or equivalent) in either an engineering or a scientific field and 3 years of appropriate, responsible nuclear experience associated with implementation of a facility chemistry program.

The Performance Manager reports to the Technical Services Manager and has the responsibility for coordinating and maintaining testing programs for the facility. This includes testing of systems and components to ensure that the systems and components are functioning as specified in design documents. This individual will have, as a minimum, a bachelor's degree (or equivalent) in either an engineering or a scientific field and 4 years of appropriate, responsible nuclear experience associated with implementation of testing programs.

The Projects Manager reports to the Technical Services Manager and has the responsibility for the implementation of facility modifications and for maintaining the configuration management system. This individual also provides engineering support, as needed, to support facility operation and maintenance, and support of performance testing of systems and equipment. The Projects Manager will have, as a minimum, a bachelor's degree (or equivalent) in an engineering or scientific field and have a minimum of 5 years of appropriate, responsible nuclear experience.

The Engineering Manager reports to the Technical Services Manager and has the responsibility for providing engineering support at the facility. This includes ensuring the safe operation of enrichment equipment and support equipment, providing maintenance support for equipment and systems, and developing operating and maintenance procedures for the facility. The individual is responsible for the development of all design changes to the plant. The Engineering Manager will have, as a minimum, a bachelor's degree (or equivalent) in an engineering or scientific field and have a minimum of 5 years of appropriate, responsible experience in implementing and supervising a nuclear engineering program.

The Maintenance Manager reports to the Technical Services Manager and has the responsibility of directing and scheduling maintenance activities to ensure proper operation of the facility, including preparation and implementation of maintenance procedures. This includes activities such as repair and preventive maintenance of facility equipment. The Maintenance Manager also has the responsibility for coordinating and maintaining testing programs for the facility. This includes testing of systems and components to ensure that the systems and components are functioning as specified in design documents. This individual will have, as a minimum, a bachelor's degree (or equivalent) in an engineering or scientific field and 4 years of responsible nuclear experience.

The Administration Manager reports to the Human Resources Manager and has the responsibility for ensuring that support functions such as accounting, word processing, and general office management are provided for the facility. This individual will have a minimum of 3 years of appropriate, responsible experience in implementing and supervising administrative responsibilities at an industrial facility.

The Community Relations Manager reports to the Human Resources Manager and has the responsibility for providing information about the facility and LES to the public and media. During an abnormal event at the facility, the Community Relations Manager ensures that the public and media receive accurate and up-to-date information. This individual will have as a minimum, a bachelor's degree in Public Relations, Political Science, or Business Administration, and 3 years of appropriate, responsible experience in implementing and supervising a community relations program.

The Security Manager reports to the Human Resources Manager and has the responsibility for directing the activities of security personnel to ensure the physical protection of the facility. This individual is also responsible for the protection of classified matter at the facility and obtaining security clearances for facility personnel and support personnel. In matters involving physical protection of the facility or classified matter, the Security Manager has direct access to the Plant Manager. This individual will have, as a minimum, a bachelor's degree in an engineering or scientific field and 5 years of experience in the responsible management of physical security at a facility requiring security capability similar to that required for the facility. No credit for academic training may be taken toward fulfilling this experience requirement.

The Document Control Manager reports to the Human Resources Manager and has the responsibility for adequately controlling documents at the facility. This individual will have a minimum of 3 years of appropriate, responsible experience in implementing and supervising a document control program.

The Training Manager reports to the Human Resources Manager and has the responsibility for conducting training and maintaining training records for personnel at the facility. This individual will have a minimum of 5 years of appropriate, responsible experience in implementing and supervising a training program.

The minimum operating shift crew consists of a Shift Manager (or Deputy Shift Manager in the absence of the Shift Manager); one Control Room operator; one Radiation Protection technician; one operator for each Cascade Hall and associated UF_6 handling systems; and security personnel. When only one Cascade Hall is in operation, a minimum of two operators are required.

At least one criticality safety engineer will be available, with appropriate ability to be contacted by the Shift Manager, to respond to any routine request or emergency condition. This availability may be offsite if adequate communication ability is provided to allow response as needed.

The applicant has a program in place to make personnel position descriptions available onsite for NRC inspections.

2.3.2   Management Control

Section 2.3 of the SAR (LES, 2005) summarizes how the activities that are essential for implementation of the management measures and other HS&E functions are documented in formally approved, written procedures, prepared in compliance with a formal document control program. The mechanism for reporting potentially unsafe conditions or activities to the HS&E organization and facility management is also summarized. This mechanism involves giving employees that feel safety or quality is being compromised the responsibility and right to initiate the "stop work" process to ensure work is returned to safe conditions. Employees also have the right to access line management, the safety organization, requirements under 10 CFR Part 19, and the Corrective Action Program to ensure their concerns are addressed.

2.3.2.1   Configuration Management

A Configuration Management program is provided to define and maintain a technical baseline for the facility and provide a formal process for making changes to that baseline. All changes made to the facility are made in accordance with the Configuration Management program. Section 11.3.1 of this Safety Evaluation Report (SER) discusses the Configuration Management program.

2.3.2.2   Maintenance

A maintenance program will be implemented, during operations, that will include planned and scheduled preventive maintenance, surveillance, and performance trending, to ensure that IROFS are available and reliable to perform their intended functions.

2.3.2.3   Training and Qualifications

The applicant will implement a formal planned training program that will include indoctrination training for all employees, addressing criticality, radiological, chemical, and industrial safety. The level of indoctrination training will depend on the specific jobs to be performed. Continued or periodic retraining will be established, when applicable, to ensure employee proficiency. Further in-depth training will occur in the area of specific job areas. Radiological and criticality safety retraining will occur annually. Training records will be maintained by the Human

Resources Manager. Additional information on the applicant's training program is provided in Section 11.3 of the SAR (LES, 2005) and in Section 11.3.3 of this SER.

2.3.2.4 Procedures

The applicant will conduct all operations involving licensed material in accordance with approved, written procedures. These procedures will generally include operating procedures, administrative procedures, maintenance procedures, and emergency procedures.

2.3.2.5 Audits and Assessments

The applicant will implement a QA Program that requires periodic audits of activities affecting quality, to ensure that these activities are being conducted in accordance with procedures and the QA Program requirements. The audits will be identified, scheduled, and performed in accordance with a written plan. The frequency of audits will depend on the safety significance, status, and work history of the activity. The SRC and the QA Department will conduct operational reviews and program audits. Further information on audits is provided in Section 11.5 of the SAR (LES, 2005).

2.3.2.6 Safety Review Committee

Section 2.2.3 of the SAR states that the SRC will report to the Plant Manager and will provide technical and administrative review and audit of operations that could affect plant worker, public safety, and environmental impacts. The scope of activities reviewed and audited by the SRC shall, as a minimum, include the following:

- Radiation protection;
- Nuclear criticality safety;
- Hazardous chemical safety;
- Industrial safety, including fire protection;
- Environmental protection;
- As Low As is Reasonably Achievable (ALARA) policy implementation; and
- Changes in facility design or operations.

The SRC will conduct at least one facility audit per year for the above areas. The SRC will be composed of at least five members, including the Chairman. Members of the SRC may be from the LES corporate office or technical staff. The five members will include experts on operations and all safety disciplines (criticality, radiological, chemical, industrial). The Chairman, members, and alternate members of the SRC will be formally appointed by the Plant Manager, will have an academic degree in an engineering or physical science field, and, in addition, will have a minimum of 5 years of technical experience, of which a minimum of 3 years will relate directly to one or more of the safety disciplines (criticality, radiological, chemical, industrial). The SRC will meet at least once per calendar quarter.

Review meetings will be held within 30 days of any incident that is reportable to NRC. These meetings may be combined with regular meetings. After a reportable incident, the SRC will review the incident's causes, the responses, and both specific and generic corrective actions to ensure resolution of the problem is implemented. A written report of each SRC meeting and audit will be forwarded to the Plant Manager and appropriate managers within 30 days, and be retained in accordance with the records management system.

2.3.2.7   Incident Investigations

A Corrective Action Program will be implemented to identify, investigate, analyze, and document abnormal events that have the potential to threaten or weaken the applicant's health, safety, and environmental protection programs. Additional detail is provided in Section 11.3.6 of this SER.

2.3.2.8   Employee Concerns

The applicant will implement a "stop-work" process for any employee who feels that safety or quality could be compromised in any work activity. This program is implemented through general employee indoctrination training and the applicant's management, the safety organization, NRC requirements in 10 CFR Part 19, and through the Corrective Action Program.

2.3.2.9   Records Management

The applicant will implement a records management program to control the preparation and issuance of applicant documents. This document control program will include a formal process for preparing, reviewing, approving, and issuing revisions to documents. Further discussion of the records management program is provided in Section 11.3.7 of this SER.

2.3.2.10   Written Agreements with Offsite Emergency Agencies

The applicant will coordinate emergency actions with appropriate State and local offsite emergency agencies through written agreements. Further discussion of Emergency Management is provided in Chapter 8 of this SER.

2.3.3   Transition from Design and Construction to Operations

LES is responsible for the design, QA, construction, testing, initial start-up, operation, and decommissioning of the facility.   .

Toward the end of construction, the focus of the organization will shift from design and construction to initial start-up and operation of the facility. As the facility nears completion, LES will staff the facility to ensure smooth transition from construction activities to operation activities. Urenco will have personnel integrated into the LES organization to provide technical support during startup of the facility and transition into the operations phase.

As the construction of systems is completed, the systems will undergo acceptance testing as required by procedure, followed by turnover from the construction organization to the operations organization by means of a detailed transition plan. The turnover will include the physical systems and corresponding design information and records. After turnover, the operating organization will be responsible for system maintenance and configuration management. The design basis for the facility is maintained during the transition from construction to operations through the configuration management system described in Chapter 11, "Management Measures," of the SAR (LES, 2005).

2.4 EVALUATION FINDINGS

The staff reviewed the organization and administration for the proposed facility according to Chapter 2 of NUREG-1520 (NRC, 2002). The staff reviewed the applicant's organization, management position summaries and qualifications, and management controls. These organizational and administrative elements describe: (1) clear responsibilities and associated resources for the design, construction, and operation of the facility; and (2) its plans for managing and operating the project. The staff reviewed these plans and commitments and concludes that they provide reasonable assurance that an acceptable organization, administrative policies, and sufficient competent resources have been established or are committed, to satisfy the applicant's commitments for the design, construction, and operation of the facility per 10 CFR 30.33, 10 CFR 40.32, 10 CFR 70.22, 10 CFR 70.23, and 10 CFR 70.62(d).

2.5 REFERENCES

(ASME, 1994) American Society of Mechanical Engineers (ASME). ASME NQA-1, "Quality Assurance Requirements for Nuclear Facility Applications," 1994.

(ASME, 1995) American Society of Mechanical Engineers (ASME). ASME NQA-1a, Addenda to "Quality Assurance Requirements for Nuclear Facility Applications," 1995.

(LES, 2005) Louisiana Energy Services (LES). "National Enrichment Facility Safety Analysis Report," Revision 6, 2005.

(NRC, 2002) U.S. Nuclear Regulatory Commission (NRC). NUREG-1520, "Standard Review Plan for the Review of a License Application for a Fuel Cycle Facility," 2002.

3.0 INTEGRATED SAFETY ANALYSIS (ISA) AND ISA SUMMARY

The purpose of this review is to ensure that the Integrated Safety Analysis (ISA) and ISA Summary meet the regulatory requirements specified in 10 CFR Part 70, Subpart H, "Additional Requirements for Certain Licensees Authorized to Possess a Critical Mass of Special Nuclear Material." The review determines whether appropriate hazards and baseline design criteria have been addressed. The review also determined whether acceptable Items Relied on for Safety (IROFS), management measures, and likelihoods and consequences have been designated for higher-risk accident sequences, and whether, with IROFS, the performance requirements of 10 CFR 70.61 have been met. For those cases involving nuclear criticality safe-by-design components, the review determines whether the performance requirements of 10 CFR 70.61 are met through demonstration that failure of those components is highly unlikely. The review also determined whether programmatic commitments to maintain the ISA and ISA Summary are acceptable.

In particular, this review considered information provided by the applicant related to:

1. The use of baseline design criteria for the design of the facility in accordance with 10 CFR 70.64(a).

2. Commitments regarding the applicant's safety program, including the ISA, pursuant to the requirements of 10 CFR 70.62; and

3. ISA summaries submitted in accordance with 10 CFR 70.62(c)(3)(ii) and 70.65.

3.1   REGULATORY REQUIREMENTS

The following regulatory requirements are applicable to the ISA and ISA Summary content:

1. 10 CFR 70.62 specifies the requirement to establish and maintain a safety program, including performance of an ISA that demonstrates compliance with the performance requirements of 10 CFR 70.61;

2. 10 CFR 70.62(c) specifies requirements for conducting an ISA, including a demonstration that credible high-consequence and intermediate-consequence events meet the safety performance requirements of 10 CFR 70.61;

3. 10 CFR 70.64 specifies requirements for baseline design criteria and facility and system design and facility layout; and

4. 10 CFR 70.65(b) specifies the contents of an ISA Summary.

The regulations, in 10 CFR 70.62, require an applicant to establish and maintain a safety program that demonstrates compliance with the performance requirements of 70.61. The safety program is required to contain three elements: (1) process safety information; (2) an integrated safety analysis (ISA); and (3) management measures. The integrated safety analysis must be conducted and maintained by the applicant and must identify the following:

- Radiological hazards related to possessing or processing licensed material;

- Chemical hazards of licensed material and hazardous chemicals produced from licensed material;

- Facility hazards that could affect the safety of licensed materials and thus present an increased radiological risk;

- Potential accident sequences caused by process deviations or other events internal to the facility and credible external events, including natural phenomena;

- The consequence and likelihood of occurrence of each potential accident sequence identified and the methods used to determine the consequences and likelihood; and

- Each item relied on for safety (IROFS) identified pursuant to 10 CFR 70.61, the characteristics of its preventive, mitigative, or other safety function and the assumptions and conditions under which the item is relied upon to support compliance with 10 CFR 70.61.

The regulations, in 10 CFR 70.61, provide that the ISA must evaluate compliance with performance requirements. Those requirements specify that the risk of each credible high-consequence event must be limited such that the likelihood of occurrence is highly unlikely and the risk of each credible intermediate-consequence event must be limited such that the likelihood of occurrence is unlikely.

The license application must include a description of the safety program under 10 CFR 70.65(a). In addition, the applicant is required to submit to the NRC an ISA Summary. The Summary is required to contain:

- A general description of the site with emphasis on those factors that could affect safety;

- A general description of the facility with emphasis on those areas that could affect safety;

- A description of each process analyzed in the ISA in sufficient detail to understand the theory of operation and, for each process, the hazards identified in the ISA and a general description of the types of accident sequences;

- Information that demonstrates compliance with the performance requirements of 10 CFR 70.61, including a description of the management measures, requirements for criticality monitoring and alarms and the information regarding the baseline design criteria and defense-in-depth practices set forth in 10 CFR 70.64;

- A description of the team, qualifications, and the methods used to perform the ISA;

- A list briefly describing each IROFS in sufficient detail to understand their functions in relation to the performance requirements of 10 CFR 70.61;

- A description of the proposed quantitative standards used to assess consequences to an individual from acute chemical exposure to licensed material or chemicals produced from licensed material;

- A descriptive list that identifies all IROFS that are the sole item preventing or mitigating an accident sequence that exceed the performance requirements of 10 CFR 70.61; and

- A description of the definitions of unlikely, highly unlikely, and credible, as used in the evaluations in the ISA.

3.2   REGULATORY ACCEPTANCE CRITERIA

The acceptance criteria used during the U.S. Nuclear Regulatory Commission's (NRC's) review of the applicant's ISA and ISA Summary are outlined in Sections 3.4.3.1 and 3.4.3.2 of NUREG-1520 (NRC, 2002).

3.3   STAFF REVIEW AND ANALYSIS

The Staff reviewed the safety program as described in the applicant's ISA Summary (LES, 2005b) in order to assess compliance with the regulatory requirements. This includes information describing the site, facility, processes, and baseline design criteria. The summary also details the method used by the applicant to identify hazards associated with the processes identified, a description of the accident sequences identified, the potential consequences for each accident, and the identification of applicable IROFS for each accident found to be credible for which consequences could be classified as intermediate or high. A description of the safety function of each IROFS was described, along with the means by which the IROFS will be implemented.

The staff reviewed the applicant's Safety Analysis Report (SAR), portions of the Integrated Safety Analysis (ISA), and ISA Summary. Since the proposed facility has not been constructed, the staff visited a similar gas centrifuge uranium enrichment facility at Almelo, The Netherlands, to become familiar with the proposed processes and plant layout. The staff also conducted an in-office review of the ISA at the AREVA engineering offices in Marlborough, Massachusetts, and two in-office reviews of criticality, chemical safety, and other related documents at the applicant's Washington, D.C., offices. The staff analyzed the applicant's proposed Safety Program that includes the elements of process safety information, integrated safety analysis, and management measures, to determine that the requirements of 10 CFR 70.62 are met. The staff also conducted detailed, vertical slice reviews of various accident scenarios, selected on a sampling basis, to confirm that the Safety Program and associated elements are adequately implemented by the applicant to achieve the performance requirements of 10 CFR 70.61.

In accordance with the guidance in Section 3.5.2.3 of NUREG-1520 (NRC, 2002), the vertical slice review examined how the ISA method was applied to a selected subset of facility processes in order to obtain reasonable assurance that ISA methods would be effective in the other processes not sampled by the staff. The staff reviewed the applicant's HAZOP methodology and confirmed that it met the guidance in NUREG-1513 and generally acceptable industry practices (AIChE, 1989 and 1992). The HAZOP technique identifies and evaluates safety hazards in process plants and the technique requires detailed information concerning the design and operation of a process, and is typically used, as in this case, during or after the detailed design phase. Implementation of the technique involves the use of an interdisciplinary team and systematic approach to identify hazard and operability problems (i.e., accident sequences). The results of the HAZOP analysis are the team's findings, which include identification of the accident sequences and items relied on for safety (IROFS). As a result of the initial staff review, the applicant added a "safe-by-design" method to the ISA for application to passive design component features related to nuclear criticality safety. The staff subsequently determined that use of the HAZOP and "safe-by-design" ISA methods provided reasonable assurance that the applicant identified all accident sequences that could exceed the performance requirements of 10 CFR 70.61.

Accident sequences related to chemical safety, nuclear criticality safety, and fire protection were selected for a detailed staff "vertical-slice" review based on gas centrifuge uranium enrichment process knowledge and professional judgement. The vertical-slice review examined how the ISA methods were applied and examined appropriate safety information not included in the ISA Summary. The vertical slice review included both high and intermediate consequence accident scenarios. The purpose of the review was to determine whether accident sequences, consequences, and likelihoods were reasonably determined, and whether appropriate IROFS and management measures were selected to limit the risk of the analyzed events (i.e., high-consequence events to "highly unlikely," and each intermediate-consequence events to "unlikely"). The results of the staff's vertical-slice review of a smart sample of accident sequences in each technical discipline will provide reasonable assurance that, if the methods described in the SAR, and discussed above, are appropriately applied by the applicant, all accident sequences and related IROFS will be identified by the applicant. For nuclear criticality safe-by-design components, the staff review determined whether those components met the criteria for highly unlikely.

3.3.1   General Information

The staff evaluated information describing the site, facility, processes, baseline design criteria, the safety program and integrated safety analysis (ISA) to determine whether the performance criteria of 10 CFR 70.61 are met. The staff's evaluation considered the applicant's implementation of the baseline design criteria for the facility under 10 CFR 70.64. The development and implementation of the safety program, including the elements of process safety information, integrated safety analysis and management measures, were reviewed to confirm that the applicant had established an acceptable methodology for conducting an ISA. The results of the applicant's ISA, contained in the ISA Summary, were further reviewed by the staff to confirm that it identified appropriate hazards and associated accident sequences that could exceed the performance requirements of 10 CFR 70.61, including likelihood and consequence levels, and appropriate items relied on for safety (IROFS) and the measures to assure that they will be available and reliable to prevent or mitigate those accident sequences.

3.3.1.1   Site Description

A description of the proposed facility and site is provided in SAR Section 1.3 (LES, 2005a) and ISA Summary Section 3.2 (LES, 2005b). The general description topics include site geography, demographics and land use, meteorology, hydrology, geology, seismology and stability of subsurface materials. External events, such as explosions and aircraft crashes, and natural phenomena, including tornados, hurricanes, floods, and earthquakes, are assessed to determine the likelihood of occurrence and their impact on the facility.

The regulations in 10 CFR 70.65(b)(1) require each application to include a general description of the site in the ISA Summary, with emphasis on those factors that could affect safety (i.e., nearby facilities, meteorology, and seismology).

3.3.1.1.1   Site Geography

The proposed site is in Southeastern New Mexico in Lea County, approximately 1.6 km (1 mi) west of the New Mexico-Texas border on the north side of New Mexico Highway 234. Andrews County, Texas, lies across the border from the site. The site is about 8 km (5 mi) east of Eunice, New Mexico, and 32 km (20 mi) south of Hobbs, New Mexico. The site is 220 ha (543 acres) in size and is located within County Section 32, Township 21 South, Range 38 East. The site is owned by Lea County.

The proposed site is relatively flat with elevations between 1033 and 1045 m (3,390 to 3,430 ft) above sea level. The site slopes to the southwest, is undeveloped, and is used for domestic livestock grazing.

3.3.1.1.2   Demographics and Land Use

Information about demographics is provided in Section 1.3.2 of the SAR (LES, 2005a) and Section 3.2.2 of the ISA Summary (LES, 2005b).

3.3.1.1.2.1   Local Population and Land Use

The proposed site is in Lea County, New Mexico, and is about 1.6 km (1 mi) from the New Mexico-Texas border. Andrews County, Texas abuts the border on the Texas side. Together, the counties have a combined population of 68,515, based on the 2000 census. In 1990, the combined population was 70,130. This decrease is counter to the trends within New Mexico and Texas, which had state-wide population increases of 20.1 percent and 22.8 percent, respectively, over that 10-year period. The population decreases in Lea and Andrews Counties are caused by decrease in petroleum industry jobs since the mid-1980s. It is expected that population growth in these two counties in the next 30 years will be at a lower rate than the overall rates in New Mexico and Texas.

Lea County covers 11,378 km^2 (4393 mi^2), which is about three times the size of the State of Rhode Island. Andrews County covers 3,895 km^2 (1504 mi^2).

Major population centers near the proposed site include the following:

> Eunice, New Mexico, about 8 km (5 mi) west of the site;
> Hobbs, New Mexico, about 32 km (20 mi) north of the site;
> Jal, New Mexico, about 37 km (23 mi) south of the site;
> Lovington, New Mexico, about 64 km (39 mi) north-northwest of the site;
> Andrews, Texas, about 51 km (32 mi) east of the site;
> Seminole, Texas (in Gaines County) about 51 km (32 mi) east-northeast of the site; and
> Denver City, Texas (in Gaines County) 65 km (40 mi) north-northeast of the site.

Outside of these population centers, population density is very low. The nearest residences are located about 4.3 km (2.6 mi) west of the proposed facility site.

Within 8 km (5 mi) of the site, land is primarily open land used for cattle grazing. Oil and gas potentials are absent within this range, although operations are widespread beyond this area. Nearby industrial activities include a quarry and a "produced-water" reclamation company. Lea County operates a county landfill on the south side of New Mexico Highway 234, and about 1.6 km (1 mi) east of the proposed site, Waste Control Specialists (WCS) operates a hazardous chemical waste disposal facility and has licenses for the treatment and storage of low-level radioactive and mixed wastes. A natural gas processing plant is located about 6 km (4 mi) from the site.

3.3.1.1.2.2   Local Public Services

Fire fighting services are provided locally by Eunice Fire and Rescue, which is located 8 km (5 mi) from the proposed site. It is staffed by a full-time fire chief and 34 volunteer firefighters. Equipment includes three pumpers, one tanker, and three grass trucks. Eunice Fire and Rescue also has agreements for mutual assistance with all Lea County fire departments.

Police and law enforcement services are provided by the Eunice Police Department, which has five full-time officers. The Lea County Sheriff's Department also has a substation in Eunice. Agreements between Lea and Andrews Counties provide mutual support when needed. The New Mexico State Police can also provide support.

Educational institutions in Eunice include an elementary school, a middle school, a high school, and a private K-12 school. The nearest other schools are in Hobbs, New Mexico, 32 km (20 mi) north of Eunice. The nearest schools in Andrews County, Texas, are in Andrews, Texas, about 51 km (32 mi) from the proposed site.

There are two hospitals in Lea County – one is located in Hobbs, New Mexico, 32 km (20 mi) north of the proposed site, and the other in Lovington, New Mexico, 64 km (39 mi) north-northwest of the site. The hospital in Hobbs is a 250-bed facility capable of handling acute and stable chronic care patients. The hospital in Lovington is a full-service, 27-bed facility. The Eunice clinic is the nearest medical center to the proposed facility site. The nearest nursing home facilities are in Hobbs, New Mexico.

There are no recreation facilities near the site. The Eunice Golf Course is located approximately 15 km (9.2 mi) west of the site. A historical marker and picnic area are located about 3.2 km (2 mi) west of the proposed site at the intersection of New Mexico Highways 234 and 18.

3.3.1.1.2.3 Water Use

Southeast New Mexico has a semi-arid climate with an average annual precipitation of 33 to 38 cm (13 to 15 in.). The proposed site has no surface water and/or drainage features. Essentially all precipitation either infiltrates the soil or is evapotranspirated. There are no significant bodies of water or navigable waterways in the vicinity of the proposed site. There is also no agricultural activity in the site vicinity although there are various crops grown in Lea and Andrews Counties. Cattle grazing does occur at the proposed site and in the nearby vicinity. Dairy farming is important in Lea County, although none takes place near the site and/or in Andrews County.

Known sources of water near the site include: a man-made pond at the quarry, adjacent to the proposed site, stocked with fish for private use; Baker Spring, which is an intermittent surface-water feature, located about 1.6 km (1 mi) northeast of the proposed site; and several cattle watering holes, where groundwater is pumped by windmills and stored in above-ground tanks.

3.3.1.1.2.4 Nearby Highways

Information concerning public roads is provided in Section 1.3.2.4 of the SAR (LES, 2005a) and Section 3.2.1.2 of the ISA Summary (LES, 2005b). The New Mexico State Highway 234 passes along the southern boundary of the proposed facility. Vehicles transporting propane travel this highway at a relatively high frequency (Snooks, 2003).

The risk associated with the potential hazard of a highway propane explosion to the proposed facility was analyzed by the applicant (Snooks, 2003). The analysis used the largest volume of propane transported at one time. This volume was determined based on discussions with propane operators (Snooks, 2003). During an accident, a large truck with a bounding gross weight of 4536 kg (10,000 lb) was assumed to be totally crashed. Also, the structures of the proposed facility for consideration of an explosion impact were assumed to be designed to withstand 6.9 kPa (1 psi) overpressure, as suggested by Regulatory Guide 1.91 (NRC, 1978b). Based on this design-basis overpressure, a safety-significant structure must be at least 0.4 km (0.24 mi) (381 m (1,251 ft)) from the point of explosion, to avoid damage. This safe-separation distance is the approximate distance from New Mexico State Highway 234 to the proposed TSB, a safety-significant structure.

The applicant calculated the likelihood of a propane truck accident, on Highway 234, that could have an effect on the proposed facility, using information including annual truck accident rate and miles, annual number of shipments passing the proposed facility, conditional probability of occurrence of significant incidents from the accidents recorded, and exposure distance of a structure in miles. The U.S. truck occupancy safety data from 1990 through 2001 were obtained from the Bureau of Transportation Statistics (Snooks, 2003). However, only data from the most recent 5-year period (1997–2001) were used in the analysis. The applicant indicated the most recent data better represent the current and future statistics because of the improvements in transportation equipment and roadway conditions (LES, 2004a). The same 5-year data for cargo tank truck incidents were used to calculate the conditional probability of significant incidents. The number of propane shipments passing the proposed facility on New

Mexico State Highway 234 was obtained by interviewing the local propane operators. The exposure distance used in the analysis was twice the safe separation distance for conservatism. The probability of an incident based on the data discussed in this paragraph was determined to be 2.07×10^{-8} (Snooks, 2003). The applicant defined not credible with a likelihood of occurrence less than 10^{-6}, which is consistent with that suggested in NUREG–1520 (NRC, 2002a), and accepted by the staff. The applicant concluded that a propane truck explosion was a credible event. This probability, however, meets the definition of highly unlikely. Therefore, the potential consequence does not have to be determined because the event sequence is highly unlikely in accordance with 10 CFR 70.61(b).

Summary

Based on review of the highway propane-explosion-hazard risk assessment, staff concludesl: (i) the data used in the analysis are from acceptable sources; and (ii) the approach used in the analysis for likelihood determination is acceptable because it relied on conservative bounding assumptions and a design-basis over-pressure suggested by Regulatory Guide 1.91 (NRC, 1978b).

3.3.1.1.2.5   Nearby Gas Pipelines

Information concerning gas pipelines passing through or located near the proposed facility site is provided in Sections 1.3.2.4 of the SAR (LES, 2005a) and 3.2.2.4 of the ISA Summary (LES, 2005b).

Natural Gas Pipeline

The applicant identified an underground natural gas pipeline located along the south property line running parallel to New Mexico State Highway 234. This 40.6-cm (16-in.) natural gas pipeline is a low-pressure line (<345 kPa (<50 psi)) and is located approximately 545 m (1800 ft) from the proposed TSB, a safety-significant structure. The 40.6-cm (16-in.) natural gas pipeline is buried approximately 0.9 m (3 ft) below the surface (Thomson, 2004). The natural gas transported by the pipeline includes 72 percent methane, 11 percent ethane, 7 percent propane, and less than 1 percent hydrogen sulfide. The pipeline gas flow is between 5.7×10^3 and 1.4×10^4 m^3/day (2×10^5 and 5×10^5 ft^3/day). A parallel 35.6-cm (14-in.) gas pipeline also is identified, but is not in use.

A hazard risk assessment for the 40.6-cm (16-in.) natural gas pipeline was performed by the applicant (Thomson, 2004). The hazards associated with a natural gas pipeline explosion may include blast overpressure, missile generation, and thermal radiation. The assessment performed by the applicant considered all these hazards, and the likelihood of a gas pipeline explosion causing damage to safety-significant structures of the proposed facility was determined by summing the probabilities of these hazards.

The applicant (Thomson, 2004) pointed out that the explosion-generated missile hazard depends on several factors. Because of insufficient information on these factors, the applicant assumed every natural gas detonation would result in a missile-impact hazard. Therefore, the probability for a missile hazard was the same as the explosion probability. Staff finds this assumption conservative and acceptable.

In addressing the potential thermal effects, the applicant stated that the potential thermal effects might be bounded by a similar analysis involving a natural gas pipeline presented by the Tennessee Valley Authority (TVA) in its Preliminary Safety Analysis Report for Hartsville Nuclear Plants (Thomson, 2004). The natural gas pipeline analyzed by TVA was 55.9 cm (22 in.) in diameter, with a 3861-kPa (560-psi) operating pressure, whereas the natural gas pipeline in question for the proposed facility is 40.6 cm (16 in.) in diameter, with a 345-kPa (50-psi) operating pressure. The distance of the natural gas pipeline to the proposed facility is less than that to the Hartsville Nuclear Plants [545 versus 808 m (1800 versus 2650 ft)]. The applicant contended, considering the conservatism in the pipeline size and operating pressure, that TVA results for radiant heat flux would bound the results for the safety-significant structures of the proposed facility if a detailed analysis were performed. For a worst-case condition, the radiant heat incidence obtained by TVA was less than 9085 kJ/m^2 (800 Btu/ft^2). To cause spontaneous ignition of wood, a radiant heat incidence of 19,874 kJ/m^2 (1750 Btu/ft^2) would be required (Thomson, 2004, Attachment 9). Therefore, a much higher radiant heat flux exposure would be necessary to cause damage to concrete structures. As indicated in the ISA Summary (LES, 2005b), the safety-significant structures of the proposed facility will be concrete structures. Therefore, the applicant stated (Thomson, 2004) the explosion-induced thermal hazards on the proposed facility might be neglected. Staff agrees with the applicant's assessment that the radiant heat flux resulting from a potential natural gas flame near the safety-significant structures of the proposed facility would be bounded by the results calculated for the TVA facility. Consequently, the associated thermal effects on the proposed facility would be negligible.

In estimating the probability of pipeline explosion-induced blast overpressure hazard, the applicant considered three parameters: (1) gas pipeline rupture incidents per mile; (2) conditional probability for significant incidents; and (3) exposure distance in miles. Data from the Office of Pipeline Safety (OPS) web site (OPS, 2004) were used by the applicant to calculate gas pipeline rupture incidents per mile. These data included the annual mileage related to natural gas transmission operations (1984–2003) and the detailed accounts, including rupture length (mid-1984–present) and telephone records (1987–2001) of the reported accidents. The data for annual mileage presented in Thomson (Thomson, 2004, Table 1) were slightly lower than the data currently posted on the OPS web site. Staff does not expect this slight difference to affect significantly the gas pipeline rupture incidents per mile. The applicant selected 4 years of data (1998–2001) to calculate gas pipeline rupture incidents per mile. The applicant (LES, 2004a) pointed out that these 4 years of data are comparable to the number of gas pipeline rupture incidences for other years and, therefore, representative of the available rupture incidence data.

Among the rupture incidents in these 4-year data, incidents with a recorded rupture length of less than 3.1 cm (0.1 ft) were not included in the analysis. The applicant indicated these incidents would not be a hazard to the safety-significant structures, based on past experience (Thomson, 2004; LES, 2004a). Fifty incidents were identified to have had a rupture length greater than 3.1 cm (0.1 ft). The gas pipeline rupture incidents per mile, the ratio of these rupture incidents to the sum of the annual mileages, was approximately 5.7×10^{-5} ruptures per mile (Thomson, 2004). Staff determined that the applicant's justification for: (I) use of the 4-year data (1998–2001) for natural gas pipeline hazard analysis is acceptable; and (ii) the use of a limiting rupture length [3.1 cm (0.1 ft)] to screen out nonconsequential rupture incidents from further consideration are reasonable.

Not all rupture incidents would involve explosions. The rupture incidents without explosions would not generate blast overpressure, or missiles, or induce thermal radiation; thus, these incidents would not be a safety concern. Of the 50-rupture incidents used in the analysis, the applicant identified seven which were explosion related (Thomson, 2004). These explosion incidents represented a fraction of 0.14 (defined as explosion probability, to facilitate discussion) of the rupture incidents. Gas cloud explosions may be classified as deflagrations or detonations (NFPA, 1995). Detonations are rapid explosions, generating supersonic pressure waves and capable of producing blast over-pressures sufficient to cause damage. Deflagrations, on the other hand, are relatively slow explosions produced by rapid chemical reactions which generate only subsonic pressure waves. Typically, 28 percent of the explosions are estimated to be detonation related (NAESC, 1999). The product of the detonation rate and the explosion probability formed the conditional probability, and this conditional probability was 0.0392. The applicant used this conditional probability to modify the probability of gas pipeline rupture incidents per mile, to reflect the fact that only large explosions could produce blast overpressures sufficient to cause damage to the structure in question.

The exposure distance is the length of a natural gas pipeline segment, with the explosion resulting from a rupture at any point of this segment posing a safety concern to the structures. This exposure distance is a function of the safe-separation distance. To determine the safe-separation distance, two aspects need to be considered. One is the distance, D_1, from a gas-release location along a direct pathway to the proposed facility to the lower flammable limit of a gas plume, and D_3 to the upper flammable limit. Within these two limits, the flames from an ignited gas cloud could propagate in a self-sustaining manner (CCPS, 2002). Another aspect is the distance, D_2, estimated from the edge of the gas plume to a safety-significant structure designed to withstand an overpressure of 6.9 kPa (1 psi).

The applicant used the computer program ALOHA (USEPA, 1999) to estimate D_1 and D_3. D_1 and D_3 were estimated to be 1248 and 665 m (4095 and 2181 ft) (Thomson, 2004). In the calculation, the gas plume was assumed to contain solely methane. This assumption is acceptable because methane is the major constituent of natural gas. The mass of the released methane within the flammable range was then determined using D_1, D_3, and the wind speed in the ALOHA calculation. After the trinitrotoluene weight of the methane mass was obtained using the equation suggested by the National Fire Protection Association and Society of Fire Protection Engineers (NFPA, 1995), the equation in Regulatory Guide 1.91 (NRC, 1978b) was then used for D_2. In calculating D_2, two important parameter values (yield and theoretical net heat of combustion) were obtained from two fire protection handbooks (NFPA, 1995; NFPA, 1991). The staff finds the values used are appropriate and from reliable sources. D_2 was determined to be approximately 448 m (1471 ft) (Thomson, 2004). The minimum distance required for a safety-significant structure to avoid damage from a pipeline explosion (safe-separation distance) was $D_1 + D_2$ (i.e., 1697 m (5566 ft)). Because the closest location of the natural gas pipeline of concern was approximately 545 m (1800 ft) from a safety-significant structure of the proposed facility, hazards associated with a substantial section of the natural gas pipeline would pose a safety concern for that structure. This pipeline section was located within a circle centered at the edge of the structure with a radius of $D_1 + D_2$ (i.e., 1697 m (5566 ft)). The applicant estimated conservatively the length of the pipeline section (exposure distance) to be the diameter of the circle (i.e., approximately 3.45 km (2.1 mi)).

Several parameters were used in the ALOHA calculation (Thomson, 2004)

- Postulated release from a hole with a diameter equal to that of the pipeline. This assumption is bounding because the amount of gas flow through the pipeline at a given time is controlled by the pipeline diameter and gas pressure and this amount is the maximum that can be released at a given time should there be a natural gas pipeline rupture.

- Release duration—1 hour. This value was the maximum expected time required to shut off the gas supply and bleed the system. According to the pipeline operator, the natural gas pipeline in question is 22.5-24.1 km (14-15 mi) in length with three manual shutoff valves. Two of the valves are located at the end and one is located in the middle of the pipeline (Thomson, 2004, Attachment 5: LES, 2004d). To reach, and manually shut off one of the valves in responding to a natural gas pipeline rupture accident, the pipeline operator estimated that approximately one hour would be required. Once the gas supply is shut off and the pipeline is bled, no more natural gas would be available for release. Furthermore, the applicant indicated that, in the event of a rupture, a steady-state concentration would be reached in less than one hour (Thomson, 2004). Consequently, the use of one hour as the release duration in the ALOHA analyses is reasonable and acceptable to the staff.

- A stable atmosphere, with minimal dispersion. The wind speed of 1 m/s (3.3 ft/s) used in the analysis was in the range of stable wind speeds of Pasquill Class F for the proposed facility site (LES, 2004a; LES, 2005c). This stable wind class occurs 2.2 percent of the time for the site (LES, 2005c). The selection of wind speed is consistent with Regulatory Guide 1.78 (NRC, 2001).

With these, and all the parameters which are known, the applicant estimated the probability of hazards associated with a pipeline explosion-induced blast overpressure to be 4.7×10^{-6}/year (Thomson, 2004). As indicated earlier, the probability of a missile hazard was assumed to be equal to the probability of a pipeline explosion. Consequently, the final probability of the natural gas pipeline hazard affecting the structures was 9.4×10^{-6}/year. The applicant concluded the natural gas pipeline explosion event was a credible event. This probability, however, meets the definition of highly unlikely. Therefore, the potential consequences do not have to be determined because the event sequence is highly unlikely, in accordance with 10 CFR 70.61(b).

Carbon Dioxide Pipeline

An underground 25.4-cm (10-in.)-diameter, high-pressure carbon dioxide pipeline currently runs across the property. The applicant (2005a) stated that the normal operating pressure of the pipeline was 13,445 kPa (1950 psi), and the maximum operating pressure was 14,479 kPa (2100 psi) (Thomson, 2004; LES, 2004a). The ISA Summary (LES, 2005b) indicated this pipeline will be relocated along the western and southern boundary of Section 32 so the pipeline will be positioned at least 396.2 m (1300 ft) west of the facility restricted area and is approximately 945 m (3100 ft) [estimated from Figure 3.2-3 of the ISA Summary (LES, 2005b)] from the Separations Building Module, a safety-significant structure that houses two cascade halls. The applicant concluded that, at this distance from the proposed facility, the pipeline was not a safety concern. Staff agrees with the applicant's assessment that the carbon dioxide pipeline is not a safety concern to the Separations Building Module.

Figure 3.2-3 of the ISA Summary (LES, 2005b) shows that a portion of the carbon dioxide pipeline, at the southern boundary of Section 32, is close to the two natural gas pipelines located south of the proposed facility site. As noted earlier, the natural gas pipeline is approximately 545 m (1800 ft) south of the TSB. Even though carbon dioxide is not flammable, pipeline ruptures resulting from the high pressure might be credible. According to the pipeline operator, a clearance requirement of 6- to 9- m (20- to 30-ft) minimum separation is normally required between parallel running pipelines and pipelines crossing one another for safety concerns (LES, 2004d). Because of this clearance requirement, the applicant indicated rupture of the high pressure carbon dioxide pipeline will not affect the hazard probability of the natural gas pipeline (LES, 2004d). The staff concludes that the applicant's assessment is acceptable.

Onsite Natural Gas Pipeline

According to the applicant (Harper, 2003a), the proposed facility will include an on-site natural gas pipeline. This pipeline will be used to provide natural gas for heating the boiler in the Central Utility Building (CUB).

The potential hazard analyzed for this pipeline was an explosion within the CUB (Harper, 2003a). By examining the rupture incidents per mile (5.7×10^{-5} ruptures per mile) for the natural gas pipeline and the likelihood of a rupture leading to an explosion, the applicant determined that a scenario of a gas explosion in the CUB would be highly unlikely. It would be even less likely for a detonation to occur with sufficient energy to damage the CUB and the nearby TSB or Separations Building Modules.

At the present time, the pipeline to be used for on-site natural gas supply is a 10-cm (4-in.) diameter, low-pressure [69-kPa (10-psi)] line (LES, 2004a). This pipeline will have an excess flow valve located at the entrance of the proposed facility site, and this excess flow valve will automatically shut the gas flow off in case of a pipeline leak. Because the diameter of the on-site natural gas pipeline is small, the operating pressure is low, and the excess flow valve is used, a potential explosion of the on-site natural gas pipeline that is of sufficient magnitude to damage the safety-significant structures of the proposed facility is considered not credible (LES, 2004a).

Staff reviewed the hazard assessments conducted for the natural gas pipelines and concludes the approach used to assess the natural gas pipeline hazards is acceptable, and the data used for the assessment were from reliable sources. Staff also concludes that the assessments performed for the carbon dioxide and the on-site natural gas pipeline hazards are based on data from reliable sources and based on an acceptable approach.

3.3.1.1.2.6   Nearby Air Transportation

Information concerning potential aircraft crash hazards is provided in Sections 1.3.2.4 of the SAR (LES, 2005a), and Section 3.2.1.2.4 of the ISA Summary (LES, 2005b). The analysis concerning the potential hazard of aircraft crash to the proposed facility site was documented in the aircraft hazard risk determination report (Yeung, 2003). The information, including the number of operations and holding patterns of six local airports, was analyzed by Yeung (2003). These airports include:

- Lea County Regional Airport—40 km (25 statute miles) northwest of the proposed facility site;

- Eunice Airport—24 km (15 statute miles) west of the proposed facility site;

- Lea County/Jal Airport—40 km (25 statute miles) south-southwest of the proposed facility site;

- Andrews County Airport—48 km (30 statute miles) east of the proposed facility;

- Gaines County Airport—48 km (30 statute miles) northeast of the proposed facility; and

- Seminole Spraying Services (private)—48 km (30 statute miles) northeast of the proposed facility.

The information about the number of operations and holding patterns of each airport was obtained from the Federal Aviation Administration (FAA) (Yeung, 2003). Military flights are operated from the Lea County Regional Airport. The number of military operations is included in the number of operations for the Lea County Regional Airport. In analyzing potential aircraft crash hazards, the applicant used the proximity criteria provided in Section 3.5.1.6 of NUREG-0800 (NRC, 1981) to screen out the hazards associated with some flight activities from further consideration. According to the proximity criteria, aircraft crash hazards do not have to be considered if (I) the distance, D, of the facility to the airport is between 5 and 10 statute miles and the projected number of annual operations is less than 500 D^2, or D is greater than 10 statute miles and the projected number of operations is less than 1000 D^2, (ii) the facility is at least 5 statute miles from the edge of military training routes, including low-level training routes, except those associated with a usage greater than 1000 flights per year, or where activities (such as practice bombing) may create an unusual stress situation, and (iii) the facility is at least two statute miles beyond the nearest edge of a federal airway, holding pattern, or approach pattern. Based on the first proximity criterion provided in Section 3.5.1.6 of NUREG-0800 (NRC, 1981), the applicant determined the presence of these airports not to be a safety concern for the proposed facility.

Based on information from the FAA (Yeung, 2003), the applicant concluded the holding patterns for four of the six airports were, in general, more than the 3.2-km (2-statute miles) proximity criterion (third criterion) provided in Section 3.5.1.6 of NUREG-0800 (NRC, 1981) for airway distance from the site of interest. The applicant pointed out no specific holding patterns existed for Eunice or Lea County/Jal Airport (LES, 2004a; LES, 2005a). For the Eunice Airport, the annual operations are small, approximately 480 flights per year. This number is substantially smaller than the threshold limit (225,000 annual operations) provided in NUREG-0800 (NRC, 1981). The applicant determined the operation of the Eunice Airport does not pose any aircraft crash hazard to the proposed facility because of the relatively small number of annual operations. For the Lea County/Jal Airport, the applicant indicated the airport is more than 32 km (20 mi) away from the proposed facility, and the landing procedure usually will not be initiated until an aircraft is within 32 km (20 mi) of the airport (LES, 2004a; LES, 2005a). Therefore, even if an aircraft is placed in a holding pattern, it will not bring the aircraft near the proposed facility. Consequently, there is no aircraft crash hazard from holding patterns to the proposed facility from the operations of the Lea County/Jal Airport. The staff concurs with the conclusion made by the applicant concerning the potential hazards associated with the random aircraft holding patterns at the Eunice and Lea County/Jal Airports.

The applicant identified a low-level federal airway passing within 9 km (~6 statute miles) northeast of the proposed facility. Using the method provided in Section 3.5.1.6 of NUREG–0800 (NRC, 1981), the probability of an aircraft on the airway crashing onto the proposed facility was estimated to be 3.4×10^{-7} (Yeung, 2003). This probability makes the aircraft crash an incredible event or hazard to be considered in either design or integrated safety analysis.

Three military training routes were identified in the region. The closest route to the proposed facility was approximately 26 km (16 statute miles) southwest of the site. Based on the second proximity criterion provided in Section 3.5.1.6 of NUREG–0800 (NRC, 1981), this military training route posed no safety threat to the proposed facility.

Based on the review of aircraft-crash risk assessment, staff concludes that: (i) the aircraft transportation information used in the analysis was obtained from a reliable source; and (ii) the risk of public exposure was evaluated using the acceptable proximity criteria and method provided in Section 3.5.1.6 of NUREG–0800 (NRC, 1981).

3.3.1.1.2.7   Demographics and Land Use Summary

The staff reviewed the site demographic information presented by the applicant and concludes that the applicant has adequately described and summarized general site demographical information related to local population, identification of population centers, schools, commercial facilities, land use, and water use. Population information is provided based on the latest census information.

3.3.1.1.3   Meteorology

3.3.1.1.3.1   Tornado Hazard

Information about the tornadoes and design-basis tornado at the proposed facility is provided in Sections 1.3.3.3 of the SAR (LES, 2005a), and Section 3.2.3.4.1 of the ISA Summary (LES, 2005b).

There is an average of nine tornadoes a year in New Mexico, and the occurrence of tornadoes in the vicinity of the proposed facility is rare. Tornadoes are classified using the Fujita Tornado Damage Scale (F-scale) with classifications ranging from F0-F5 (NOAA, 2005). Eighty-seven tornadoes of low magnitude (F0 to F2) were reported in Lea County, New Mexico, between January 1, 1950, and December 31, 2004. Only one additional tornado was reported as F3 on May 17, 1954. Two tornadoes, one in 1998 and the second in 1999, had a magnitude of F0 and were located near Eunice. All the reported tornadoes were associated with very light damage (NCDC, 2005).

The Modified IDR tornado hazard assessment model (McDonald, 1995) is used to quantify the tornado risks at the facility. The tornado hazard assessment (1) defines the local region that surrounds the site, (2) determines the occurrence rate and associated confidence limits, (3) determines the number of tornados per F-scale category, (4) estimates the damage path area of each F-scale tornado and calculates damage areas with confidence limits, and (5) calculates the tornado hazard probabilities for each F-scale wind speed category. Due to insufficient damage path area data for the one-degree square area (4034 square miles) surrounding the proposed facility, a six-degree area was defined. A linear regression analysis was performed to obtain a continuous area-intensity function. The total number of recorded tornados in the local region for the years of 1954 through 1999 were broken down into the respective F-scale and the mean damage path was determined from the historical tornado records. The results of the tornado hazard assessment identified an expected annual probability of an F-3 tornado with 260 km/h (162 mph) wind speed occurring within the one-degree square area where the plant site is located as 1.21 E-05. When both straight winds and tornado hazards are considered, the expected wind speed with an annual probability of occurrence of 1.0 E-05 is 302 km/h (188 mph).

The design parameters for the design-basis tornado are listed in ISA Summary Section 3.3.2.2.2.1 (LES 2005b). All safety-significant structures will be designed to withstand the design-basis tornado. The development of these tornado design-basis-related parameters was discussed in SAR Section 1.3.3.3 (LES 2005a), and ISA Summary Section 3.2.3.4 (LES, 2005b). The staff review and acceptance of the development of the design-basis tornado are discussed in Section 1.3.3.3.1 of this SER. The values of design-basis tornado-related parameters are listed as follows:

Design-Basis Wind Speed:	302 km/h (188 mph)
Radius of Damaging Winds:	130 m (425 ft)
Atmospheric Pressure Change:	−3.83 kg/m^2 (−80 lbs/ft^2)
Rate of Atmospheric Pressure Change:	−1.44 kg/m^2/s (−30 lbs/ft^2/s)

The tornado design-basis is characterized as a 100,000-year tornado, which is the equivalent of a "F-3" tornado on the Fujita Tornado Scale.

To estimate the tornado hazard, the assessment model proposed by McDonald and Lu (McDonald, 1995) was used. The local region selected for the assessment was the 1-degree square area between latitudes 31.94° and 32.94° and longitudes 102.58° and 103.58°. According to the applicant (Harper, 2003b), tornadoes in this region were determined to most likely affect the site.

The historical tornado records from 1954 to 1999 for the selected region were used for the assessment. Linear regression analysis was performed by the applicant to obtain the occurrence-intensity relationship. The applicant defined the design-basis tornado for the proposed facility as the tornado with a return period of 100,000 years. Based on the occurrence-intensity relationship, the wind speed for this design-basis tornado was determined to be 302 km/h (188 mph). The design parameters, including atmospheric pressure change, rate of atmospheric pressure change, and radius of damaging winds, were calculated for the design-basis tornado using the tornado wind field model proposed by McDonald (NRC, 1983). Based on this model, the atmospheric pressure change for the design-basis tornado was 3.83 kPa (80 psf), the rate of atmospheric pressure change was 1.44 kPa/s (30 psf/s), and the radius of damaging wind was 130 m (425 ft).

The applicant also established the damage area-intensity relationship in terms of the F-Scale for the proposed facility site. Because the data for the damage path areas associated with the tornadoes for the selected 1-degree square region were not sufficient to establish a damage area-intensity relationship, the applicant used the damage path area data related to tornadoes in a 6-degree square region (bounded by latitude 30°–35° and longitude 100°–105°), containing the proposed facility site, to determine the damage area-intensity relationship.

The staff concludes that the design-basis tornado is acceptable because it was estimated based on an acceptable method and was based on a 100,000-year return period, sufficient to make the frequency of more damaging tornados highly unlikely.

3.3.1.1.3.2   High Winds and Hurricanes

Information about high winds at the proposed facility is provided in Sections 1.3.3.1 of the SAR (LES, 2005a); Section 3.2.3.4.2 of the ISA Summary (LES, 2005b), and Section 3.6.1.4 of the Environmental Report (ER) (LES, 2005c).

According to the SAR (LES, 2005a), no meteorological data were available for the proposed facility site. Although the measured wind data at Midland–Odessa, Texas, and Roswell, New Mexico, were discussed in the SAR (LES, 2005a) and ER (LES, 2005c), the Midland–Odessa annual extreme wind data were used exclusively to estimate the high-wind hazard at the proposed facility site (LES, 2005b; Harper, 2003b). The annual extreme data used range from 1973 through 1999. The wind speeds were 3-second gust speeds measured at 10 m (32.8 ft) above ground. The Midland–Odessa weather station is located at the regional airport approximately 103 km (64 mi) east-southeast of the proposed site, whereas the Roswell station is approximately 161 km (100 mi) northwest of the proposed site. The climate data for both locations were collected by the National Oceanic and Atmospheric Administration (NOAA) (LES, 2005c).

The largest wind speed for the annual extreme straight-line winds from 1973 through 1999 at Midland–Odessa was 140 km/h (87 mph) and the smallest annual extreme straight-line wind speed was 84 km/h (52 mph) (Harper, 2003b). The mean and standard deviation wind speeds were 111.5 and 16.6 km/h (69.3 and 10.3 mph).

The high-wind hazard assessment was performed by fitting the annual extreme wind data using the Fisher–Tippett Type I distribution model. The applicant chose the speed of a wind with an annual probability of 1.0×10^{-5} for the design-basis straight-line wind speed for the proposed facility. This design-basis straight-line wind speed was 252 km/h (157 mph) (LES, 2005b). All safety significant structures will be designed for this wind speed.

Because the proposed facility is not located near the coastal area (805 km (500 mi) from the coast), hurricanes affecting the coastal area will have no effect on the performance of the proposed facility. Consequently, consideration of hurricane hazards on the design of the proposed facility is not needed.

Based on the review of the information concerning high winds, the staff concludes that high-wind hazards and the associated design-basis straight-line winds have been addressed acceptably because the data used for assessment were from a recognized source and the method used for assessing high-wind hazards is an accepted method.

3.3.1.1.3.3   Temperature Extremes

Information about the temperature at the proposed facility site is provided in Section 3.6.1.2 of the ER (LES, 2005c).

The regional temperatures in Hobbs, New Mexico [32 km (20 mi) north of the proposed facility site]; Midland–Odessa, Texas; and Roswell, New Mexico, are discussed in the ER. The discussions are based on 30-year records (from 1971 through 2000). As indicated previously, NOAA collected the climate data for Midland–Odessa and Roswell. However, the Western Regional Climate Center collected the climate data for Hobbs (LES, 2005c).

The highest recorded monthly mean maximum temperature was 38.9 °C (102.1 °F), and the lowest recorded monthly mean minimum temperature was -5.1 °C (22.8 °F) for Hobbs, New Mexico. No such data were presented in the ER for Midland–Odessa or Roswell. The highest daily maximum and lowest daily minimum temperatures were 46.7 °C (116.0 °F) and -23.9 °C (-11.0 °F) for Midland–Odessa, and 45.6 °C (114.0 °F) and -22.8 °C (-9.0 °F) for Roswell. No such data were presented for Hobbs. As indicated, the highest daily maximum and the lowest daily minimum temperatures for Midland–Odessa and Roswell were similar.

The staff reviewed the temperature information and concludes that the information is acceptable because recognized data sources were used and the temperature extremes are properly determined.

3.3.1.1.3.4   Extreme Precipitation

Section 1.3.3.2 of the SAR (LES, 2005a); and Sections 3.2.3.2 and 3.2.3.4.4 of the ISA Summary (LES, 2005b) discuss the rainfall precipitation at the proposed facility site. The precipitation data for Hobbs, Midland–Odessa, and Roswell were listed in Tables 3.2-14 through 3.2-16 of the ISA Summary (LES, 2005b). These data were collected from the Western Regional Climate Center and NOAA and are based on data from 1971 through 2000 (LES, 2005c). The maximum monthly totals were 35.13 cm (13.83 in.) for Hobbs; 24.6 cm (9.7 in.) for Midland–Odessa; and 17.5 cm (6.88 in.) for Roswell. The minimum monthly totals were zero for all locations. The highest 24-hour precipitation was 15.2 cm (5.99 in.) for Midland–Odessa and 12.5 cm (4.91 in.) for Roswell.

According to the SAR (LES, 2005a), the local intense probable maximum precipitation was estimated from NOAA data (NOAA, 1982). The local intense probable maximum precipitation was approximately 43.9 cm (17.3 in.) in 1 hour, over a 2.6-km² (1-mi²) area.

The staff reviewed the information concerning regional precipitation and local intense probable maximum precipitation presented in the SAR (LES, 2005a) and the ISA Summary (LES, 2005b) and concludes that the information is acceptable because recognized data sources, such as NOAA, were used.

3.3.1.1.3.5   Snow

Section 1.3.3.2 of the SAR (LES, 2005a), and Section 3.2.3.3 of the ISA Summary (LES, 2005b) discuss the regional snowfall. NOAA collected the snowfall data. The maximum monthly snowfall/ice pellets were 24.9 cm (9.8 in.) for Midland–Odessa and 53.3 cm (21.0 in.)

for Roswell. The maximum snowfall/ice pellets during a 24-hour period were 12.47 cm (4.91 in.) for Midland–Odessa and 41.91 cm (16.5 in.) for Roswell. No snowfall information was available for Hobbs, New Mexico.

The staff reviewed the information concerning snow precipitation presented in the SAR (LES, 2005a) and the ISA Summary (LES, 2005b) and concludes that the information is acceptable because recognized data sources, such as NOAA, were used.

3.3.1.1.3.6   Lightning and Thunderstorms

Section 1.3.3.3 of the SAR (LES, 2005a), and Section 3.2.3.4.5 of the ISA Summary (LES, 2005b) describe the potential of thunderstorms and lightning strikes at the proposed facility site. The applicant indicated thunderstorms occur every month and are most common in spring and summer at the proposed facility site.

The applicant estimated the lightning strike frequency at the proposed facility site to be 1.36 flashes per year. The applicant stated in the ISA Summary (LES, 2005b) that the proposed facility will be designed for lightning protection.

The staff reviewed the information about lightning and concludes that the lightning strike frequency determined for the site is acceptable and appropriate. Staff further concludes that the design approach proposed by the applicant to protect the proposed facility from lightning effects is acceptable.

3.3.1.1.3.7   Sandstorms

Section 1.3.3.3 of the SAR (LES, 2005a) describes the potential of sandstorms at the proposed facility site. Blowing sand and dust may occur occasionally. Large dust storms with the potential of covering a large region are rare (DOE, 2003). Staff reviewed the information about sandstorms presented in the SAR (LES, 2005a) and finds the information sufficient and acceptable.

Staff concludes that, based on the individual parameter assessments discussed above, the applicant's material regarding meteorology presented in the SAR and ISA Summary is sufficient and acceptable to use in determining appropriate mechanical and thermal loads for the safe design of principal structures, systems, and equipment.

3.3.1.1.4   Hydrology

Site surface water and groundwater hydrology is discussed in Sections 1.3.4 of the SAR (LES, 2005a) and 3.2.4 of the ISA Summary (LES, 2005b). The applicant obtained hydrological data principally from previous investigations conducted by WCS, which is located 1.6 km (1 mi) east of the proposed site. WCS operates a hazardous chemical treatment and disposal facility. The applicant performed a limited number of geotechnical studies that demonstrate that the WCS data are applicable to the proposed site.

The proposed site contains no surface water and/or surface water drainage features, with essentially all precipitation subject to either infiltration or evapotranspiration.

The applicant performed subsurface studies of the alluvial material that overlies the Chinle red bed clays. These alluvial deposits are 9 to 15 m (30 to 60 ft) thick. The Chinle formation consists of a low-permeability clay unit having a thickness of 323 to 333 m (1060 to 1092 ft). No perched water systems in the alluvial deposits were found, although one well produced water samples, because of a limited groundwater occurrence.

The low permeability Chinle formation essentially isolates the deep and shallow groundwater systems. Within the Chinle formation are three distinct groundwater systems, with no interconnections. The first is a siltstone or silty sandstone unit with some saturation at 65 to 68 m (214 to 222 ft) below the surface. This unit is a low-permeability formation that does not yield groundwater easily. The second unit is a saturated siltstone layer approximately 30.5 m (100 ft) thick, at an elevation of 183 m (600 ft) below the surface. The third unit is the Santa Rosa formation at 340 m (1115 ft) below the surface. The Santa Rosa unit is the first occurrence of a well-defined aquifer system. However, this system is considered non-potable because of high concentrations of dissolved solids.

At the quarry site, north of the proposed site, there are shallow groundwater occurrences. These shallow perched systems, however, are intermittent and limited and caused by a layer caliche or caprock at the surface that in places is fractured and can lead to rapid infiltration of precipitation forming the perched water system. Caprock, however, is not present at the proposed site, and, therefore, it is not expected that significant perched water systems would be produced.

Baker Spring is located about 1.6 km (1 mi) northeast of the proposed site. However, this spring is intermittent and flows only after precipitation events.

Several localized shallow perched groundwater systems exist to the east of the proposed site and are used to supply water pumped by windmills to tanks for grazing livestock. These perched systems are located above the Chinle clays, but the volume of water produced is limited.

Because of the lack of sufficient surface and groundwater supplies, the applicant will not make withdrawals of groundwater at the site. Instead, the applicant is proposing to obtain water for plant use from Eunice and Hobbs municipal supplies. These water supplies are obtained from well fields near Hobbs, New Mexico. The applicant is also not proposing to inject water into groundwater systems at the site.

Since there are no surface water bodies in the immediate vicinity of the site, flooding is not a design objective. The only potential flooding at the plant would occur from intense local precipitation events. Flood protection is provided by establishing building floor levels above the calculated depth of ponded water caused by intense precipitation events (see Sections 3.3.2.1.3.4 and 3.3.2.1.3.5 of this SER).

The staff reviewed the applicant's hydrological data and finds that it provides sufficient information to assess site flooding hazards and ground- and surface water impacts.

3.3.1.1.5   Geology

3.3.1.1.5.1   Seismic Hazard - General

Seismic hazards are discussed in Section 1.3.5 of the SAR (LES, 2005a); and Sections 3.2.5 and 3.2.6 of the ISA Summary (LES, 2005b).

The following areas concerning the seismic hazard applicable to the safety analysis and design of the proposed facility were reviewed:

- Seismic source characterization;
- Ground motion attenuation;
- Seismic hazard calculation;
- Development of site-specific spectra; and
- Surface faulting.

Seismic Source Characterization

Geological and Tectonic Settings

Section 3.2.5 of the ISA Summary (LES, 2005b) provides a description of the regional and local geological and tectonic settings. The proposed facility site is located within the Central Basin Platform area. The Central Basin Platform Area is situated between the Midland and Delaware Basins, all of which are part of the Permian Basin, a 250-million-year-old structure. The Permian Basin is a downward flexure of a large thickness of originally flat-lying bedded, sedimentary rock. The base of the Permian Basin sediments extends to approximately 1525 m (5000 ft) beneath the proposed facility site. The top of the Permian section is approximately 434 m (1425 ft) below ground surface. These sediments are overlain by sedimentary strata of the Triassic Age Dockum Group. The upper formation of the Dockum Group is the Chinle Formation, locally overlain by either the Tertiary Ogallala, Gatuña, or Antlers Formations, or Quaternary alluvium. At the proposed facility site, geotechnical borings identified up to 0.6 m (2 ft) of loose eolian sand underlain by dense to very dense, fine- to medium-grained sand and silty sand of the Gatuña Formation. The sands of the Gatuña Formation are locally cemented with caliche. Beneath the Gatuña Formation, the Chinle claystone, a hard and highly plastic clay, was encountered in geotechnical borings at depths from 10.7 to 12.2 m (35 to 40 ft).

As noted in the ISA Summary (LES, 2005b), the Southeast New Mexico–West Texas area is presently structurally stable. The Laramide Orogeny (late Cretaceous to Early Tertiary time) uplifted the region to its present elevation, and there has been no substantial tectonic activity since this early Tertiary deformation. The Permian Basin has subsided slightly since the Laramide Orogeny. However, this subsidence is believed to be a result of dissolution of the Permian evaporite layers by groundwater or possibly compaction from oil and gas extraction. As stated in the ISA Summary (LES, 2005b), no active faults have been identified at the site. Faulting consists of geologically older subsurface faults in the Permian Basin subregion related to the development of the Permian Basin and the Laramide Orogeny. The nearest evidence of Quaternary faulting is 161 km (100 mi) west of the site, in the Basin and Range tectonic province.

Historical Seismicity

Section 3.2.6.1 of the ISA Summary (LES, 2005b) summarizes the historical seismicity at the proposed facility site. As stated in the ISA Summary (LES, 2005b), the assessment of historical seismicity included earthquakes in the region of interest known from felt or damage records and from more recent instrumental records (since the early 1960s). The largest earthquake known to occur within 322 km (200 mi) of the site was the August 16, 1931, earthquake near Valentine, Texas. This earthquake had an estimated M_L (Local Magnitude) of 6.0 to 6.4 and produced a maximum epicentral intensity of VIII on the Modified Mercalli intensity scale. This earthquake occurred approximately 237 km (147 mi) from the proposed site location. Within 80 km (50 mi) of the site, the largest historical earthquake was a M_L 5.0 event in 1992, approximately 16 km (10 mi) southwest of the site. Other significant events between 322 km (200 mi) and 80 km (50 mi) of the proposed facility site ranged in M_L from 4.0 to 5.7.

Earthquakes in the region of the proposed facility site include isolated and small clusters of low-to-moderate-magnitude events toward the Rio Grande Valley of New Mexico and in Texas, southeast of the proposed site. According to the ISA Summary (LES, 2005b), no earthquakes in the site region are known to be correlated to specific faults. An earthquake catalog based on the historic seismicity in the region [322-km (200-mi) radius] was presented in the ISA Summary (LES, 2005b). This catalog was composed of data from: the Advanced National Seismic System (NCEDC, 2004); University of Texas Institute for Geophysics (UTIG, 2002); New Mexico Tech Historical Catalog (NMIMT, 2003); and New Mexico Tech Regional catalogs. The catalog identified a substantial cluster of seismic activity that has occurred on and near the Central Basin Platform since the mid-1960s. It was suggested by DOE (DOE, 2003) and noted in the ISA Summary (LES, 2005b) that Central Basin Platform earthquakes are not tectonic in origin but instead are related to water injection and withdrawal resulting from secondary recovery operations in oil fields in the Central Basin Platform area. The ISA Summary (LES, 2005b) noted, however, the January 2, 1992, event was attributed to a tectonic origin because of its determined focal depth of approximately 12 km (7 mi). This event was likely associated with pre-existing zones of weakness within the crust that formed in the distant geologic past. These zones of weakness are characterized by deeply buried and poorly characterized faults, some of which accomplish a periodic release of strain that builds up continually in the North American continental plate. At the proposed facility site, postulated earthquakes that could impact safe operation of the proposed facility are associated with zones of crustal weakness in the Central Basin Platform and the Basin and Range tectonic province.

The staff concludes that information concerning seismic source characterization presented in the ISA Summary (LES, 2005b) is acceptable. The information provides a complete summary of seismicity and potential fault and tectonic sources.

Ground Motion Attenuation

Details of ground motion attenuation functions used to compute the hazard are described in Section 3.2.6.4.1 of the ISA Summary (LES, 2005b). Several attenuation models were used in the ISA Summary. The Nuttli attenuation model developed by the U.S. Department of the Army, Waterways Experiment Station (USDA, 1973) was primarily selected because it was used in the DOE (DOE, 2003) seismic hazard assessment. The Toro, et al. (Toro, 1997) attenuation model also was used in the hazard calculations for comparison.

The attenuation models used in the ISA Summary were applicable to locations within the Central U.S.. The proposed facility site is located at 103° west longitude, slightly east of the 105° west longitude cutoff for Central and Eastern U.S. sites, as specified in Regulatory Guide 1.165 (NRC, 1997). In addition, Frankel, et al. (Frankel, 1996) specified attenuation zones for the U.S. in its hazard mapping project. The U.S. Geological Survey (USGS) boundary separating the Western U.S. and the Central and Eastern U.S. attenuation zones also is located at approximately 105° west longitude and slightly to the west of the proposed facility site. The proposed facility site is thus situated within the area in which both the Central and Eastern U.S. attenuation models are applicable, and the staff concludes that applicant's evaluation of ground motion attenuation is acceptable.

Seismic Hazard Calculation

A probabilistic seismic hazard analysis performed for the proposed facility site is discussed in Section 3.2.6.4.2 of the ISA Summary (LES, 2005b). The method used to calculate the probabilistic seismic hazard was that of McGuire (McGuire, 1976). The probabilistic seismic hazard analysis incorporated seismic source zones from both local and distant seismic sources. Several alternative distant source zones were used in the hazard calculations. These distant source zones included the Rio Grande rift and Basin and Range area sources and were taken directly from DOE (DOE, 2003). Similarly, a suite of alternative local area source zones was used to account for local seismicity, including the Central Basin Platform area source zone.

The total seismic hazard at the proposed facility site was the sum of ground motion effects from all distant and local seismic sources. Twelve seismic hazard curves were developed for a combination of various seismic source zones, attenuation models mentioned previously, b-values, and upper bound magnitudes. As described in the ISA Summary (LES, 2005b), the resultant seismic hazard curve was developed through a weighted average of the individual curves. A total weight of 0.75 was given to the first seismicity model, which includes the Rio Grande Rift and Basin and Range seismic source zones and the Central Basin Platform, embedded within a seismic source zone defined by a 161-km (100-mi) radius from the site. A combined weight of 0.25 was assigned to a second seismicity model that incorporated a seismic source zone defined by a 322-km (200-mi) radius from the site and a third seismicity model that incorporated a 161-km (100-mi) radius from the site. These two models did not incorporate the Central Basin Platform and Rio Grande Rift and Basin and Range seismic source zones and, therefore, were not as strongly weighted as the first seismicity model. The resulting 250- and 475-year return period peak horizontal ground accelerations were estimated at 0.024 and 0.036g. The respective 10,000- and 100,000-year return period peak horizontal ground accelerations were estimated at 0.15 and 0.31g.

Comparison of the hazard results for the proposed facility site with USGS National Seismic Hazard Mapping Project (USGS, 2004) suggests the 10,000-year return period hazard results may underestimate ground motions at the site. For the 10,000-year return period, the USGS estimates a peak ground acceleration of 0.40g, with the 0.2–5 second spectral acceleration at 0.97g. These values are 2.6 to 2.8 times higher than the respective values calculated by the applicant. The applicant provided an evaluation of the USGS results (LES, 2004b). The applicant concluded that USGS used overly conservative maximum magnitude and activity rate information in the background source zone.

The staff agrees with the applicant's assessment that the USGS hazard calculation used larger maximum magnitudes and seismicity rates and is therefore more conservative. Additionally, the USGS study was also not site specific because the USGS hazard calculation used a site spacing of 0.1 degrees latitude and longitude for the Western U.S. and 0.2 degrees for the Central and Eastern U.S., which represents a much larger area. The applicant's calculation is, on the other hand, site specific and more representative of the site.

In addition, the seismic hazard calculated for facility site is similar to that calculated for the nearby Waste Isolation Pilot Plant (DOE, 2003). The calculated 10,000-year return period peak ground acceleration at the Waste Isolation Pilot Plant is slightly less than 0.15 g. Based on all the information available, the staff concludes that the seismic hazard described in the ISA Summary (LES, 2005b) is acceptable because it is based on a method that follows current industry practice and includes available data.

Development of Site-Specific Spectra

Information on the rock conditions hazard was used to obtain the response spectra at the proposed facility site. The attenuation model (USDA, 1973) was considered to predict ground motions at "firm rock" conditions, which are attributed to the Triassic Age claystones (Dockum Ground including the Chinle Formation) underlying the proposed facility site. As determined by 5 geotechnical borings, above the Triassic claystones is a thickness of 10.7 to 12.2 m (35 to 40 ft) of sands of the Gatuña Formation and loose eolian sands that belong to a soil classification of C (Dobry, 2000). As outlined in the ISA Summary (LES, 2005b), the firm rock uniform hazard response spectra were transformed to soil conditions by multiplying the firm rock uniform hazard response spectra by the appropriate soil-amplification factors specified by Dobry, et al (Dobry, 2000). Vertical component uniform hazard response spectra were determined to be a function of frequency and were determined using the formulation outlined in NRC Regulatory Guide 1.60 (NRC, 1973).

The design-basis earthquake for the proposed facility site was selected as the 10,000-year (1.0 $\times$ 10^{-4} mean annual probability) earthquake. As indicated in the ISA Summary (LES, 2005b), the applicant proposed the method outlined in DOE–STD–1020 (DOE, 2002) or ASCE/Structural Engineering Institute (SEI) 43-05, "Seismic Design Criteria for Structures, Systems, and Components in Nuclear Facilities" (ASCE/SEI, 2005) to demonstrate compliance to a target performance goal of 1.0×10^{-5} annual probability by designing to a seismic hazard of 1.0×10^{-4} annual probability [for further discussion, see Section 3.3.3(C)(v) of this Safety Evaluation Report (SER)]. The design response spectra for the proposed facility were based on the 10,000-year uniform hazard response spectra described in the previous paragraph. The horizontal and vertical design response spectra have standard response spectral shapes (at 5 percent damping) based on the Newmark and Hall amplification factors (NRC, 1978a). The spectrum is anchored at the 10^{-4} per year peak horizontal ground acceleration of 0.151g determined from the weighted final seismic hazard curve. Amplification factors were then applied using Dobry, et al. (Dobry, 2000) soil class C definition to account for the local soil column.

The staff considers the method and results used to develop site-specific spectra acceptable because they follow current industry practice and are consistent with methods used at other nuclear facilities (e.g., DOE, 2002). Thus, staff conclude that the site-specific hazard and response spectra are technically sound.

Surface Faulting

There is no geologic, geophysical, or seismological evidence of active surface faulting in the vicinity of the proposed facility site. As stated in the ISA Summary (LES, 2005b), the nearest recent faulting is located more than 161 km (100 mi) west of the site. Therefore, surface faulting was not considered a credible disruptive event for the proposed facility.

Recently, a fault was discovered at the nearby Waste Control Specialists (WCS) site. However, subsequent fault investigations revealed that the faulting is inactive because no faults exist in formations younger than Triassic age (205 to 240 million years old) (LES, 2005a).

Geotechnical investigation indicated the soil beneath the proposed facility site is a layer of loose eolian sand underlain by the Gatuña Formation (dense to very dense sand and silty sand). Below the Gatuña Formation is the Chinle claystone, a very hard, highly plastic clay. The Chinle claystone was encountered at depths approximately 10.7 to 12.2 m (35 to 40 ft). For the top 7.6 m (25 ft) of soils, the blow-count values ranged from 20 to 76. Beneath the 7.6-m (25-ft) horizon, typical blow-count values were more than 60, with even larger blow-count values for the Chinle claystone.

Based on the staff's review of the applicant's submittals on seismic hazard potential discussed in the above sections, staff concludes that the potential hazards have been acceptably addressed by the applicant.

3.3.1.1.5.2   Slope Stability

Section 1.3.1.2 of the SAR (LES, 2005a), and Section 3.2.1.1 of the ISA Summary (LES, 2005b) describe the topography at the proposed facility site. The SAR (LES, 2005a) and ISA Summary (LES, 2005b) indicated the site topography is relatively flat, with a gradual elevation increase from southwest to northeast. The staff site visit on May 27–28, 2004 (NRC, 2004b), confirmed the area at the proposed facility is relatively flat. Consequently, slope stability is not a safety concern for this proposed facility.

3.3.1.1.5.3   Liquefaction

Liquefaction potential of soils beneath the proposed facility is discussed in Section 3.2.7.1 of the ISA Summary (LES, 2005b). According to the ISA Summary (LES, 2005b), except a top layer of loose sand (up to 0.6 m (2 ft)), the soils at the proposed site are dense to very dense and the groundwater level is at least 30 m (98 ft) below ground surface. Consequently, the applicant concluded the potential for liquefaction was remote (LES, 2005b, c).

The staff reviewed the geotechnical investigation information presented in the ISA Summary (LES, 2005b, c) and concurs with the applicant that the potential for liquefaction of soils at the site may not be a safety concern for the proposed facility. The applicant committed to perform additional geotechnical investigations at the site to confirm that liquefaction is not a safety concern for the proposed facility. Additional site testing will be evaluated in accordance with NRC Regulatory Guide 1.198, "Procedures and Criteria for Assessing Seismic Soil Liquefaction at Nuclear Power Plant Sites" (NRC, 2003b).

3.3.1.1.5.4   Settlement

Settlement of foundations for the proposed facility is discussed in Section 3.2.7 of the ISA
Summary (LES, 2005b). In its ISA Summary, the applicant stated that only five borings were
drilled at the proposed facility site to determine the suitability of the site. The applicant
recognized the geotechnical results obtained from the five borings were not sufficient for final
design purposes. The applicant states that due consideration will be given to settlement and
differential settlement during final design, and final design details will be based on a more
comprehensive geotechnical investigation to be undertaken when additional project details
become available. Allowable soil bearing pressures will be evaluated in accordance with Naval
Facilities Engineering Command Design Manual NAVFAC DM-7.02, "Foundations and Earth
Structures" (NAVFAC, 1996). Building settlement analyses will be performed in accordance
with Naval Facilities Engineering Command Design Manual NAVFAC DM-7.01, "Soil
Mechanics" (NAVFAC, 1986) and Winterkorn and Fang, "Foundation Engineering Handbook"
(Winterkorn, 1975).

The staff reviewed the information presented concerning differential settlements and find the
applicant's commitment to perform additional geotechnical investigations using acceptable
geotechnical standards for final facility design to be acceptable.

Conclusion

Staff finds that, based on the individual parameter assessments discussed above, the
applicant's material regarding the geological information presented in the SAR and ISA
Summary is sufficient and acceptable to use in determining appropriate criteria for the safe
design of principal structures, systems, and equipment.

3.3.1.1.6   Site Description Conclusion

Based on the above, the Staff concludes that sufficient site information has been provided to
support identification of those factors that could affect safety, including related internal and
external hazards and associated accident sequences that could exceed the performance
requirements of 10 CFR 70.61. The staff concludes that ISA Summary site description meets
the requirements of 10 CFR 70.65(b)(1).

3.3.1.2   Facility Description

A description of the proposed facility arrangement, buildings and major components, and
structural design criteria is provided in SAR Chapter 1 (LES, 2005a) and ISA Summary Section
3.3 (LES, 2005b). The facility location and the distance from the boundary in all directions are
provided in ISA Summary Table 3.3-1 (LES 2005b). The distance to the nearest resident is
identified in ISA Summary Section 3.2 (LES 2005b), as about 4.3 km (2.6 mi) west of the site.
The arrangement and location of buildings on the facility site, and identification of the site
boundary (controlled area) and controlled access area boundary, are shown in ISA Summary
Figure 3.3-1 (LES 2005b).

3.3.1.2.1 Buildings and Major Components

This section contains a review of the civil structures of the proposed facility. According to the SAR (LES, 2005a), the proposed facility consists of the following structures:

- Three Separations Building Modules
- Technical Services Building
- Cylinder Receipt and Dispatch Building
- Centrifuge Assembly Building
- Blending and Liquid Sampling Area
- Uranium Byproduct Cylinder Storage Pad
- Central Utilities Building
- Administration Building
- Visitor Center
- Site Security Building

Among these structures, the Separations Building Modules, Technical Services Building, Cylinder Receipt and Dispatch Building, the centrifuge test facility in the Centrifuge Assembly Building, and the Blending and Liquid Sampling Area are determined to be safety significant, based on the ISA performed by the applicant (LES, 2005b). These structures are required to be designed to withstand design-basis natural phenomena hazards and external hazards as required by the baseline design criteria of 10 CFR 70.64(a)(2).

Each Separations Building Module includes two cascade halls. Each Separations Building Module has a uranium hexafluoride handling area and a process services area. A Separations Building Module is 170 m (557.75 ft) long, 67.9 m (222.75 ft) wide, and 13 m (42.7 ft) high, and has 12,703 m² (137,025 ft²) of space. The Technical Services Building is a two-story building with 9,192 m² (98,942 ft²) of space. The Cylinder Receipt and Dispatch Building is 246.2 m (807.75 ft) long, 45.9 m (150.6 ft) wide, and 13 m (42.7 ft) high, and has 11,300 m² (121,638 ft²) of space. The Centrifuge Assembly Building is 195.5 m (641.4 ft) long, 50.9 m (167 ft) wide, and 11 to 16 m (36.1 to 52.5 ft) high, and has 11,364 m² (122,322 ft²) of space. The Blending and Liquid Sampling Area is 33.5 m (109.9 ft) long, 45.9 m (150.6 ft) wide, and 10 m (32.8 ft) high, and has 1,538 m² (16,555 ft²) of space.

3.3.1.2.2 Structural Design Criteria

Structural design criteria for the proposed facility is discussed in Section 3.3 of the SAR (LES, 2005a) and Section 3.3.2.1 of the ISA Summary (LES, 2005b). Specifically, structural design loads for the proposed facility are discussed in Section 3.3.2.2 of the ISA Summary (LES, 2005b).

3.3.1.2.2.1 Codes and Standards

A list of the codes and standards for the structural design of the proposed facility are provided in Section 3.3.2.1 of the SAR (LES, 2005a). These codes and standards included guidance for general structural design, concrete design, precast concrete design, steel construction, and testing and material selections.

Staff reviewed the cited codes and standards to be used for the design and construction of the proposed facility structures and conclude that the cited codes and standards conform with standard engineering practice and are reasonable and acceptable.

3.3.1.2.2.2   Structural Design Loads

Straight Wind Load

ISA Summary Section 3.3.2.2.1, (LES, 2005b), identifies the design-basis straight-line wind as 252 km/h (157 mph) for all safety-significant structures. This design-basis wind is characterized as a 100,000-year return period wind. The applicant will use American Society of Civil Engineers (ASCE) ASCE 7-98 (ASCE, 1998) for the determination of wind loads and the design for wind loads for all safety-significant structures and components exposed to winds. For structures that are not safety significant, the design-basis wind speed [130 km/h (80 mph)] will be that of a 50-year return period, and the Uniform Building Code (ICBO, 1997) will be followed. Staff concludes that the method used to calculate wind loads is based on an acceptable standard.

Because the design-basis wind is characterized as a 100,000-year wind, the event sequences associated with high winds are highly unlikely events based on the definition given in ISA Summary Section 3.1, "General ISA Information" (LES 2005b) (see SER Section 3.3.3.c.1.3 for the definition of highly unlikely).

Tornado Wind Load

The values of design-basis tornado-related parameters are listed as follows:

Design-Basis Wind Speed:	302 km/h (188 mph)
Radius of Damaging Winds:	130 m (425 ft)
Atmospheric Pressure Change:	−3.83 kg/m^2 (−80 lbs/ft^2)
Rate of Atmospheric Pressure Change:	−1.44 kg/m^2/s (−30 lbs/ft^2/s)

The tornado design-basis is characterized as a 100,000-year tornado, which is the equivalent of a "F-3" tornado on the Fujita Tornado Scale. The applicant concluded a tornado was a credible event. This probability, however, meets the definition of highly unlikely. Therefore, the potential consequences for an event of this magnitude or greater do not have to be determined because the event sequence is highly unlikely, in accordance with 10 CFR 70.61(b).

In ISA Summary Sections 3.3.2.1 and 3.3.2.2, the applicant indicated that ASCE 7-98 (ASCE, 1998) will be used for the determination of tornado wind loads and the design for tornado wind loads for all safety-significant structures and components exposed to tornado winds. Use of ASCE 7-98 (ASCE, 1998) for tornado load calculation and design of structures and components for tornado loads is acceptable to the staff.

Tornado-Generated Missile Loads

Three types of missiles associated with the design-basis tornado were selected for design of safety-significant structures. As indicated in ISA Summary Sections 3.2.3.4.1 and 3.3.2.2.3.2 (LES, 2005b), these missiles included (I) a 6.8-kg (15-lb), 10.2 × 30.5-cm (2 × 4-in) timber plank, (ii) 34-kg (75-lb), 7.6-cm (3-in)-diameter steel pipe, and (iii) 1,361-kg (3,000-lb) automobile. The associated vertical and horizontal impact velocities of each missile also were provided in the same sections of the ISA Summary.

The tornado-generated missile effects on structures will be considered in two aspects. For the localized impact, depth of penetration and scabbing thickness will be determined for reinforced concrete targets. The applicant will use the formulas provided in "Structural Analysis and Design of Nuclear Plant Facilities" (ASCE, 19080) for depth of penetration and scabbing thickness calculations. Staff finds the use of these formulas acceptable because these formulas are commonly used for assessing local penetration and scabbing effects. The applicant will use the requirements in Section C.7.2.2 of the American Concrete Institute (ACI) ACI 349-90 (ACI, 1990) for determining minimum concrete thickness to resist hard missiles such as steel pipes and the requirements in Section C.7.2.3 of ACI 349-90 (ACI, 1990) to check against punching shear resulting from soft missiles such as timbers. The use of ACI 349-90 (ACI, 1990) is acceptable to staff because this standard is developed specifically for nuclear safety-related concrete structures. For steel targets, the formula provided in "Structural Analysis and Design of Nuclear Plant Facilities" (ASCE, 1980) will be used to estimate the perforation thickness. The requirement in this document will be used to establish the required steel thickness. Staff considers the use of this ASCE document acceptable because it is a design code for nuclear plant facilities.

For design consideration of tornado-generated missile effects on overall structural response, all missile momentum is assumed to transfer into the target during impact, and the target is assumed to behave elasto-plastically. An equivalent static load acting at the impact area will be calculated using the procedure outlined in "Structural Analysis and Design of Nuclear Plant Facilities" (ASCE, 1980). This equivalent will be used in combination with other design loads using conventional design methods. The applicant may also employ other formulations such as presented in ACI 349-90 (ACI, 1990) for design to resist the overall effects of missile impact. Staff find the use of either of these methods acceptable because they represent standards and codes for nuclear facilities.

Flood Design

The grade level of the proposed facility is above the maximum foreseeable flood level (LES, 2005a). The only potential flood may result from intense local precipitation. Two flood-protection design features will be used. First, the floor level of the proposed facility will be 0.15 m (0.5 ft) higher than the finished grade elevation, which will be at least 0.45 m (1.5 ft) above all roads. Second, an earth berm and intercept trench will be constructed uphill of the building structures to divert water flow. NRC staff has concluded that the flood-protection design features are adequate to protect building structures from water damage.

Seismic Loads

Development of a design-basis earthquake for the proposed facility site is discussed in Section 3.2.6 of the ISA Summary (LES, 2005b). The site-specific, design-basis earthquake is discussed in ISA Summary Sections 3.2.6.5 and 3.3.2.2.5.2 (LES, 2005b). The staff review and acceptance of the selected design-basis ground acceleration are discussed in Section 3.3.1.1.7.a of this SER.

The design-basis earthquake for the proposed facility is based on an earthquake with a return period of 10,000 years (10^{-4} annual probability). The corresponding peak horizontal and vertical ground accelerations were 0.15g. In ISA Summary Section 3.3.2.2.5.1 (LES, 2005b), the applicant indicated the seismic design loads will be calculated using the method outlined in Chapter 16, Division IV, of the Uniform Building Code (ICBO, 1997) for all buildings and structures, including items such as equipment supports, for the proposed facility. The proposed facility site is located in seismic zone 1 based on International Conference of Building Officials (ICBO, 1997). In considering soil amplification, suggested soil amplification factors for Soil Type C defined by Dobry, et al. (Dobry, 2000) will be used. These amplification factors are listed in Section 3.2.6.4.3 of the ISA Summary (LES, 2005b). The applicant will verify the appropriateness of the soil amplification factors for the proposed facility during final design. The staff considers this method acceptable because it follows modern practices and is consistent with methods used at other nuclear facilities [e.g., DOE–STD–1020–2002 (DOE, 2002)].

The applicant states that structures designed to a seismic hazard of 10^{-4} annual probability can meet the performance goal of 10^{-5} annual probability so the seismic hazards for structures relied on for safety satisfy the definition of highly unlikely. As a result, according to 10 CFR 70.61(b), the potential consequences resulting from seismic hazards do not have to be assessed in the ISA. This additional performance may be achieved because of conservatism in the design including conservative specifications of material strength and elastic design approach. The applicant will use either the method outlined in DOE–STD–1020–2002 (DOE, 2002) or ASCE Standard Seismic Design Criteria (ASCE, 2003) for such demonstration. Specifically, the applicant intends to show that even though the design code allowables are exceeded for the targeted structural performance goal, the ultimate capability of the design is not exceeded. Confirmatory seismic performance calculations for structures relied on for safety will be conducted during detailed design.

Staff reviewed the approach proposed for seismic design-load calculation and concludes that the method presented in the Uniform Building Code (UBC) (ICBO, 1997) is acceptable for determining seismic design loads because the UBC (ICBO, 1997) was developed with emphasis on earthquake effects. Staff also concludes that the applicant's approach in demonstrating that they meet the performance goal of 10^{-5} annual probability is reasonable because of the conservatism in the conventional design approach.

Snow Loads

ISA Summary Section 3.2.3 (LES, 2005b) describes the design basis values for snow or ice loads. According to the SAR (LES, 2005a), the design-basis snow load for the proposed facility was developed based on the method provided in the NRC "Site Analysis Branch Position for Winter Precipitation Loads" (NRC, 1975). The design-basis snow load was a combination of 100-year snowfall load at the site and the weight of the 48-hour probable maximum winter precipitation for the month corresponding to the selected snowfall.

According to Figure 7-1 of American Society of Civil Engineers (ASCE) ASCE 7-98 (ASCE, 1998), the ground snow load with a 2-percent annual probability of being exceeded (i.e., a 50-year mean recurrence interval) at the proposed facility site was approximately 0.48 kPa (10 psf). The applicant used this value to determine the 100-year return period snow load using the method described in ASCE 7-98 (ASCE, 1998). The 100-year return period snow load was determined to be 0.56 kPa (12 psf). The applicant (Harper, 2003c) estimated the 48-hour probable maximum winter precipitation using the method described in Hydrometeorological Report No. 33 (USWB, 1956). The 48-hour probable maximum winter precipitation was determined to be 48.3 cm (19.0 in.), which corresponds to a loading of 0.95 kPa (19.8 psf). The design-basis ground snow load was 1.53 kPa (32 psf). The design-basis snow load for buildings not safety significant will be determined in accordance with the UBC (ICBO, 1997) based on a mean return period of 50 years.

The staff reviewed the information about design-basis snow load presented in the SAR (LES, 2005a) and the ISA Summary (LES, 2005b) and concludes that the information is acceptable and appropriate because acceptable standards and acceptable methods were used to determine the 100-year return period snow load and the 48-hour probable maximum winter precipitation.

Rain Loads

In Section 3.3.2.2.6.2 of the ISA Summary (LES 2005b), the applicant states the rain loads on roofs may be related to a 100-year return period rainfall and a localized intense rainfall. This 100-year return period rainfall was established based on a storm with a duration of 1 hour. According to the International Plumbing Code (ICC, 2000), the rainfall rate of this storm in Hobbs, New Mexico, was 7.6 cm/h (3 in/h). Staff concludes that use of the International Plumbing Code (ICC, 2000) in determining the 100-year return period rainfall for the site is acceptable. For the 100-year return period rainfall, the roofs will be designed, assuming the primary roof drains are blocked, to prevent water from accumulating in excess of the normal roof design live load.

For the localized intense rainfall, the rain loads on roofs assuming a total blockage of the roof drainage system will be accounted for as an individual load. This rain load is for safety-significant structures only. Including a rain load determined assuming loss of roof drainage capacity in the design of roofs for safety-significant structures is reasonable and acceptable to the staff.

Process and Equipment-Derived Loads

The process and equipment-derived loads for the proposed facility include equipment; piping; heating, ventilation, and air conditioning; and electric tray and conduit loads. Equipment heavier than 454 kg (1,000 lb) will be treated as individual dead loads. Weights for other equipment will be accounted for as appropriate uniform dead loads in the particular area.

Piping loads were estimated at 2.39 kg/m² (50 lbs/ft²) based on combined dead and live loads. Heating, ventilation, and air conditioning duct loads were 1.44 kg/m² (30 lbs/ft²) and were estimated based on combined dead and live loads. Electric tray and conduit loads were 104 kg/m (70 lb/ft) of tray and conduit and a 91-kg (200-lb) concentrated load at midspan of the tray. The electric tray and conduit loads were estimated based on combined dead and live loads.

The applicant states (LES, 2005a) that the equipment; piping; heating, ventilation, and air conditioning; and electric tray and conduit loads provided in the ISA Summary represent the minimum loads expected to be used. These minimum loads will be assumed as live loads and treated as such in load combinations. During final design, individual loads (dead versus live loads) will be identified and accounted for in the design accordingly by applying appropriate load factors defined in the load combinations. The resulting actual load will be compared to the load obtained using the minimum load value treated as a live load. The larger of the two loads will be used for design. The staff concludes that the approach used for considering loads created by equipment; piping; heating, ventilation, and air conditioning; and electric trays and conduit in the load combination is conservative and acceptable.

Settlement-Induced Loads

Settlement of foundations for the proposed facility is discussed in Section 3.2.7 of the ISA Summary (LES, 2005b). In its ISA Summary, the applicant stated that only five borings were drilled at the proposed facility site to determine the suitability of the site. The applicant recognized the geotechnical results obtained from the five borings were not sufficient for final design purposes. The applicant states in SAR Chapter 3 (LES, 2005a) that additional geotechnical investigations will be performed to obtain further data for the final design of the proposed facility. Allowable soil bearing pressures will be evaluated in accordance with Naval Facilities Engineering Command Design Manual NAVFAC DM-7.02, "Foundations and Earth Structures" (NAVFAC, 1996). Building settlement analyses will be performed in accordance with Naval Facilities Engineering Command Design Manual NAVFAC DM-7.01, "Soil Mechanics" (NAVFAC, 1986) and Winterkorn and Fang, "Foundation Engineering Handbook," (Winterkorn, 1975).

NRC reviewed the information presented concerning differential settlements and concludes that the applicant's commitment to perform additional geotechnical investigations using acceptable geotechnical standards for final facility design is acceptable.

Explosion Overpressure

In assessing potential hazards of propane truck accidents and a nearby natural gas pipeline explosion, the applicant (Snooks, 2003; Thomson, 2004) used the assumption that the safety-significant structures of the proposed facility would be designed to withstand 6.9 kPa (1 psi) overpressure. This design-basis overpressure was included in Section 3.3.2.2.8.2 of the ISA Summary (LES, 2005a) for design of safety-significant structures. Staff concludes that treating the explosion overpressure as an extreme environmental load in the load combination is acceptable.

Load Combinations

The design loads considered in the load combination include two parts: normal and extreme environmental loads. The normal loads were listed in Section 3.3.2.2.8.1 of the ISA Summary and the extreme environmental loads were listed in Section 3.3.2.2.8.2 (LES, 2005b).

In ISA Summary Section 3.3.2.2.8 (LES, 2005b), the applicant states that all concrete structures will be designed using the ACI strength method (ACI, 1999). Load combinations for the design of safety-significant concrete structures will follow those provided in ACI 349-90 (ACI, 1990), while load combinations for other structures will follow ASCE 7-98 (ASCE, 1998). Specific load combinations for both safety-significant and non-safety-significant concrete structures and components are listed in Section 3.3.2.2.8.3 of the ISA Summary (LES, 2005b).

The rain load caused by localized intense rainfall will be evaluated using the load combination A.10 in Section 3.3.2.2.8.3 by replacing the design-basis tornado with this rain load (LES, 2005b).

All structural steel will be designed using the American Institute of Steel Construction (AISC) Allowable Stress Method (AISC, 1989). Specific load combinations for design of structure steel are listed in Section 3.3.2.2.8.3 of the ISA Summary (LES, 2005b).

The load combinations for foundations are provided in Section 3.3.2.2.8.3, "Combined Load Applications," of the ISA Summary (LES, 2005b). The minimum factor of safety of foundations against overturning was 1.5 and sliding was 2.0.

Staff reviewed the information presented in the ISA Summary regarding the design loads and approaches used to develop seismic design loads for the safety-significant structures and components and find these design loads were developed based on site characteristics. Staff concludes that the approaches for determining the design loads followed acceptable codes and standards. Staff also concludes that the approach proposed by the applicant to demonstrate the structures designed to a seismic hazard of 10^{-4} annual probability, meeting the performance goal of 10^{-5} annual probability, is acceptable because the conventional design approach contains adequate conservatism. The design loads and the approaches described satisfy the requirements of 10 CFR 70.64(a)(2), 70.64(a)(4), and 70.65(b)(2).

3.3.1.2.2.3   Foundations

Foundations will be shallow concrete footings. The applicant anticipated the footings for all safety-significant buildings will be founded on firm and dense sandy soils. The allowable bearing pressure is 335 kPa (7,000 lbs/ft^2). In areas where the footings bear in existing or new fill areas, the allowable bearing pressure is 144 kPa (3,000 lbs/ft^2). The allowable pressure may be higher in areas where the fill material is entirely rock.

3.3.1.2.3   Facility Description and Design Basis Conclusion

The staff concludes that sufficient information has been provided describing those areas that could affect safety, including identification of the controlled area boundaries. The facility description is adequate to allow identification of the general features, including internal and external hazards and associated accident sequences that could exceed the performance criteria of 10 CFR 70.61. The ISA Summary facility description meets the requirements of 10 CFR 70.65(b)(2). The staff also concludes that appropriate design loads and associated codes and standards were identified to meet 10 CFR 70.64(a)(2) and (4).

3.3.1.3   Processes

The enrichment processes and equipment analyzed as part of the ISA process are described in the ISA Summary, Section 3.4 (LES, 2005b). The enrichment systems are comprised of the Uranium Hexafluoride (UF$_6$) Feed System, Cascade System, Product Take-off System, and Tails Take-off System. Major support systems include the Product Blending System, Product Liquid Sampling System, and Contingency Dump System. Systems used to support the enrichment process and the handling of UF$_6$ are the Gaseous Effluent Vent System (GEVS), Centrifuge Test Facility and Centrifuge Post Mortem Facility, and Material Handling. A subsection containing information pertaining to the functional description, major components, design description, interfaces, design and safety features, operating limits, instrumentation, and IROFS is provided for each of the 10 enrichment and supporting systems.

The process function is to enrich the amount of the ^{235}U isotope in UF_6 from 0.711 weight percent up to a maximum of 5.0 weight percent through a mechanical centrifuge separation process. Naturally occurring uranium will be received from a conversion facility in the form of UF_6 shipped in Type 48Y or 48X cylinders qualified to American National Standard Institute (ANSI) Standard N14.1 (ANSI, 1995). The UF_6 will be in a solid state under vacuum. The UF_6 feed system will heat the cylinder to 53°C (127°F) to sublime the UF_6 into a gas. The feed purification system removes light gas components from the feed to a specified level prior to admittance to the cascade in order to protect the centrifuges and enhance efficiency. At the feed purification Low Temperature Take-off Stations, Type 48X or 48Y cylinders are cooled to -25°C (-13°F). Gaseous UF_6 enters the cylinders and desublimes into the solid phase. UF_6 cold traps, aluminum oxide (Al_2O_3) traps, and vacuum pumps are used to transfer residual light gas to the GEVS. The traps remove any UF_6 or hydrogen fluoride (HF) from the effluent stream. From purification, UF_6 is transferred to the cascades where enrichment occurs. The cascades are operated under a significant vacuum (about 65 mbar or 0.09 psia) to assure the UF_6 does not desublime back into a solid state at ambient temperatures.

The cascade systems separates the UF_6 feed stream into a product stream and a tails stream. The product stream is routed to the Product Take-off System, where it is transported by vacuum pumps to the Product Low Temperature Take-off Station. These stations are operated at -25°C (-13°F) and the UF_6 desublimes into the solid form inside a Type 30B or 48Y cylinder. The Type 30B cylinder is used for final product, while the Type 48Y cylinder is used for future blending operations. Any light gas impurities are purged through cold traps followed by product vent vacuum pump/chemical trap sets.

The Tails Take-off System withdraws the depleted UF_6 stream and provides a means to withdraw UF_6 from the centrifuge cascades under abnormal conditions. The UF_6 is routed to a Low Temperature Take-off Station operated at -25°C (-13°F). The gaseous UF_6 is desublimed into a solid form inside a Type 48Y cylinder.

The Product Blending system is used to provide a specific enrichment of ^{235}U by blending UF_6 at two different enrichment levels. The donor stations can handle Type 30B and 48Y product cylinders. The UF_6 is sublimed back into a gas and transported to a Blending Receiver Station containing an empty Type 30B cylinder operated at -25°C (-13°F), where the UF_6 desublimes back into a solid form.

The only system at the facility that changes solid UF_6 into liquid UF_6 is the Product Liquid Sampling System. A filled Type 30B cylinder is placed in an autoclave that is heated to 70°C (158°F) by electric heaters. When the pressure reaches +2.5 bar (36.3 psia), the temperature is stabilized for about 16 hours to allow homogenization prior to sampling. The sample bottles are connected to the cylinder via a header, all located within the confines of the autoclave. The main safety feature of the autoclave is to provide a secondary confinement barrier in the event of a UF_6 leak. All Type 30B cylinders are required to meet ANSI N14.1 (ANSI, 1995) requirements, which include a cylinder design pressure of 1380 kPa (200 psi) and testing to 2760 kPa (400 psi).

Section 3.5 (LES, 2005b) of the ISA Summary emphasizes capacities, redundancies, and other provisions for coping with routine and non-routine events. The system descriptions include functional requirements, design capacities, system interfaces, and descriptions of major components. Operational characteristics and safety considerations are also described. The applicant asserts that the health and safety of the public are protected such that a failure or inadvertent operation of any utility system would not result in the release of hazardous quantities of chemicals or radiation. These systems include the Building Ventilation System, Electrical System, Compressed Air System, Water Supply, Cooling Water Systems, Septic Systems, Communication and Alarm Annunciation Systems, Fire Protection, Control Systems, Standby Diesel Generator System, Nitrogen System, Liquid Effluent Collection and Treatment System, Solid Waste Collection System, Decontamination Workshop, Fomblin Oil Recovery System, Laundry System, Ventilated Room, Chemical Laboratory, and the Cylinder Preparation Room. Many of these systems do not contain IROFS. However, codes and standards to which these systems will be designed are identified in Table 3.3-8 of the SAR (LES, 2005a) and Table 3.5-1 of the ISA Summary (LES, 2005b).

The applicant did not identify any essential utility service systems that are required to support the IROFS safety function. However, upon completion of the design, the IROFS boundaries will be defined by using LES procedure, "IROFS Boundary Definition" (LES, 2005a). Any essential utility service systems that are required to support the safety function of the IROFS will fall within the boundary of the IROFS per above procedure. The applicant states that if any essential utility service systems are within the IROFS boundary per final design of the IROFS, then these essential utility service systems will be designed to withstand environmental stresses caused by environmental and dynamic service conditions under which failure could prevent the satisfactory accomplishment of the IROFS safety function. Also, if an essential utility is identified in the future, its continued operation will be addressed at that time through the applicant's design change process.

Instrumentation and Control (I&C)

Process schematics (LES, 2005a) show instrumentation component functions in sufficient detail to understand the I&C systems at a functional level, and the results of the ISA based on those functional descriptions. The detailed information on the physical components will be developed when the final design is complete. Once the final design is complete, the applicant will maintain I&C system design documentation onsite as part of the ISA (see Section 3.8 of the ISA Summary (LES, 2005b)). The design documentation will include hardware design details, identification of essential support utilities, operating ranges and limits for measured process variables, and safety limits and safety margins, as applicable. The applicant has also committed to following good engineering design practices established by the ISA and current applicable codes and standards (NRC, 2002a and LES, 2005a).

Measured process variable operating ranges and limits (i.e., trip setpoints) are discussed in Section 3.4 of the ISA Summary (LES, 2005b). Section 3.8.1 of the ISA Summary addresses IROFS. For hardware IROFS with instrumentation providing automatic prevention or mitigation actions, setpoint calculations will be performed in accordance with a setpoint method consistent with the applicable guidance provided in Regulatory Guide 1.105, "Setpoints for Safety-Related Instrumentation," Revision 3, dated December 1999 (NRC, 1999a).

The Process Control System (PCS) is essentially a distributed control system interconnecting the various major components over a plant control network. The PCS and its software will not have any direct, hardwired interface to systems or components responsible for the actuation, control, or reset of safety systems and components (LES, 2005a and NRC, 2002a).

Since the proposed design, which is not complete at the time of this review, does not include IROFS that use software, firmware, microcode, PLCs, and/or any digital device, including hardware devices which implement data communication protocols, the staff finds the current applicant commitments satisfactory. However, should the applicant choose to implement design changes to include any of the preceding features, prior NRC approval will be necessary (see License Condition No.1 in Section 3.4 of this SER).

In the facility SAR, the applicant provided preliminary design basis information for I&C systems that it identified as IROFS for the facility. The design information is at the system functional level. Individual components and vendors had not yet been selected. Based on the staff's review of the SAR, supporting information provided by the applicant, and the applicant's commitments to the industry standards and guidance cited in the sections above for I&C systems, the staff finds that the preliminary design meets the requirements of 10 CFR 70.61 and 70.64(a)(10).

Given that these conclusions are based on preliminary design information and the possibility that the applicant may choose to implement design changes as discussed in the previous section on I&C, the staff is imposing a license condition to ensure that the final design is adequate and acceptable to the staff. Specifically, the following condition will be included in the license:

> "Currently, there are no IROFS that have been specified as using software, firmware, microcode, PLCs, and/or any digital device, including hardware devices which implement data communication protocols (such as fieldbus devices and Local Area Network controllers), etc. Should the design of any IROFS be changed to include any of the preceding features, the licensee shall obtain Commission approval prior to implementing the change(s). The licensee's design change(s) shall comply with accepted best practices in software and hardware engineering, including software quality assurance controls as discussed in the Quality Assurance Program Description (LES, 2005a) throughout the development process and the applicable guidance of the following industry standards and regulatory guides as specified in SAR Chapter 3 (LES, 2005a):
>
> 1. American Society of Mechanical Engineers (ASME) NQA-1-1994, Part II, subpart Part 2.7, "Quality Assurance Requirements of Computer Software for Nuclear Facility Applications," as revised by NQA-1a-1995 Addenda of NQA-1-1994 and ASME NQA-1-1994, Part 1, Supplement 11S-2, "Supplementary Requirements for Computer Program Testing." (Refer to SAR Chapter 11, Appendix A, Section 3.)
>
> 2. Electric Power Research Institute (EPRI) NP-5652, "Guideline for the Utilization of Commercial Grade Items in Nuclear Safety Grade Applications," June 1988.

3.   EPRI Topical Report (TR) -102323, "Guidelines for Electromagnetic Interference Testing in Power Plants," Revision 1, December 1996.

4.   EPRI TR-106439, "Guideline on Evaluation and Acceptance of Commercial Grade Digital Equipment for Nuclear Safety Applications," October 1996.

5.   Regulatory Guide 1.152, "Criteria for Digital Computers in Safety Systems in Nuclear Power Plants," Revision 1, January 1996.

6.   Regulatory Guide 1.168, "Verification, Validation, Reviews, and Audits for Digital Software Used in Safety Systems of Nuclear Power Plants," Revision 1, February 2004.

7.   Regulatory Guide 1.169, "Configuration Management Plans for Digital Computer Software Used in Safety Systems of Nuclear Power Plants," September 1997.

8.   Regulatory Guide 1.170, "Software Test Documentation for Digital Computer Software Used in Safety Systems of Nuclear Power Plants," September 1997.

9.   Regulatory Guide 1.172, "Software Requirements Specifications for Digital Computer Software Used in Safety Systems of Nuclear Power Plants," September 1997.

10.   Regulatory Guide 1.173, "Developing Software Life Cycle Processes for Digital Computer Software Used in Safety Systems of Nuclear Power Plants," September 1997.

If any above changes result in IROFS requiring operator actions, the licensee shall conduct a human factors engineering review of the human-system interfaces using the applicable guidance in NUREG-0700, "Human-System Interface Design Review Guidelines," Revision 2, dated May 2002 (NRC, 2002d), and NUREG-0711, "Human Factors Engineering Program Review Model," Revision 2, dated February 2004."

Contingent on the above provision, the staff finds the above I&C safety-related commitments acceptable.

Electrical

In ISA Summary, Section 3.5.2 (LES, 2005b), the applicant states that the overall electrical power system is designed with a high level of redundancy to maintain a reliable power supply to the process equipment for investment protection. Total loss of electrical power does not have any safety implications. Furthermore, the applicant states in ISA Summary Section 3.8.1 (LES, 2005b), that IROFS systems will be designed to be fail-safe.

The applicant also describes in Section 3.5 of this ISA Summary (LES, 2005b) how the various support functions would be performed during a loss of power. Included in the discussion were descriptions of the Standby Diesel Generators, Uninterruptible Power Supplies, and defense-in-depth approach. The applicant will provide two standby diesel generators to power short break loads and uninterruptible power to process instruments, plant control system, emergency lighting, site communications system, and environmental monitoring to allow safe shutdown of the facility. The applicant will mitigate the effects of hydrogen ignition in areas containing batteries as part of final design through implementation of the National Fire Protection Association (NFPA) 70 E, "Standard for Electrical Safety Requirements for Employee Workplaces," and American National Standards Institute (ANSI) C2, "National Electrical Safety Code," if ventilation control of hydrogen gas is required.

The staff concludes that the descriptions of the overall electrical power system is adequate to allow evaluation of the completeness of the hazard and accident identification tasks and the likelihood and consequences of accidents. The descriptions covered basic process function and theory, and the general arrangement, function, and operation of the major components in the process. Process design, equipment, and instrumentation were sufficiently discussed.

Human Factors

The staff's analysis included the process used to conduct the human factors engineering review of the Communication and Alarm Annunciation System, the Central Control Room, and the Criticality Accident Alarm System (CAAS) as it applies to IROFS requiring operator actions. The applicant describes the Control Room in ISA Summary, Section 3.3.1.2.2.17 (LES, 2005b), the Central Control System in ISA Summary Section 3.5.9.2.1 (LES, 2005b), the Communication and Alarm Annunciation System in ISA Summary, Section 3.5.7 (LES, 2005b), and the CAAS in SAR, Section 5.3 (LES, 2005a) and ISA Summary, Section 3.1.5 (LES, 2005b).

The applicant described (LES, 2005a) how it used the design, maintenance, and operating experience of Urenco to derive best practices for all elements of the plant. Incorporated implicitly in the Core Plant Design (CPD) are all the human factors engineering enhancements from over 30 years of experience. The Urenco SP5 plant design, on which the proposed facility is based, is based on this CPD. Additionally, every element of the design is subject to HAZOP analysis and design review. Operations experienced personnel are mandatory members during such HAZOPs and design reviews. Urenco design review guidelines include design review topics of functionality, operability, maintenance, layout, and orientation. Urenco engineering design safety principles also address the human factors engineering issues. Specifically these principles state that the design of all interfaces between operating personnel and the plant should follow good human factors and ergonomics practice. These principles also note that analysis of the safety function tasks requires determination of the demands on personnel in order to evaluate the feasibility of the tasks and provide input to the design interfaces. The design of tasks and equipment are fully compatible with training arrangements for operations personnel, proposing staffing levels, and the development of operating procedures.

In ISA Summary Section 3.8.1 (LES, 2005b), the applicant states that for those IROFS requiring operator actions, a human factors engineering review of the human-system interfaces will be conducted using the applicable guidance in NUREG-0700, "Human-System Interface Design Review Guidelines," Revision 2, dated May 2002 (NRC, 2002d), and NUREG-0711, "Human Factors Engineering Program Review Model," Revision 2, Dated February 2004 (NRC, 2004a).

The staff concludes that appropriate human factors guidance has been identified for conducting reviews of the human-system interfaces for those IROFS requiring operator actions, and the engineering review of the Communication and Alarm Annunciation System, the Central Control Room, and the CAAS.

Process Conclusion

The operating ranges and limits for measured variables for processes that are controlled by IROFS are included in the ISA Summary or the classified portion of the ISA. The staff concludes that this information is sufficient to permit an understanding of the theory of operation and assess compliance with the performance requirements of 10 CFR 70.61. The staff concludes that an adequate description was provided for each process analyzed in the ISA to understand the theory of operation; and for each process, to allow evaluation of the completeness of the hazards that were identified and the accident sequences identified.

3.3.2   Baseline Design Criteria and Defense-In-Depth

ISA Summary, Section 3.1.7 (LES, 2005b), addresses compliance with 10 CFR 70.64. The ISA accident sequences for credible high and intermediate consequence events define the design basis events. The IROFS and safety parameter limits for these events are available in the ISA documentation. Together, the IROFS and safety parameter limits ensure that the associated baseline design criteria (BDC) are met. Specific details on how the facility design or operation conforms to the BDC are located in the individual sections of the SAR. Staff review of external events, including natural phenomena hazards, is in SER Section 3.3.1.1.

The applicant will ensure that the structures, systems, and components (SSCs) that are determined to have safety significance are designed, fabricated, erected, and tested in accordance with the criteria set forth in 10 CFR Part 50, Appendix B, "Quality Assurance Criteria for Nuclear Power Plants and Fuel Reprocessing Plants." Appropriate records of the design, fabrication, erection, procurement and testing of SSCs which are determined to have safety significance are maintained throughout the life of the plant.

The design addresses natural phenomena hazards, fire protection, environmental and dynamic effects, chemical protection, emergency capability, utility services, inspection, testing and maintenance, criticality safety, instrumentation and controls, and defense-in-depth. For a more detailed review of the individual BDC, see SER Sections 3.3.1.1 and 3.3.1.2 for natural phenomena hazards, environmental and dynamic effects; Section 3.3.1.3 for utility services; Chapter 5 for nuclear criticality safety; Chapter 7 for fire protection; Section 3.3.1.3 and Chapter 6 for chemical safety; Chapter 8 for emergency management; and Chapter 11 for management measures for surveillance, maintenance and testing.

The applicant states that IROFS components and systems will be qualified using the applicable guidance in the Institute of Electrical and Electronics Engineers, Inc. (IEEE) Standard IEEE-323, 1983, "IEEE Standard for Qualifying Class 1E Equipment for Nuclear Power Generating Stations." The qualification of each IROFS component and system will demonstrate that the IROFS will perform their safety function under the environmental and dynamic service conditions in which they will be required to function and for the length of time the function is required. Additionally, non-IROFS components and systems will be qualified to withstand environmental stress caused by environmental and dynamic service conditions under which their failure could prevent satisfactory accomplishment of the IROFS safety functions. Furthermore, IROFS components and systems will be designed, procured, installed, tested, and maintained using the applicable guidance in Regulatory Guide 1.180, "Guidelines for Evaluating Electromagnetic and Radio-Frequency Interference in Safety-Related Instrumentation and Control Systems," Revision 1, dated October 2003 (NRC,2003).

In Section 3.1.7.J of the ISA Summary (LES, 2005b), the applicant states that I&C systems will be provided to monitor variables and operating systems that are significant to safety over anticipated ranges for normal operation, for abnormal operation, for accident conditions, and for safe shutdown. These systems will ensure adequate safety of process and utility service operations in connection with their safety function. Controls will be provided to maintain these variables and systems within the prescribed operating ranges under all normal conditions. I&C systems will be designed to fail into a safe state or to assume a state demonstrated to be acceptable if conditions such as disconnection, loss of energy or motive power, or adverse environments are experienced.

The status and operation of hardware IROFS involving instrumentation that provides automatic prevention or mitigation of events will be monitored by the plant control system (PCS) by means of an alarm. This alarm will be provided by an isolated, hardwired digital signal from the associated IROFS to the PCS programmatic logic controller (PLC). This signal will only be directed from the associated IROFS to the PCS PLC. The required separation and isolation will be provided at the IROFS hardware interface in the process equipment for the connections to the PCS PLC. Consistent with ANSI/IEEE-279-1971, "Criteria for Protection Systems for Nuclear Power Generating Stations" (ANSI/IEEE, 1971), the isolation devices will be classified as part of the IROFS boundary and will be designed such that no credible failure at the output of the isolation device will prevent the associated IROFS from meeting its specified safety function. Additionally, the applicant has committed to the criteria contained in IEEE Standard 603-1998, "IEEE Standard Criteria for Safety Systems for Nuclear Power Generating Stations" (IEEE, 1998), for separation and isolation of IROFS systems and components.

The staff conducted an on-site review of both the classified and non-classified portions of the ISA and concludes that the applicant has appropriately addressed the baseline design criteria for new facilities, as required by 10 CFR 70.64.

Defense-in-Depth

The applicant based the facility and system designs on defense-in-depth practices. The design incorporates a preference for engineered controls over administrative controls, and to incorporate features that enhance safety by reducing challenges to IROFS. Facility and system IROFS are identified in ISA Summary Section 3.8 (LES, 2005b), and are described in ISA Summary Section 3.4 (LES, 2005b). The utility and support systems are described in ISA Summary Section 3.5 (LES, 2005b). The system descriptions identify their associated IROFS and additional design and safety features that provide defense-in-depth. The staff reviewed the identified safety features for each of the process systems described in the ISA Summary and determined that the applicant had established a preference for the selection of engineered controls over administrative controls to increase overall system reliability and reduce challenges to IROFS.

The staff concludes that the applicant has appropriately addressed the defense-in-depth practices for new facilities, as required by 10 CFR 70.64(b).

<u>Baseline Design Criteria and Defense-In-Depth Conclusion</u>

Based on the above, the staff concludes that the applicant has provided sufficient information in the ISA Summary to demonstrate that the baseline design criteria was applied in an acceptable manner to the design of the gas centrifuge uranium enrichment facility, and the facility and system design and facility layout will be based on defense-in-depth practices.

3.3.3   Safety Program and Integrated Safety Analysis

3.3.3.1   Safety Program

The staff reviewed the applicant's proposed safety program commitments identified in SAR Section 3.0 (LES, 2005a) to determine that the three elements of process safety information, integrated safety assessment, and management measures demonstrates compliance with the performance requirements of 10 CFR 70.61, that records are established and maintained to demonstrate compliance with 10 CFR 70.62(b) through (d), and that records are established and maintained documenting each discovery that an IROFS or management measure has failed to perform on demand or has degraded such that the performance requirements of 10 CFR 70.61 are not satisfied.

3.3.3.1.1   Process Safety Information

The applicant has developed process safety information program requirements in Section 3.0.1 of the Safety Analysis Report (SAR) (LES, 2005a). Program commitments require that up-to-date documentation of process safety information pertaining to the hazards of all materials used or produced in the process, the technology of the process, and the process equipment are compiled. The program commitments also require that process safety information be maintained up-to-date in accordance with the program elements described in SAR Section 11.1 (LES 2005a). SER Chapter 11 discusses the staff review and evaluation of the proposed program elements needed to maintain the process safety information documentation. Therefore, the staff concludes that if the program elements are followed, there is reasonable assurance that the applicant will compile and maintain process safety information up-to-date.

The applicant has committed to develop procedures and criteria for changing the ISA that meet the requirements of 10 CFR 70.72 in accordance with procedure development and implementation requirements contained in SAR Section 11.4 (LES, 2005a). Also, the ISA will be maintained by personnel with the appropriate experience and expertise in engineering and process operations. The ISA Team for the various processes consists of individuals who are knowledgeable in the ISA method(s) and the operation, hazards, and safety design criteria of the particular process. Training and qualifications of individuals responsible for maintaining the ISA are described in SAR Sections 2.2, 3.2, and 11.3 (LES, 2005a).

3.3.3.1.2   Integrated Safety Analysis

In SAR Section 3.0.2 (LES, 2005a), the applicant identifies ISA program commitments that were used to establish the ISA process. Those commitments include the performance of an ISA for each process that identifies the radiological hazards, chemical hazards that could increase radiological risk, facility hazards that could increase radiological risk, potential accident sequences, consequences and likelihood of each accident sequence, and IROFS including the assumptions and conditions under which they support compliance with the performance requirements of 10 CFR 70.61. Implementation of the Configuration Management Program described in SAR Section 11.1 (LES, 2005a) will maintain the ISA and supporting documentation accurate and up-to-date. The ISA update process will account for any changes to the facility or its processes. The update process will also verify that initiating event frequencies and IROFS reliability values assumed in the ISA remain valid. Management policies, organizational responsibilities, revision time frame, and procedures to perform and approve revisions to the ISA are outlined in SAR Section 11.0 (LES, 2005a). Personnel used to update and maintain the ISA and ISA Summary will be trained and qualified, as described in SAR Section 11.3 (LES 2005a). Proposed changes to the facility will be evaluated using the ISA method(s). All IROFS will be maintained available and reliable, and unacceptable performance deficiencies are addressed. The adequacy of existing IROFS and associated management measures are promptly evaluated to determine if they are impacted by changes to the facility. Unacceptable performance deficiencies associated with IROFS are addressed that are identified through updates to the ISA. Written procedures will be maintained onsite in accordance with the requirements of SAR Section 11.4 (LES, 2005a).

3.3.3.1.3 Management Measures

In SAR Section 3.0.3 (LES 2005a), the applicant describes management measures that comprise the principal mechanism by which the reliability and availability of each IROFS is ensured. General requirements applicable to each IROFS for configuration management, maintenance, training and qualifications, procedures, audits and assessments, incident investigation, records management, and other quality assurance elements, are discussed. Any management measures deviating from these general requirements, which are consistent with the performance requirements assumed in the ISA documentation, are discussed in the ISA Summary (LES, 2005b). Incident investigations are conducted within the applicant's Corrective Action Program. Incidents associated with IROFS, and any items that may affect the function of IROFS, include: processes that behave in unexpected ways; procedural activities not performed in accordance with an approved procedure; discovered deficiency, degradation, or non-conformance of an IROFS; or, any items that may affect the function of IROFS. Feedback from the results of incident investigations are used, as appropriate, to modify management measures to provide continued assurance of the availability and reliability of IROFS, in order to meet the performance requirements of 10 CFR 70.61. All records associated with IROFS, and any item that may affect the function of IROFS, will be managed and controlled in a systematic manner in order to provide identifiable and retrievable documentation. The management measures are further detailed in SAR Chapter 11 (LES 2005a), and evaluated in Chapter 11 of this Safety Evaluation Report (SER). All IROFS are Quality Level 1 items per the applicant's approved Quality Assurance Program Description.

3.3.3.1.4 Safety Program Conclusions

Based on the above information, the staff concludes that the applicant meets the requirements of 10 CFR 70.62(a)(1) through (3) to establish and maintain a safety program that includes process safety information, integrated safety analysis and management measures, and appropriate safety program records. The staff also concludes that the applicant has an appropriate program to establish and maintain records of IROFS failures that will be retrievable for NRC inspection.

3.3.3.2 Integrated Safety Analysis and ISA Summary

The staff reviewed the ISA Summary to confirm that it contained appropriate information pertinent to: (1) a general description of the site, (2) a general description of the facility, (3) a description of each process in sufficient detail to understand the theory of operation and the hazards that were identified in the ISA and a general description of the accident sequences; (4) information demonstrating compliance with the performance requirements of 10 CFR 70.61, including a description of management measures, the requirements for criticality monitoring and alarms, and the baseline design criteria; (5) a description of the team, qualifications, and the methods used to perform the ISA; (6) a list briefly describing each IROFS in sufficient detail to understand its function in relation to the performance requirements of 10 CFR 70.61; (7) a description of the proposed quantitative standards used to assess the consequences of chemical exposures; (8) a list of sole IROFS, and (9) a description of the definitions of "not unlikely," "unlikely," and "highly-unlikely," as used in the evaluations in the ISA.

The staff also reviewed selected portions of the applicant's Integrated Safety Analysis (ISA) to confirm that the applicant performed an ISA of appropriate detail for the proposed gas centrifuge uranium enrichment process, that identifies: radiological hazards, chemical hazards of licensed material and hazardous chemicals produced from licensed material; facility hazards; potential accident sequences caused by credible internal and external events, including natural phenomena; consequences and likelihood of occurrence of each potential accident sequence; each item relied on for safety (IROFS), as required by 10 CFR 70.62(c).

3.3.3.2.1 ISA Summary Content

ISA Summary Section 3.2 (LES, 2005b) contains the site description, and includes information on site geography, demographics and land use, meteorology, hydrology, geology, seismology, and stability of subsurface materials. Based on the review conducted in Section 3.3.1.1 of this SER, the staff has determined that the information provided is sufficient to comply with 10 CFR 70.65(b)(1).

ISA Summary Section 3.3 (LES, 2005b) contains the facility description, and includes information on the buildings and major components, and structural design criteria. Based on the review conducted in Section 3.3.1.2 of this SER, the staff has determined that the information provided is sufficient to comply with 10 CFR 70.65(b)(2).

ISA Summary Sections 3.4 and 3.5 (LES, 2005b) contain descriptions of the process and utility and support systems, Sections 3.3 describes the facility hazards and Sections 3.1 and 3.7 describe the radiological, and chemical hazards and accident sequences. Based on the review conducted in Section 3.3.1.3 of this SER, the staff has determined that the information provided is sufficient to comply with the requirements of 10 CFR 70.65(b)(3).

ISA Summary Section 3.1 (LES, 2005b) contains information regarding the ISA methods used to demonstrate compliance with the performance requirements of 10 CFR 70.61, a description of the management measures, the requirements for criticality monitoring and alarms, and baseline design criteria. The demonstration of meeting the performance requirements is contained in Sections 3.7 and 3.5 of the ISA Summary (LES, 2005b). Based on the review conducted in Section 3.3.1.1 of this SER and additional information in Chapters 4, 5, 6, and 7 of this SER, the staff has determined that this information is sufficient to comply with 10 CFR 70.65(b)(4).

ISA Summary Section 3.1 (LES, 2005b) contains information regarding the description of the ISA Team, qualifications, and the methods used to perform the ISA. Based on the review conducted in Section 3.3.3.2.2 of this SER, the staff has determined that this information is sufficient to comply with 10 CFR 70.65(b)(5).

ISA Summary Section 3.8, Table 3.8-1 (LES, 2005b) contains a list briefly describing each IROFS in sufficient detail to understand their functions in relationship to the performance criteria of 10 CFR 70.61. Based on the review conducted in Section 3.3.3.2.4 of this SER, the staff has determined that this information is sufficient to comply with 10 CFR 70.65(b)(6).

ISA Summary Section 3.1, and Tables 3.1-3 and 4 (LES, 2005b) contains information regarding the selection of quantitative standards used to assess the consequences to an individual from acute chemical exposure to licensed material or chemicals produced from licensed material. Based on the review conducted in Section 3.3.3.2.2.2 of this SER, the staff has determined that this information is sufficient to comply with 10 CFR 70.65(b)(7).

ISA Summary Section 3.8, Table 3.8-2 (LES, 2005b) contains a list of sole IROFS used for preventing or mitigating an accident sequence that exceeds the performance requirements of 10 CFR 70.61. Based on the review conducted in Section 3.3.3.2.5 of this SER, the staff has determined that this information is sufficient to comply with 10 CFR 70.65(b)(8).

ISA Summary Section 3.1 and Table 3.1-5 (LES, 2005b) contains a description of the definitions of unlikely, highly unlikely and credible as used in the evaluations in the ISA. Based on the review conducted in Section 3.3.3.2.2.2 of this SER, the staff has determined that this information is sufficient to comply with 10 CFR 70.65(b)(9).

ISA Summary Content Conclusion

Based on the staff's review of the ISA Summary, the staff concludes that the ISA Summary content complies with the requirements of 10 CFR 70.65(b).

3.3.3.2.2   ISA Introduction

The staff reviewed information concerning the ISA methodology located in SAR Section 3.1 (LES, 2005a) and ISA Summary Section 3.1.1 (LES, 2005b). Two ISA teams were employed, a classified ISA team to deal with the sensitive classified processes and a non-classified ISA team to analyze the remaining processes. The ISA was performed in accordance with a procedure developed to guide the conduct of the ISA, and several participants were employed on both ISA teams to ensure consistency of the results. The applicant used a semi-quantitative risk method consistent with "Example Procedure for Accident Sequence Evaluation," Appendix A to Chapter 3 of NUREG-1520 (NRC, 2002a). An integration checklist is used as a guide to facilitate the integrated review process. Included in the applicant's review is the identification of IROFS functions that may be simultaneously beneficial and harmful with respect to different hazards, and interactions that might not have been considered in previous subsets. The review is intended to ensure that the designation of one IROFS does not negate the preventive or mitigation function of another IROFS.

3-43

3.3.3.2.2.1   Integrated Safety Analysis Team Qualifications

In SAR Section 3.2 (LES, 2005a), the applicant describes the two ISA teams (classified and non-classified). To facilitate consistency of results, several members were employed for each team. The ISA was performed by teams with expertise in various technical disciplines and included personnel with experience and knowledge specific to each process or system being evaluated. The applicant states that the ISA Team Leader was trained and knowledgeable in the ISA method used. The ISA Manager is responsible for the overall direction of the ISA. The ISA Team Leader has an adequate understanding of the process, but is not the responsible cognizant engineer or enrichment process expert. Staff review of the team's credentials confirmed that they are consistent with NUREG-1520 (NRC, 2002a). The staff also finds that the above responsibilities and authorities are consistent with the guidance provided in NUREG-1520 (NRC, 2002a), and are, therefore, acceptable.

Based upon the above, the staff concludes that the applicant's ISA team composition and membership meet the requirements of 10 CFR 70.62(c)(2).

3.3.3.2.2.2   Hazard and Accident Sequence Identification

The applicant used a process hazard method (HAZOP) to identify potential hazards related to UF_6 process systems and Technical Services Building Systems and potential accident sequences. The HAZOP method is a structured technique commonly used in the chemical industry that is well suited to analyze processes during or after a detailed design stage. The method uses an interdisciplinary team to identify hazards and operability problems resulting from deviations from the process's design intent that could lead to undesirable consequences. In SAR Section 3.1.2 (LES, 2005a), the applicant states that implementation of the HAZOP method was accomplished by either validating the Urenco HAZOPs for the NEF design or performing a new HAZOP for systems where there were no existing HAZOPs. For cases where there was an existing HAZOP, the ISA team developed a new HAZOP through the validation process. The format of the HAZOP tables used in the ISA documentation is provided. The "Fire" and "External Event" guidewords are handled on a facility-wide basis. The applicant states that the HAZOPs were revised/updated as necessary to be consistent with the requirements of 10 CFR Part 70 and as described in NUREG-1513 (NRC, 2001) and NUREG-1520 (NRC, 2002a). The process hazards are described in the ISA Summary Section 3.6 (LES, 2005b), and the associated accident sequences are identified in ISA Summary Section 3.7 (LES, 2005b).

The HAZOP method is widely used in the chemical processing industry because it is suitable for performing a detailed analysis of a wide range of hazards to identify potential accident sequences. NUREG-1513, "Integrated Safety Analysis Guidance Document," Appendix A, "Flowchart for Selecting a Hazards Analysis Technique" (NRC, 2001), identifies the HAZOP technique as an acceptable approach. Therefore, the staff concludes that the process hazard analysis method used by the applicant is acceptable for the identification of potential radiological, chemical and facility hazards, and potential accident sequences caused by process deviations or other events internal to the facility and credible external events, including natural phenomena that could lead to a loss of UF_6 confinement or a criticality.

The hazard identification process used by the applicant documented materials that are radioactive, fissile, flammable, explosive, toxic, and reactive, and identified potentially hazardous conditions. Hazards were assessed individually for the potential impact on discrete components. However, hazards related to fires and external events were assessed on a facility-wide basis. The Fire Hazard Analysis (discussed in SAR Section 7.2 (LES, 2005a)) was consulted in order to place reasonable and conservative bounds on the fire scenarios. External events evaluated included seismic, tornado, tornado missile and high wind, snow and ice, flooding, local precipitation, transportation and nearby facility accidents, aircraft, pipelines, highway, railroad, and internal flooding from above ground storage tanks.

The applicant considered common mode failures and common cause situations, support system failures, divergent impacts of IROFS, non-IROFS impacts on system performance, multiple impact scenarios, system interactions and interdependence, and major hazards or events which tend to be common cause situations that could lead to interactions between processes, systems, buildings, etc.

SAR, Chapter 6, Tables 6.1-1 through 6.1-6 (LES, 2005a), identifies the hazardous properties of all chemicals used on site, and the inventories and locations of chemicals of concern (including UF_6). Potential interactions involving UF_6 and any reaction products are identified in SAR, Section 6.1.2 (LES, 2005a). These chemicals include HF and UO_2F_2. This section identifies the physical properties, reactivity, toxicological properties, and flammability of these chemicals. Chemical reactions and interactions involving UF_6 and water, Fomblin oil, chemical trap materials, and other materials used in the process are described in SAR, Section 6.2.1 (LES, 2005a). SAR Section 7.2 describes the fire hazards analysis for those facility areas containing licensed material. ISA Summary Chapter 3 identifies the external hazards. Other general types of process hazards are briefly discussed in ISA Summary, Section 3.6 (LES, 2005b). The applicant identified either a loss of confinement (of UF_6) or a criticality as the hazard of concern in ISA Summary, Section 3.1.4 (LES, 2005b). The staff agrees that all accident sequences at a gas centrifuge uranium enrichment facility that could exceed the performance requirements of 10 CFR 70.61 are the direct result of either a loss of confinement (of UF_6) or a criticality. Potential accident sequences that could result in a UF_6 release or criticality of high or intermediate consequence are discussed in detail in ISA Summary Section 3.7 (LES, 2005b).

Based on the process hazard analysis method used to identify hazards, a review of the information in the application and ISA Summary and the staff's visit to a similar gas centrifuge uranium enrichment facility in Almelo, The Netherlands, the AREVA engineering offices in Marlborough, Massachusetts, and the applicant's offices in Washington, D.C., the staff has reasonable assurance that the applicant has identified all of the potential internal and external hazards and accident sequences that could affect the safety of licensed materials, and has identified the materials and conditions that could result in a hazardous condition, including the expected inventory amounts and locations. Potential interactions between materials or conditions that could result in hazardous conditions were appropriately considered.

Risk Matrix Development

The applicant utilized a risk index method in order to categorize accident sequences in terms of likelihood of occurrence and consequences. This evaluation identifies the accident sequences which could exceed the performance requirements of 10 CFR 70.61 and, therefore, require designation of Items Relied on for Safety (IROFS). The staff reviewed the methods used by the applicant to determine the consequences and likelihood of each potential accident sequence identified through the HAZOP process. In SAR Section 3.1.3.3 (LES, 2005a), the applicant describes the consequence verses likelihood risk matrix. The risk matrix and computed index values are shown in SAR Table 3.1-6 (LES, 2005a). Use of the risk matrix is demonstrated in ISA Summary Section 3.1 (LES, 2005b).

Consequence Analysis Method

In SAR Section 3.1.3.1 (LES, 2005a), the applicant used the dispersion method discussed in SAR Section 6.3.2 (LES, 2005a), for determining the chemotoxic exposure to HF and UO_2F_2. The radiological and chemical consequence severity levels are provided in SAR Table 3.1-3 (LES, 2005a). Information on the chemical dose limits specific to NEF are found in SAR Table 3.1-4 (LES, 2005a). The applicant developed credible accident scenarios, and the dispersion analysis and chemical/radiological dose assessment associated with each accident sequence in accordance with NUREG/CR-6410, "Nuclear Fuel Cycle Facility Accident Analysis Handbook" (NRC, 2002b). The consequences of an inadvertent criticality were conservatively assumed to be high for both the public and the workers. For accident sequences postulated to result in an inadvertent criticality, IROFS are specified to ensure subcriticality under all normal and credible abnormal conditions.

Based on the above, the staff finds that applicant's method for consequence determination is consistent with the guidance described in NUREG-1513 (NRC, 2001) and NUREG-1520 (NRC, 2002a), and therefore, acceptable.

Consequence Categories

Accident sequences identified by the applicant as a result of the process hazards analysis are categorized as either a "high" consequence event, an "intermediate" consequence event, or are considered a low consequence event in accordance with the performance criteria of 10 CFR 70.61. ISA Summary Table 3.1-3 (LES, 2005b) identifies the consequence severity categories for workers, offsite public and the environment for defined radiation and chemical doses. The values proposed by the applicant for both radiological dose and chemical dose are consistent with the guidance in NUREG-1520, Appendix A, Table A-5 (NRC, 2002a). All of the identified high-consequence and intermediate-consequence events are listed in the ISA Summary (LES, 2005b).

Quantitative Standards for Chemical Consequences

The applicant identifies the radiological and chemical dose consequence categories for workers, offsite public, and environment in ISA Summary Table 3.1-3 (LES, 2005b), and SAR Table 6.3-2 (LES, 2005a). The chemical dose consequence categories are consistent with the guidance contained in NUREG-1520, Table A-1 (NRC, 2002a). SAR Tables 6.3-3 and 6.3-4 (LES, 2005b) contain the AEGL values for HF and UF_6, respectively. The applicant proposed a worker-exposure strategy that incorporates 10-minute AEGL values for HF, as used in NUREG-1391 (NRC, 1991). The staff finds the use of AEGL standards to be acceptable because these are unambiguous quantitative standards developed by the National Advisory Committee for Acute Guideline Levels for Hazardous Substances (AEGLs) that are used nationally in a broad application for emergency planning, response, and prevention in the community, the workplace, transportation and remedial action. Application of the consequence categories is discussed in Section 6.3.2.1 of the SAR. The quantitative standards that the consequence categories are based on are in accordance with 10 CFR 70.65(b)(7) and are consistent with the standards in 10 CFR 70.61.

Likelihood Determination

The likelihood evaluation method is discussed in SAR Section 3.1.3.2, (LES, 2005a). The definitions of "not unlikely," "unlikely," and "highly unlikely" are in SAR Table 3.1-5 (LES, 2005a). "Not unlikely" is defined as "Likelihood Category 3" with a probability of occurrence of more than 10^{-4} per event per year. "Unlikely" is defined as "Likelihood Category 2" with a probability of occurrence of between 10^{-4} and 10^{-5} per event per year. "Highly unlikely" is defined as "Likelihood Category 1" with a probability of occurrence of less than 10^{-5} per event per year. The staff concludes the definitions are acceptable because they are consistent with the guidance provided in NUREG-1520 (NRC, 2002a).

Not-Credible Determination

In SAR Section 3.1.3.2, (LES, 2005a) the applicant states that the definition of "not credible" is taken from NUREG-1520 (NRC, 2002a), and any one of the following independent acceptable sets of qualities could define an event as not credible:

a. An external event for which the frequency of occurrence can conservatively be estimated as less than once in a million years;

b. A process deviation that consists of a sequence of many unlikely human actions or errors for which there is no reason or motive (In determining that there is no reason for such actions, a wide range of possible motives, short of intent to cause harm, must be considered. Necessarily, no such sequence of events can ever have actually happened in any fuel cycle facility.)

c. Process deviations for which there is a convincing argument, given physical laws that they are not possible, or are unquestioningly extremely unlikely.

The staff concludes that the above definition of "not credible" is consistent with the regulatory guidance and suitable for use in determining the likelihood performance requirements of 10 CFR 70.61.

Likelihood Evaluation Method

ISA Summary Section 3.1.1.3.2 (LES, 2005b) discuss the definitions of "Not Unlikely," "Unlikely," and "Highly Unlikely," and ISA Summary, Table 3.1-5 (LES, 2005b), cross references the three likelihood categories with a probability of occurrence based on approximate order of magnitude ranges. The proposed values are consistent with NUREG-1520, Chapter 3, Appendix A, Table A-6 (NRC, 2002a). Implementation of the evaluation method is discussed in ISA Summary Section 3.1.1.3 (LES, 2005b). ISA Summary Tables 3.7-1 through 3.7-4 (LES, 2005b), show how each designated IROFS acts to prevent or mitigate the consequences of an accident sequence. When multiple IROFS are designated, IROFS interactions are considered. The likelihood of failure was qualitatively evaluated for each IROFS, often based on the operational history of similar facilities. Each sequence was evaluated as either "Not Unlikely," "Unlikely," or "Highly Unlikely."

Safe-By-Design Evaluation Method for Nuclear Criticality Safety

In SAR Section 3.1.3.2 (LES, 2005a), the applicant described a safe-by-design ISA method for selected equipment for nuclear criticality safety (NCS) used to identify safe-by-design components, the failure of which would be highly unlikely. The applicant described the connection between subcriticality and the safe-by-design ISA process for NCS. Using the safe-by-design ISA process, there are no accident sequences and, hence, IROFS are not identified because it is highly unlikely these components would fail. Those safe-by-design components are considered items which may affect IROFS.

A qualitative determination of highly unlikely can apply to passive design component features of the facility that do not rely on human interface to perform the NCS function. Safe-by-design components are those components that by their physical size or arrangement have been shown to have a $k_{eff} < 0.95$. The definition of safe-by-design components encompasses two different categories of components. The first category includes those components that are safe-by-volume, safe-by-diameter, or safe-by-slab thickness (i.e., favorable geometry components). A set of generic, conservative NCS calculations has determined the maximum volume, diameter, or slab thickness that would result in a $k_{eff} < 0.95$. A favorable geometry component has a volume, diameter, or slab thickness that is less than the associated value for $k_{eff} < 0.95$. The components in the second category (i.e., non-favorable geometry components) require a more detailed NCS Analysis to demonstrate $k_{eff} < 0.95$. For the non-favorable geometry components, the design configuration is not bounded by the results of the generic, conservative NCS calculations for maximum volume, diameter, or slab thickness that would result in a $k_{eff} < 0.95$.

For failures of these passive safe-by-design components (i.e., both favorable geometry components and non-favorable geometry components) to be considered highly unlikely, those components must also meet the criterion that the only potential means to effect a change that might result in a failure to function would be to implement a design change (i.e., no potential failure mode exists). The evaluation of the potential to adversely impact the safety function of these design features includes consideration of potential mechanisms to cause bulging, corrosion, or breach of confinement/leakage and the subsequent accumulation of material. The evaluation further includes consideration of adequate controls to ensure that the double contingency principle is met. For each of these passive design components (i.e., both favorable geometry components and non-favorable geometry components), it must be concluded that there is no credible means to effect a geometry change that might result in a failure of the safety function and that significant margin exists.

For favorable geometry components, significant margin is defined as a margin of at least 10 percent, during both normal and upset conditions, between the actual design parameter value of the component and the value of the corresponding critical design attribute. For non-favorable geometry components, significant margin is defined as $k_{eff} < 0.95$, where $k_{eff} = k_{calc} + 3\sigma_{calc}$. This calculation of k_{eff} conservatively assumes the components are full of uranic breakdown material at maximum enrichment, with worst credible moderation, and with worst credible reflection.

These passive, safe-by-design features (i.e., both favorable geometry components and non-favorable geometry components) are considered items that may affect IROFS. As a result, Quality Level 1 requirements apply to these features. Also, the configuration management program required by 10 CFR 70.72 ensures the maintenance of the safety function of these features and assures compliance with both the double contingency principle and the defense-in-depth criterion of 10 CFR 70.64(b).

Additionally, the applicant commits to apply the guidelines of the ASME QA standard NQA-1-1994, "Quality Assurance Program Requirements for Nuclear Plants," including supplements as revised by the ASME NQA-1-a-1995 Addenda (ASME, 1994), to all IROFS and any item, structures, systems, controls, and administrative controls that are determined to affect the functions of the IROFS, and, in general, to items required to satisfy regulatory requirements. All IROFS are Quality Level 1 components.

Safe-By-Design Highly-Unlikely Determination for NCS

SER Chapter 5 provides the staff's review of the applicant's demonstration for meeting highly unlikely for safe-by-design components.

Risk Index Evaluation

The staff reviewed the applicant's risk index evaluation description in SAR Section 3.1.4 (LES, 2005a). The results of the ISA are summarized in tabular form. The risk index evaluation process cross references an accident sequence consequence with its likelihood to first determine the total risk, and then to demonstrate that with the application of the selected IROFS, the performance requirements of 10 CFR 70.61 are met. The ISA results are summarized in tabular form, with columns for the initiating event and IROFS. For sequences that place the system in a vulnerable state, the duration of the vulnerable state is considered, and a duration index assigned. The values of all index numbers in a sequence are added to obtain a total likelihood index, T. Accident sequences are then assigned to one of three likelihood categories, depending on the value of the index in accordance with SAR Table 3.1-8 (LES, 2005a). The criteria of SAR Tables 3.1-9 through 3.1-11 are used to assign index numbers to accident sequences. The staff reviewed the risk index evaluation for selected accident scenarios in the fire, chemical, and criticality safety areas as part of the "vertical-slice" review and determined that the applicant's evaluation process was appropriately applied. The staff concludes that the risk evaluation methodology described by the applicant is consistent with the guidance in NUREG-1520 (NRC, 2002a) and suitable for determining which accident sequences require IROFS, and the level of risk reduction provided the IROFS to comply with 10 CFR 70.61.

3.3.3.2.3   Compliance Item Commitments

In SAR Sections 3.3.1 through 3.3.5 (LES, 2005a), the applicant committed to: (1) provide justification for having Initiating Event Frequency (IEF) index numbers for certain accident sequences or else revise the ISA Summary; (2) provide justification for having Failure Probability Index Numbers (FPINs) when using Administrative Controls with 'use of' a component or device or else revise the ISA Summary; (3) provide justification for having FPINs when using Administrative Controls with 'verification of' a state or condition or else revise the ISA Summary; (4) use three of four criteria when using Administrative Controls with 'independent sampling;' and (5) meet all criteria when using Enhanced Administrative Controls, which require 'independent verification' of a safety function.

3.3.3.2.4   Descriptive List of IROFS

ISA Summary, Table 3.8-1 (LES, 2005b), lists all IROFS identified in ISA Summary Section 3.7 (LES, 2005b). The IROFS address hazards that could result in a loss of UF$_6$ confinement and/or possible criticality events. Safety functions addressed include radiological safety, chemical safety, criticality safety, and fire protection. The staff notes that no IROFS are identified as a result of using the alternate "safe-by-design" ISA method because "highly unlikely" is achieved with a significant margin. The staff concludes that the applicant has provided IROFS for all identified high and intermediate consequence accident sequences.

ISA Summary Table 3.8-1 (LES, 2005b) also describes the safety function of each IROFS and the reliability management measures. More detailed descriptions of each IROFS are contained in ISA Summary Sections 3.4 and 3.5 (LES, 2005b). Hardware control descriptions contain sufficient information to permit evaluation of its reliability. Safety limits on key parameters are identified. Administrative control IROFS are adequately described to permit an understanding that adherence to it should be reliable. Additionally, the applicant commits to apply the guidelines of the ASME QA standard NQA-1-1994, "Quality Assurance Program Requirements for Nuclear Plants," including supplements as revised by the ASME NQA-1-a-1995 Addenda (ASME, 1994), to all IROFS and any item, structures, systems, controls, and administrative controls that are determined to affect the functions of the IROFS, and, in general, to items required to satisfy regulatory requirements. All IROFS are Quality Level 1 components.

The applicant states that it will ensure that IROFS components and systems will be able to perform their required safety functions under normal and accident conditions (e.g., pressure, temperature, humidity, seismic motion, electromagnetic interference, and radiofrequency interference) as required by the ISA. The applicant states that IROFS will be qualified using the applicable guidance in Institute of Electrical and Electronics Engineers (IEEE) Standard IEEE-323, 1983, "IEEE Standard for Qualifying Class 1E Equipment for Nuclear Power Generating Stations" (IEEE, 1983). Furthermore, IROFS components and systems will be designed, procured, installed, tested, and maintained using the applicable guidance in Regulatory Guide 1.180, "Guidelines for Evaluating Electromagnetic and Radio-Frequency Interference in Safety-Related Instrumentation and Control Systems," Revision 1, dated October 2003 (NRC, 2003a). Redundant IROFS systems will be separate and independent from each other. IROFS systems will be designed to be fail-safe. Process control system failures will not affect the ability of the IROFS systems to perform their required safety functions. Required testing and calibration of IROFS will be consistent with the assumptions of the ISA and setpoint calculations, as applicable. For hardware IROFS involving instrumentation that provides automatic prevention or mitigation of events, setpoint calculations are performed in accordance with a setpoint method, which is consistent with the applicable guidance provided in Regulatory Guide 1.105, "Setpoints for Safety-Related Instrumentation," Revision 3 (NRC, 1999a).

In SAR Section 3.3.6 (LES, 2005a), the applicant states that upon completion of the final design, the IROFS boundaries will be defined. In defining the boundaries of each IROFS, LES procedure, "IROFS Boundary Definitions," will be used. The procedure requires identification of each support system and component necessary to ensure the IROFS is capable of performing its specified safety function. The procedure also requires identification of the management measures necessary to support the IROFS availability and reliability ISA assumptions. Because historical failure data was not used to derive all of the indices in ISA Summary Table 3.7-1, once the facility is operating, failure data will be trended and the impact of this failure data on the values assumed in the ISA will be evaluated to validate those assumptions. By completing this procedure for each IROFS, the applicant will more specifically: (1) identify and establish engineered and administrative IROFS to protect against the credible accident sequences; (2) identify and establish management measures for specific IROFS to protect against the credible accident sequences; and (3) confirm that the credible accident sequences are highly unlikely, and that credible intermediate accident sequences are unlikely. In conjunction with the applicant's configuration management program, the IROFS boundary provides assurance that IROFS will be maintained available and reliable over the life of the facility.

SER Chapter 11 discusses the management measures that the applicant will apply to IROFS and items that could affect IROFS to assure that they are reliable and available when called upon to perform their intended safety functions.

The staff reviewed the IROFS for selected accident sequences involving chemical hazards, criticality safety and fire protection, as described in SER Chapters 5 through 7.

Under 10 CFR 70.72(d)(3), for all changes that affect the ISA Summary, the licensee must submit to NRC annually revised pages of the ISA Summary. Any changes to IROFS would be included in the annual revisions to the ISA Summary.

The staff concludes, pursuant to 10 CFR 70.23(a), that the preliminary design bases of the IROFS evaluated in this SER section will provide reasonable assurance of protection against natural phenomena and the consequences of potential accidents. Given that these conclusions are based on preliminary design information, the staff is imposing a license condition to ensure that the final design is adequate and acceptable to the staff. Specifically, the following condition will be included in the license:

"The applicant shall utilize its procedure, "IROFS Boundary Definitions," to define the boundaries of each IROFS. Completed IROFS boundaries for all IROFS shall be available for inspection at the time of the operational readiness review."

The applicant has committed to conforming to the applicable guidance of the codes and standards and regulatory guidance listed in SAR Section 3.3.7 (LES, 2005a), for the design, procurement, installation, testing, and maintenance of I&C related IROFS to comply with the performance requirements of 10 CFR 70.61.

The staff concludes that the applicant has adequately described the management measures applied to IROFS (including safety grading), characteristics of its preventive, mitigative, or other safety function, and assumptions and conditions, under which the item is relied on to support compliance with the performance requirements of 10 CFR 70.61.

3.3.3.2.5   List of Sole IROFS

ISA Summary, Table 3.8-2 (LES, 2005b), identifies each sole IROFS and cross references it to the specific accident sequences for which it is applicable. The accident sequences in turn provide a clear and unambiguous reference to the process to which the item applies. The list includes a descriptive title of the IROFS, and a clear and traceable reference to the description of the item as it appears in the full list of all IROFS. Therefore, the staff concludes that the applicant has adequately described all IROFS that are the sole item for preventing or mitigating an accident sequence identified by the applicant.

3.3.3.3 Summary of ISA Results

3.3.3.3.1 Hazards

The applicant's ISA uses the HAZOP method for identifying the hazards for UF_6 process systems, the Technical Services Building systems, the Centrifuge Assembly Building systems, and the Uranium Byproduct Storage Pad. The HAZOP analysis was applied to discrete process components. Radiological hazards identification considered the characteristics of uranium enriched to 5 percent (see Chapter 4 of this SER). Criticality hazards identification was performed for areas of the facility where fissile material is expected to be present (see Chapter 5 of this SER). Chemical hazards identification included those from licensed material and chemicals produced from licensed material, including chemical interactions (see Chapter 6 of this SER). Fire hazards identification considered in-situ and transient combustible sources and the use of fire barriers (see Chapter 7 of this SER). External hazards were considered at the site and facility level (see Chapter 3 of the SER). The applicant identified and listed hazards that could result in accident sequences exceeding the performance requirements of 10 CFR 70.61 in ISA Summary Tables 3.7-1 through 3.7-4. The HAZOP method is an acceptable hazard identification method per the guidance provided in NUREG-1513 (NRC, 2001).

Based on the above, the staff has reasonable assurance that the applicant used an acceptable hazard identification technique and identified all of the radiological hazards relating to possessing or processing licensed material; the chemical hazards of licensed material and hazardous chemicals produced from licensed material; facility hazards that could affect the safety of licensed material, and thus, present an increased radiological risk; and hazards related to process deviations or other events internal to the facility and credible external events, including natural phenomena.

3.3.3.3.2 Accident Sequences

In ISA Summary Tables 3.7-1 through 3.7-4, the applicant lists and describes the identified accident sequences and external and fire events for which the consequences could exceed the 10 CFR 70.61 performance requirements. ISA Summary Table 3.8-1 identifies each IROFS listed in the ISA Summary and how it protects against each accident sequence. The information is sufficient to determine how each accident sequence that could exceed the performance requirements of 10 CFR 70.61 is protected against by IROFS. The staff performed a review of selected high-consequence and intermediate-consequence events in the areas of chemical safety, criticality safety, and fire protection to confirm that the applicant had properly identified and analyzed accident sequences and the related consequences. The staff identified one unanalyzed accident sequence with consequences that exceeded the performance requirements that were overlooked by applicant. The accident sequence involves either a feed or depleted uranium cylinder shipment truck fire, as discussed in SER Section 7.3.2.2. The applicant addressed the accident sequence and identified additional IROFS.

Based on the above, the staff has reasonable assurance that the applicant has identified and analyzed all accident sequences that could potentially exceed the 10 CFR 70.61 performance requirements by resulting in high or intermediate consequences.

3.3.3.3.3 IROFS and Management Measures

The only accident sequence types that can potentially result in intermediate or high consequences at this facility are loss of confinement events (i.e., caused by process upsets, human error, natural phenomena, fires, and external events) and criticality accidents (i.e., which are assumed to have a high consequence to the worker). Management measures are identified to ensure the IROFS are available and reliable when needed to perform their safety function (see SER Chapter 11). All IROFS are designated Quality Level 1 items under the applicant's approved Quality Assurance Program Description. Safe-by-design components do not have IROFS identified because a failure is considered highly unlikely (i.e., there is no credible way to effect a geometry change that might result in a failure of the safety function). However, safe-by-design components are considered items which may affect IROFS, and the configuration management system required by 10 CFR 70.72, which is implemented by the facility Configuration Management Program, ensures the maintenance of the safety function of these features. See SER Chapter 5 for a discussion of IROFS and safe-by-design components related to criticality safety, Chapter 6 for chemical safety IROFS, and Chapter 7 for fire safety IROFS. The staff conducted a sample review of selected IROFS related to chemical safety, criticality safety and fire protection for selected accident sequences. The staff review identified the need for additional IROFS related to a cylinder shipment truck fire that were omitted by the applicant, as discussed in SER Section 7.3.2.2. The applicant subsequently identified appropriate IROFS for this accident scenario.

Based on the above, the staff concludes that the applicant has identified suitable IROFS to make intermediate consequence events unlikely and high consequence events highly unlikely, and suitable management measures are identified to make the IROFS available and reliable when called upon to perform their safety function.

3.3.3.3.4 Meeting the Performance Requirements of 10 CFR 70.61

Based on the above information regarding hazards, accident sequences, IROFS, and management measures, the staff concludes that the applicant has: (1) identified hazards related to this type of facility; (2) identified credible events that could exceed the performance requirements of 10 CFR 70.61 through the application of appropriate accident identification method in accordance with NUREG-1513 (see SER Chapter 3); (3) identified appropriate chemical dose and radiological dose values for determining intermediate consequence and high consequence events (see SER Section 3.3.3.b.2); (4) determined the consequences in accordance with the fuel facility accident analysis guidance in NUREG-6410 (see SER Chapters 3 and 6)(NRC, 2002b); (5) established appropriate definitions for likelihood; and (6) applied those definitions in an acceptable manner to demonstrate that intermediate consequence events are unlikely and high consequence events are highly unlikely. For safe-by-design components, the staff concludes that the applicant has identified hazards and demonstrated that failure of safe-by-design components will be highly unlikely.

3.4 EVALUATION FINDINGS

Many hazards and potential accidents can result in unintended exposure of persons to radiation, radioactive materials, or toxic chemicals incident to the processing of licensed materials. The NRC staff finds that the applicant has performed an ISA to identify and evaluate those hazards and potential accidents as required by the regulations. The NRC staff has reviewed the ISA Summary and other information, and finds that it provides reasonable assurance that the applicant has identified IROFS and established engineered and administrative controls to ensure compliance with the performance requirements of 10 CFR 70.61. Specifically, the NRC staff finds that the ISA results, as documented in the ISA Summary, provide reasonable assurance that failure of the safe-by-design components will be highly unlikely and IROFS, management measures, and the applicant's programs will, if properly implemented, make all credible intermediate consequence accidents unlikely, and all credible high consequence accidents highly unlikely. NRC staff finds that the applicant has adequately demonstrated that safe-by-design components for NCS are highly unlikely, as required by the regulations.

In the facility SAR, the applicant provided preliminary design basis information for I&C systems that it identified as IROFS for the facility. The design information is at the system functional level. Individual components and vendors had not yet been selected. Based on the staff's review of the SAR, supporting information provided by the applicant, and the applicant's commitments to the industry standards and guidance cited in the sections above for I&C systems, the staff finds that the preliminary design meets the requirements of 10 CFR 70.61 and 70.64(a)(10).

Given that these conclusions are based on preliminary design information and the possibility that the applicant may choose to implement design changes as discussed in the previous section on I&C, the staff is imposing a license condition to ensure that the final design is adequate and acceptable to the staff. Specifically, the following condition will be included in the license:

> "Currently, there are no IROFS that have been specified as using software, firmware, microcode, PLCs, and/or any digital device, including hardware devices which implement data communication protocols (such as fieldbus devices and Local Area Network controllers), etc. Should the design of any IROFS be changed to include any of the preceding features, the licensee shall obtain Commission approval prior to implementing the change(s). The licensee's design change(s) shall adhere to accepted best practices in software and hardware engineering, including software quality assurance controls as discussed in the in the Quality Assurance Program Description (LES, 2005a) throughout the development process and the applicable guidance of the following industry standards and regulatory guides as specified in SAR Chapter 3 (LES, 2005a):
>
> a.   American Society of Mechanical Engineers (ASME) NQA-1-1994, Part II, subpart Part 2.7, "Quality Assurance Requirements of Computer Software for Nuclear Facility Applications," as revised by NQA-1a-1995 Addenda of NQA-1-1994 and ASME NQA-1-1994, Part 1, Supplement 11S-2, "Supplementary Requirements for Computer Program Testing." (Refer to SAR Chapter 11, Appendix A, Section 3.)
>
> b.   Electric Power Research Institute (EPRI) NP-5652, "Guideline for the Utilization of Commercial Grade Items in Nuclear Safety Grade Applications," June 1988.

c. EPRI Topical Report (TR) -102323, "Guidelines for Electromagnetic Interference Testing in Power Plants," Revision 1, December 1996.

d. EPRI TR-106439, "Guideline on Evaluation and Acceptance of Commercial Grade Digital Equipment for Nuclear Safety Applications," October 1996.

e. Regulatory Guide 1.152, "Criteria for Digital Computers in Safety Systems in Nuclear Power Plants," Revision 1, January 1996.

f. Regulatory Guide 1.168, "Verification, Validation, Reviews, and Audits for Digital Software Used in Safety Systems of Nuclear Power Plants," Revision 1, February 2004.

g. Regulatory Guide 1.169, "Configuration Management Plans for Digital Computer Software Used in Safety Systems of Nuclear Power Plants," September 1997.

h. Regulatory Guide 1.170, "Software Test Documentation for Digital Computer Software Used in Safety Systems of Nuclear Power Plants," September 1997.

I. Regulatory Guide 1.172, "Software Requirements Specifications for Digital Computer Software Used in Safety Systems of Nuclear Power Plants," September 1997.

j. Regulatory Guide 1.173, "Developing Software Life Cycle Processes for Digital Computer Software Used in Safety Systems of Nuclear Power Plants," September 1997.

If any above changes result in IROFS requiring operator actions, the licensee shall conduct a human factors engineering review of the human-system interfaces using the applicable guidance in NUREG-0700, "Human-System Interface Design Review Guidelines," Revision 2, dated May 2002 (NRC, 2002d), and NUREG-0711, "Human Factors Engineering Program Review Model," Revision 2, dated February 2004."

The staff concludes, pursuant to 10 CFR 70.23(a), that the preliminary design bases of the IROFS evaluated in SER Section 3.3.3.2.4 will provide reasonable assurance of protection against natural phenomena and the consequences of potential accidents. Given that these conclusions are based on preliminary design information, the staff is imposing a license condition to ensure that the final design is adequate and acceptable to the staff. Specifically, the following condition will be included in the license:

"The applicant shall utilize its procedure, "IROFS Boundary Definitions," to define the boundaries of each IROFS. Completed IROFS boundaries for all IROFS shall be available for inspection at the time of the operational readiness review."

3.5 REFERENCES

(ACI, 1990) American Concrete Institute (ACI). ACI 349, "Code Requirements for Nuclear Safety Related Concrete Structures." Farmington Hills, Michigan: American Concrete Institute. 1990.

(ACI, 1999) American Concrete Institute (ACI). ACI 318, "Building Code Requirements for Structural Concrete." Farmington Hills, Michigan: American Concrete Institute. 1999.

(AIChE, 1989) American Institute of Chemical Engineers (AIChE). "Guidelines for Chemical Process Quantitative Risk Analysis." Center for Chemical Process Safety, New York. 1989

(AIChE, 1992) American Institute of Chemical Engineers (IChE). "Guidelines for Hazard Evaluation Procedures, Second Edition with Worked Examples." Center for Chemical Process Safety, New York. 1992.

(AISC, 1989) American Institute of Steel Construction (AISC). "Allowable Stress Method," 1989.

(ANSI, 1994) American National Standard Institute/American Nuclear Society (ANSI/ANS). ANSI/ANS-3.2, "Administrative Controls and Quality Assurance for the Operational Phase of Nuclear Power Plants," 1994.

(ANSI, 1995) American National Standard Institute (ANSI). ANSI N14.1, "American National Standard for Nuclear Materials - Uranium Hexafluoride - Packaging for Transport," 1995.

(ANSI/IEEE, 1971) American National Standard Institute/Institute of Electrical and Electronics Engineers (ANSI/IEEE). ANSI/IEEE 279, "Criteria for Protection Systems for Nuclear Power Generating Stations," 1971.

(ASCE, 1980) American Society of Civil Engineers (ASCE). ASCE Manuals and Reports on Engineering Practice No. 58, "Structural Analysis and Design of Nuclear Plant Facilities." Reston, Virginia, 1980.

(ASCE, 1998) American Society of Civil Engineers (ASCE). ASCE 7, "Minimum Design Loads for Buildings and Other Structures." 1998.

(ASCE/SEI, 2005) American Society of Civil Engineers (ASCE)/Structural Engineering Institute (SEI). ASCE/SEI 43-05, "Seismic Design Criteria for Structures, Systems, and Components in Nuclear Facilities," 2005.

(ASME, 1994) American Society of Mechanical Engineers (ASME). Quality Assurance (QA) NQA-1-1994, "Quality Assurance Program Requirements for Nuclear Plants," 1994.

(CCPS, 2002) Center for Chemical Process Safety. "Deflagration and Detonation Arresters," 2002. (August 7, 2004) <http://www.knovel.com/knovel2/Toc.jsp?BookID=787?>

(Dobry, 2000) Dobry, R., et al. "New Site Coefficients and Site Classification System Used Recent Building Seismic Code Provisions." *Earthquake Spectra*. Vol. 16, No. 1. 2000.

(DOE, 2002) DOE–STD–1020–2002, "Natural Phenomena Hazards Design and Evaluation Criteria for Department of Energy Facilities," 2002.

(DOE, 2003) U.S. Department of Energy (DOE). DOE/WIPP–95–2065, Rev. 7, "Waste Isolation Pilot Plant Contact-Handled (CH) Waste Safety Analysis Report," 2003.

(Fisher, 1991) Fisher, et al. "Evaluation of Health Effects in Sequoyah Fuels Workers from Accidental Exposure to Uranium Hexafluoride," PNL-7328 (NUREG/CR 5566) Pacific Northwest Laboratory, Richland, WA, 1990.

(Frankel, 1996) Frankel, A., et al. "National Seismic Hazards Maps, June 1995 Documentation." U.S. Geological Survey. 1996.

(Harper, 2003a) Harper, G.A. "Assessment of Other External Event Hazards at NEF for ISA and Design Basis." Document Identifier 51-2400552-00, Framatome ANP, Inc., 2003.

(Harper, 2003b) Harper, G.A. "Assessment of Tornado, Tornado Missiles and High Wind Loads at NEF for ISA and Design Basis—Attachment 2." Document Identifier 51-2400548-00, Framatome ANP, Inc., 2003.

(Harper, 2003c) Harper, G.A. "Assessment of Winter Precipitation Loads at NEF for ISA and Design Basis." Document Identifier 51-2400547-00, Framatome ANP, Inc., 2003.

(ICBO, 1997) International Conference of Building Officials, Uniform Building Code, 1997.

(ICC, 2000) International Code Council, International Plumbing Code, Falls Church, Virginia, 2000.

(IEEE, 1983) Institute of Electrical and Electronics Engineers (IEEE). IEEE 323, "IEEE Standard for Qualifying Class 1E Equipment for Nuclear Power Generating Stations," 1983.

(IEEE, 1998) Institute of Electrical and Electronics Engineers (IEEE). IEEE 603, "IEEE Standard Criteria for Safety Systems for Nuclear Power Generating Stations," 1998.

(LES, 1993) Louisiana Energy Services (LES). "Safety Analysis Report for Claiborne Enrichment Center (CFC)" (1993)

(LES, 2004a) Louisiana Energy Services (LES), letter to U.S. Nuclear Regulatory Commission, "Clarifying Information Related to External Hazards Analysis," August 31, 2004.

(LES, 2004b) Louisiana Energy Services (LES). Letter to U.S. Nuclear Regulatory Commission, "Clarifying Information Related to Probabilistic Seismic Hazard Analysis," November 22, 2004.

(LES, 2005a) Louisiana Energy Services (LES). "National Enrichment Facility Safety Analysis Report," Revision 6, 2005.

(LES, 2005b) Louisiana Energy Services (LES). "Integrated Safety Analysis Summary," Revision 4, 2005.

(LES, 2005c) Louisiana Energy Services (LES). "National Enrichment Facility Environmental Report," Revision 5, 2005.

(McDonald, 1995) McDonald, J.R. and D. Lu. "A Methodology for Tornado Hazard Probability Assessment." Proceeding for the 5th Department of Energy Natural Phenomena Hazard Mitigation Conference, Denver, Colorado, November 13–14, 1995. 1995.

(McGuire, 1976) McGuire, R.K., "EQRISK, Evaluation of Earthquake Risk to a Site," U.S. Geologic Survey Open File Report 76-67, 1976.

(NAESC, 1999) North Atlantic Energy Service Corporation (NAESC). "Seabrook Station Updated Final Safety Analysis Report," Docket No. 50-443, 1999.

(NAVFAC, 1986) Naval Facilities Engineering Command Design Manual NAVFAC DM-7.01, "Soil Mechanics," 1986.

(NAVFAC, 1996). Naval Facilities Engineering Command Design Manual NAVFAC DM-7.02, "Foundations and Earth Structures," 1996.

(NCDC, 2005) National Climatic Data Center (NCDC). "Storm Events." <http://www4.ncdc.noaa.gov/cgi-win/wwcgi.dll?wwevent~storms> (Accessed 3/17/05).

(NCEDC, 2004) Northern California Earthquake Data Center (NCEDC). "Composite Earthquake Catalog," 2004, <http://quake.geo.berkeley.edu/anss/>

(NFPA, 1991) National Fire Protection Association (NFPA). *Fire Protection Handbook*[TM], 7th Edition, 1991.

(NFPA, 1995) National Fire Protection Association (NFPA) and Society of Fire Protection Engineers. *Society of Fire Protection Engineers Handbook of Fire Protection Engineering*, 2nd Edition, 1995.

(NMIMT, 2003) New Mexico Institute of Mining and Technology (NMIMT). "Earthquake Catalogs for New Mexico and Bordering Areas: 1869–1998," 2003.

(NOAA, 1982) National Oceanic and Atmospheric Administration (NOAA). "Application of Probable Maximum Precipitation Estimates—United States East of the 105th Meridian," Hydrometeorological Report No. 52, 1982.

(NOAA, 2005) National Oceanic and Atomspheric Administration. "Tornadoes." <http://www.noaa.gov/tornadoes.html> (Accessed 3/17/05).

(NRC, 1973) U.S. Nuclear Regulatory Commission (NRC). Regulatory Guide 1.60, "Design Response Spectra for Seismic Design of Nuclear Power Plants," 1973.

(NRC, 1975) U.S. Nuclear Regulatory Commission (NRC). "Site Analysis Branch Position—Winter Precipitation Loads," 1975.

(NRC, 1978a) U.S. Nuclear Regulatory Commission (NRC). NUREG/CR–0098, "Development of Criteria for Seismic Review of Selected Nuclear Power Plants," 1978.

(NRC, 1978b) U.S. Nuclear Regulatory Commission (NRC). Regulatory Guide 1.91, "Evaluation of Explosions Postulated to Occur on Transportation Routes Near Nuclear Power Plants," Rev. 1, 1978.

(NRC, 1981) U.S. Nuclear Regulatory Commission (NRC). NUREG–0800, "Standard Review Plan for the Review of Safety Analysis Reports for Nuclear Power Plants," 1981.

(NRC, 1983) U.S. Nuclear Regulatory Commission (NRC). NUREG/CR–3058, "A Methodology for Tornado Hazard Probability Assessment," 1983.

(NRC, 1991) U.S. Nuclear Regulatory Commission (NRC). NUREG-1391, "Chemical Toxicity of Uranium Hexafluoride Compared to Acute Effects of Radiation." NRC: Washington, D.C. February 1991.

(NRC, 1994) U.S. Nuclear Regulatory Commission (NRC). NUREG-1491, "Safety Evaluation Report for the Claiborne Enrichment Center, Homer, Louisiana." NRC: Washington, D.C. January 1994.

(NRC, 1997) U.S. Nuclear Regulatory Commission (NRC). Regulatory Guide 1.165, "Identification and Characterization of Seismic Sources and Determination of Safe Shutdown Earthquake Ground Motion," 1997.

(NRC, 1999a) U.S. Nuclear Regulatory Commission (NRC). Regulatory Guide 1.105, "Setpoints for Safety-Related Instrumentation," Revision 3, dated December 1999.

(NRC, 1999b) U.S. Nuclear Regulatory Commission (NRC). Regulatory Guide 1.145, "Atmospheric Dispersion Models for Potential Accident Consequence Assessments at Nuclear Power Plants," Revision 1, dated November 1982.

(NRC, 2001) U.S. Nuclear Regulatory Commission (NRC). NUREG-1513, "Integrated Safety Analysis Guidance Document." NRC: Washington, D.C. May 2001.

(NRC, 2002a) U.S. Nuclear Regulatory Commission (NRC). NUREG-1520, "Standard Review Plan for the Review of a License Application for a Fuel Cycle Facility." NRC: Washington, D.C. March 2002.

(NRC, 2002b) U.S. Nuclear Regulatory Commission (NRC). NUREG-6410, "Nuclear Fuel Cycle Facility Accident Analysis Handbook." NRC: Washington, D.C. March 1998.

(NRC, 2002c) U.S. Nuclear Regulatory Commission (NRC). NUREG-0700, "Human-System Interface Design Review Guidelines," NRC: Washington, D.C. Revision 2, 2002.

(NRC, 2003a) U.S. Nuclear Regulatory Commission (NRC). Regulatory Guide 1.180, "Guidelines for Evaluating Electromagnetic and Radio-Frequency Interference in Safety-Related Instrumentation and Control Systems," Revision 1, 2003.

(NRC, 2003b) U.S. Nuclear Regulatory Commission (NRC). Regulatory Guide 1.198, "Procedures and Criteria for Assessing Seismic Soil Liquefaction at Nuclear Power Plant Sites," 2003.

(NRC, 2004a) NUREG-0711, "Human Factors Engineering Program Review Model," Revision 2, 2004.

(NRC, 2004b) Graves, H.L., U.S. Nuclear Regulatory Commission (NRC). Summary of Meeting with Louisiana Energy Services on Structural, Seismic, High Winds, and Tornado Safety Analysis. June 29, 2004.

(OPS, 2004) Office of Pipeline Safety. "Natural Gas Transmission Pipeline Annual Mileage," 2004, <http://ops.dot.gov/stats/GTANNUAL2.HTM>

(Snooks, 2003) Snooks, J.H. "Highway Propane Explosion Hazard Risk Determination." Document Identifier 32-2400573-00, Framatome ANP, Inc., 2003.

(Thompson, 2004) Thomson, S.T. "Natural Gas Pipeline Hazard Risk Determination." Document Identifier 32-400572-00, Framatome ANP, Inc., 2004.

(Toro, 1997) Toro, G.R., N.A. Abrahamson, and J.H. Schneider, "Model of Strong Ground Motions from Earthquakes in Central and Eastern North America: Best Estimates and Uncertainties," *Seismological Research Letters*, Vol. 68, No. 1, pp. 41–5, 1997.

(USDA, 1973) U.S. Department of Army, Waterways Experiment Station (USDA). "Design Earthquake for the Central United States," Miscellaneous Paper S–731–1, 1973.

(USEPA, 1999) U.S. Environmental Protection Agency (EPA) and National Oceanic Atmospheric Administration. "ALOHA—(Areal Locations of Hazardous Atmospheres) User's Manual," 1999.

(USGS, 2004) U.S. Geological Survey (USGS). "National Seismic Hazard Mapping Project," 2004. <http://eqhazmaps.usgs.gov/>

(USWB, 1956) U.S. Weather Bureau (USWB). "Seasonal Variation of the Probable Maximum Precipitation East of the 105th Meridian for Areas from 10 to 1,000 Square Miles and Duration of 6, 12, 24, and 48 Hours," Hydrometeorological Report No. 33, 1956.

(UTIG, 2002) The University of Texas Institute for Geophysics. "Compendium of Texas Earthquakes," 2002.

(Winterkorn, 1975) Winterkorn and Fang, "Foundation Engineering Handbook," 1975.

4.0 RADIATION PROTECTION

The purpose of this review is to determine whether the applicant's radiation protection (RP) program is adequate to protect the radiological health and safety of workers and to comply with the associated regulatory requirements in 10 CFR Parts 19, 20, 30, 40, and 70. Public and environmental protection is discussed in Chapter 9 of this Safety Evaluation Report.

4.1 REGULATORY REQUIREMENTS

4.1.1 Radiation Protection Program Implementation

Regulations applicable to establishment of an RP program are presented in Part 20, Subpart B, "Radiation protection programs."

4.1.2 As Low As is Reasonably Achievable (ALARA) Program

Regulations applicable to the ALARA program are presented in 10 CFR 20.1101, "Radiation protection programs."

4.1.3 Organization and Personnel Qualifications

Regulations applicable to the organization and qualifications of the radiological protection staff are presented in 10 CFR 30.33, 10 CFR 40.32, and 10 CFR 70.22, "Contents of applications."

4.1.4 Written Procedures

The regulations applicable to RP procedures and Radiation Work Permits (RWPs) are presented in 10 CFR 30.33, 10 CFR 40.32, and 10 CFR 70.22, "Contents of applications."

4.1.5 Training

The following regulations apply to the radiation safety training program:

1. 10 CFR 19.12   "Instructions to workers"

2. 10 CFR 20.2110   "Form of records"

4.1.6 Ventilation and Respiratory Protection Programs

Regulations applicable to the ventilation and respiratory protection programs are presented in Part 20, Subpart H, "Respiratory protection and controls to restrict internal exposure in restricted areas."

4.1.7   Radiation Survey and Monitoring Programs

The following NRC regulations in Part 20 are applicable to radiation surveys and monitoring programs:

1.   Subpart F   "Surveys and Monitoring"

2.   Subpart C   "Occupational Dose Limits"

3.   Subpart L   "Records"

4.   Subpart M   "Reports"

4.1.8   Additional Program Requirements

The following Part 20 regulations are applicable to the additional program requirements:

1.   Subpart L   "Records"

2.   Subpart M   "Reports"

3.   Section 70.61   "Performance requirements"

4.   Section 70.74   "Additional reporting requirements"

4.2   REGULATORY ACCEPTANCE CRITERIA

The acceptance criteria for NRC's review of the RP program are outlined in Sections 4.4.1.3; 4.4.2.3; 4.4.3.3; 4.4.4.3; 4.4.5.3; 4.4.6.3; 4.4.7.3; and 4.4.8.3 of "Standard Review Plan for the Review of a License Application for a Fuel Cycle Facility," NUREG-1520 (NRC, 2002).

4.3   STAFF REVIEW AND ANALYSIS

4.3.1   Radiation Protection Program Implementation

In Section 4.1 of the Louisiana Energy Services' (LES) Safety Analysis Report (SAR) (LES, 2005a), the applicant describes the proposed RP program for the proposed facility. The RP program is developed, documented, and will be implemented commensurate with the risk posed by a uranium enrichment operation that will meet the requirements of 10 CFR Part 20, Subpart B.

The RP program's organizational structure and the responsibilities of key program personnel are outlined in Section 4.1.1 of the SAR (LES, 2005a). The Plant Manager will be responsible for the protection of all persons against radiation exposure resulting from facility operations and material, and for compliance with applicable NRC regulations and the facility license. The RP Manager will be responsible for implementing the RP program, and in matters involving RP, will have direct access to the Plant Manager. The RP Manager and his staff will also be responsible for:

- Establishing the RP program;

- Generating and maintaining procedures associated with the program;

- Assuring that ALARA is practiced by all personnel;

- Reviewing and auditing the efficacy of the program in complying with NRC and other governmental regulations and applicable Regulatory Guides;

- Modifying the program, based on experience and facility history;

- Adequately staffing the RP group to implement the RP program;

- Establishing and maintaining the ALARA program;

- Establishing and maintaining a respirator usage program;

- Monitoring worker doses, both internal and external;

- Complying with the radioactive materials possession limits for the facility;

- Handling of radioactive wastes when disposal is needed;

- Calibration and quality assurance of all radiological instrumentation, including verification of required Lower Limits of Detection or alarm levels;

- Establishing and maintaining a radiation safety training program for personnel working in Restricted Areas;

- Performing audits of the RP program on an annual basis;

- Establishing and maintaining the radiological environmental monitoring program; and

- Posting the Restricted Areas, and within these areas, posting: Radiation, Airborne Radioactivity, High Radiation, and Contaminated Areas, as appropriate; and developing occupancy guidelines for these areas as needed.

The responsibilities of the Health, Safety, and Engineering (HS&E) Manager, the Operations Manager, and the facility personnel are outlined in Sections 4.1.1.2, 4.1.1.4, and 4.1.1.5 of the SAR (LES, 2005a), respectively.

The applicant will staff the facility with suitably trained RP personnel. The staff will be trained and qualified consistent with the guidance provided in American National Standards Institute/American Nuclear Society (ANSI/ANS) Standard 3.1, "Selection, Qualification and Training of Personnel for Nuclear Power Plants" (ANSI/ANS, 1993). The applicant will provide sufficient resources in terms of staffing and equipment to implement an effective RP program.

The applicant will ensure that the RP program will remain independent of the facility's routine operations, and that it maintains its objectivity and is focused only on implementing sound RP principles necessary to achieve ALARA goals.

The applicant will review the content and implementation of the RP program at least annually, in accordance with 10 CFR 20.1101(c). In addition, constraints on atmospheric releases are established to ensure compliance with 10 CFR 20.1101(d).

As described above, the applicant will maintain the RP program in accordance with the acceptance criteria in NUREG-1520 (NRC, 2002).

4.3.2   ALARA Program

The ALARA program will be implemented using written policies and procedures, to ensure occupational radiation exposures are maintained ALARA and that such exposures are consistent with the requirements of 10 CFR 20.1101.

The applicant states that the RP Manager will be responsible for implementing the ALARA program and preparing an ALARA program evaluation report annually, to review:

- Radiological exposure and effluent release data for trends;

- Audits and inspections;

- Use, maintenance, and surveillance of equipment used for exposure and effluent control; and

- Other issues, as appropriate, that may influence the effectiveness of the radiation protection/ALARA programs.

This report will be submitted to the Plant Manager and the Safety Review Committee (SRC).

Radiological zones, such as "Radiation Area," "Airborne Radioactivity Area," and "Contaminated Area," will be established, as necessary, to minimize the spread of contamination and reduce unnecessary radiation exposure to personnel.

Goals of the ALARA program include maintaining occupational exposures, as well as environmental releases, as far below regulatory limits as is reasonably achievable. This is accomplished by minimizing the size and number of areas with higher dose rates and designing areas to maintain the lowest dose rates reasonably achievable, particularly where facility personnel spend significant amounts of time.

The applicant will establish an SRC, which meets at least quarterly and within 30 days of any NRC-reportable incident, and will fulfill the duties of the ALARA Committee. The SRC will have at least five members, to include experts in operations, criticality safety, radiological safety, chemical safety, and industrial safety.

Responsibilities of the SRC include: (1) reviewing the effectiveness of the ALARA program; (2) determining if exposures, releases, and contamination levels are in accordance with the ALARA concept; (3) ensuring that the occupational radiation exposure dose limits of Part 20 are not exceeded under normal operations; and (4) identifying any upward trends in personnel exposures, environmental releases, and facility contamination levels.

The applicant will use the ALARA program to facilitate interaction between RP and operations personnel. The SRC reviews and revises the program goals and objectives to incorporate, when appropriate, new approaches, technologies, operating procedures, or changes that could cost-effectively reduce potential radiation exposure.

The applicant will maintain an ALARA program in accordance with the acceptance criteria as described above.

4.3.3 Organization and Personnel Qualifications

The applicant will employ only suitably trained RP personnel at the facility, by following the guidance in Regulatory Guides 8.2 and 8.10 (NRC, 1973 and NRC, 1997). Further information on personnel qualifications and training is provided in Sections 2.24, 2.3.3, and 11.3 of the SAR (LES, 2005a). The applicant will assign the responsibility of implementing the RP program functions to the RP program staff.

The RP Manager will be responsible for establishing and implementing the RP program, which includes training personnel in use of equipment, control of radiation exposure of personnel, continuous determination and evaluation of the radiological status of the facility, and conducting the radiological environmental monitoring program. The RP Manager will have direct access to the Plant Manager regarding all matters involving RP; will be skilled in the interpretation of RP data and regulations; will be familiar with the operation of the facility and RP concerns of the site; and will be a resource in radiation safety management decisions.

The RP Manager will have, as a minimum, a bachelor's degree (or equivalent) in an engineering or scientific field and 3 years of responsible nuclear experience associated with the implementation of an RP program, to include at least 2 years experience at a facility that processes uranium, including uranium in soluble form. The RP staff will be trained and qualified consistent with guidance provided in ANSI/ANS Standard 3.1 (ANSI/ANS, 1993).

The applicant will organize and staff an RP program in accordance with the acceptance criteria.

4.3.4 Written Procedures

Written procedures will be used for all operations involving licensed materials. The procedures are prepared, reviewed, and approved to carry out activities related to the RP program, and are used to ensure RP activities are conducted in a safe, effective, and consistent manner. The RP procedures are reviewed and revised as necessary, to incorporate any facility or operational changes to the facility's Integrated Safety Analysis (ISA).

RP procedures are prepared by RP staff; reviewed by members of the facility staff, personnel with enrichment plant operating experience, and other staff members, as appropriate; and approved by the RP Manager, or designee.

4-5

The applicant will perform all work in Restricted Areas in accordance with an RWP. The applicant will also issue RWPs for activities involving licensed materials not covered by operating procedures, and where radioactivity levels are likely to exceed airborne radioactivity limits, or whenever deemed as necessary by the RP Manager.

RWPs provide a description of the work or authorized activities; summary results of dose rate, contamination, and airborne radioactivity surveys, etc.; and precautions, which may include personnel protective equipment, stay-times or dose limits, record-keeping requirements, and coverage by an RP technician.

RWPs require approval by the RP Manager or designee, and have a predetermined period of validity with a specified expiration or termination time. The designee must meet the requirements of Section 4.1.2 of the SAR (LES, 2005a).

The following criteria are used for RWPs:

- Planned activities or changes to activities inside Restricted Areas or work with licensed materials are reviewed by the RP Manager or designee, for the potential to cause radiation exposures to exceed action levels or to produce radioactive contamination;

- RWPs include requirements for any necessary safety controls, personnel monitoring devices, protective clothing, respiratory protective equipment, and air sampling equipment, and the attendance of RP technicians at the work location;

- RWPs are posted at access points to Restricted Areas, with copies of current RWPs posted at the work area location;

- RWPs clearly define and limit the work activities to which they apply. An RWP is closed out when the applicable work activity for which it was written is completed and terminated; and

- RWPs are retained as a record at least for the life of the facility.

The applicant will prepare written procedures and RWPs in accordance with these acceptance criteria.

4.3.5 Training

An RP training program is designed and implemented to provide training to all personnel and visitors, unless provided with trained escorts, who enter Restricted Areas or Controlled Areas, commensurate with the radiological hazard to which they may be exposed. The level of training is based on the potential radiological health risks associated with the individual's work responsibilities.

The applicant has incorporated the provisions of 10 CFR 19.12 into the radiation training program, as outlined in Section 4.5.1 of the SAR (LES, 2005a). The requirements in 10 CFR 19.12 address required health physics information the applicant must make available to workers likely to receive exposures greater than 1 mSv (100 mrem) per year.

Retraining is performed for personnel requiring unescorted access to Restricted Areas on an annual basis, and as necessary, to address changes in policies, procedures, requirements, and the facility ISA. The HS&E Manager and RP Manager review and update the content of the formal RP training program at least every 2 years, to ensure that the programs are current and adequate.

The applicant will evaluate the effectiveness and adequacy of the training program curriculum and instructors through audits performed by operational area personnel responsible for criticality safety and RP.

The applicant will train its employees in RP in accordance with the acceptance criteria.

4.3.6   Ventilation and Respiratory Protection Programs

The design criteria, including flow velocity at openings, for the ventilation systems, are described in Sections 3.4.9 and 3.5.1 of the ISA Summary (LES, 2005b). Filters to be used in the systems include pre-filters for dust removal, High-Efficiency Particulate Air (HEPA) filters for removal of uranyl fluoride (UO_2F_2) aerosols, and activated carbon filters for hydrogen fluoride (HF) removal. Differential pressure across HEPA filters in potentially contaminated exhaust systems is either checked monthly or automatically monitored and alarmed, and filters are replaced when differential pressure exceeds the manufacturers' ratings or fail to function properly.

The containment of uranium hexafluoride (UF_6) is maintained by the process equipment, and the enrichment process, with the exception of liquid sampling, is maintained under a partial vacuum, to ensure any leaks will not result in a release of UF_6 into the work areas.

Air-flow rates at system openings (hoods, exhausted enclosures, close-capture point equipment, or ventilation systems serving these barriers) will be sufficient to preclude escape of airborne uranium and minimize the potential for worker intake, and will be checked monthly when in use and after any modifications.

To meet the respiratory protection requirements in 10 CFR Part 20, Subpart H, the applicant will prepare written procedures for the selection, fitting, issuance, maintenance, testing, training of personnel, monitoring, and record-keeping for individual respiratory protection equipment, in accordance with 10 CFR 20.1703(c)(4). The applicant has established a respiratory protection program that meets the requirements of Part 20, Subpart H.

The applicant will revise respiratory protection procedures as necessary, whenever changes are made to the facility, process, or equipment. The records of the respiratory protection program, including training for respirator use and maintenance, are maintained in accordance with the facility records management program, which is described in Section 11.7 of the SAR (LES, 2005a).

The applicant has established ventilation and respiratory protection programs in accordance with the acceptance criteria, and satisfies the regulatory requirements of Part 20, Subpart H.

4.3.7 Radiation Survey and Monitoring Programs

The applicant has a radiation survey and monitoring program using prepared written procedures that will include an outline of the program objectives; sampling procedures; data analysis methods; types of equipment and instrumentation to be used; frequency of measurements; record-keeping and reporting requirements; and actions to be taken when measurements exceed Part 20 occupational dose limits, or the administrative levels established by the applicant.

Thermoluminescent dosimeters (TLDs) that are sensitive to beta, gamma, and neutron radiation - supplied by a vendor holding dosimetry accreditation from the National Voluntary Laboratory Accreditation Program - will be required to be worn by all personnel who enter Restricted Areas. The TLDs are evaluated at least quarterly.

The applicant has established an annual administrative limit for Total Effective Dose Equivalent of 10 millisieverts (mSv) [1000 millirem (mrem)]. If 2.5 mSv (250 mrem) are exceeded in any quarter, an investigation will be performed and documented to determine what factors contributed to that exposure, including procedural reviews, efficiency studies of the air-handling system, cylinder storage protocol, and work practices.

All personnel wearing external dosimetry devices will be evaluated for internal exposures via direct bioassay, indirect bioassay, or an equivalent technique. These doses will be evaluated at least annually. Bioassay will be performed for all personnel who are likely to have had an intake of 1 milligram (mg) of uranium during a week (10 percent of the 10 mg in-a-week NRC limit) based on air-sample monitoring data. Bioassay will be performed using urinalysis having a sensitivity of 5 micrograms per liter (μg/l), assuming proper timing and amount of the sample. Workers will be restricted from activities that could result in internal exposures to soluble uranium, as long as uranalysis results are $\geq 15\ \mu$g/l.

The applicant will sum the internal and external exposure values in accordance with 10 CFR 20.1202, using procedures based on the guidance in Regulatory Guides 8.7 (NRC, 1992a) and 8.34 (NRC, 1992c).

The applicant will perform air sampling consistent with the guidance provided in Regulatory Guide 8.25 (NRC, 1992b); NUREG-1400 (NRC, 1991); and ANSI/Health Physics Society (HPS) Standard 13.1 (ANSI/HPS, 1999). Regulatory Guide 8.25 (NRC, 1992b) and NUREG-1400 (NRC, 1991) provide guidance on evaluating the need for air sampling; locating air samplers; demonstrating that air sampling is representative of inhaled air; adjustments to derived air concentrations; measuring the volume of air sampled; and evaluating sample results, including the calculation of minimum detectable activity for samples. ANSI/HPS 13.1 (ANSI/HPS, 1999) provides guidance on sampling location, sampling system design, and quality assurance and quality control.

Airborne activity levels in the Restricted Areas of the facility will be continuously monitored, with permanent air monitors designed to detect alpha emitters. Monitors located in Airborne Radioactivity Areas will have alarms, which will be set with consideration of both toxicity and radioactivity. Portable air samplers will also be available for use.

Airflow and radioactivity measurement instruments will be calibrated at least annually or according to manufacturers' recommendations, after failing an operability check, after modifications or repairs that could affect proper response, or when believed to have been damaged. The applicant will use calibration sources traceable to the National Institute of Standards and Technology, or equivalent, and which are ±5 percent of the stated values.

The applicant will provide routine contamination survey monitoring in all UF_6 process areas, with routine, periodic checks of non-UF_6 process areas, including those areas that are normally free of contamination. Restricted Areas will be surveyed at least weekly, and lunch rooms and change rooms will be surveyed daily. Monitoring will include measurements of fixed and removable surface contamination, with extent and frequency based on the potential for contamination in each area and operational experience. Removable surface contamination will be considered to be uranium that can be transferred to a dry smear paper with moderate pressure.

The applicant has defined a contaminated area as an area where removable contamination levels are above 0.33 Bq/100 cm^2 (20 dpm/100 cm^2) alpha, or 16.7 Bq/100 cm^2 (1000 dpm/100 cm^2) beta/gamma, and will initiate clean-up of contamination within 24 hours of detection, if levels exceed 83.3 Bq/100 cm^2 (5000 dpm/100 cm^2) alpha or beta/gamma for removable contamination, and 4.2 kBq/100 cm^2 (250,000 dpm/100 cm^2) alpha or beta/gamma for fixed contamination.

The facility corrective action process, which is described in Section 11.6 of the SAR (LES, 2005a), will be implemented if:

1.   Personnel dose-monitoring results or personnel contamination levels exceed the administrative personnel limits;

2.   An incident results in airborne occupational exposures exceeding the administrative limits; or

3.   The dose limits in Part 20, Appendix B, or 10 CFR 70.61 are exceeded.

The applicant has a radiation survey program consistent with the guidance contained in Regulatory Guide 8.24, "Health Physics Surveys During Enriched Uranium-235 Processing and Fuel Fabrication" (NRC, 1979), which includes the selection of instruments that should be operable and capable of measuring, at or below the required level, the types of radiation that will be encountered. These instruments will be calibrated as discussed previously in this section.

For transfer of material and equipment to unrestricted areas and release from the facility for unrestricted use, the applicant will meet surface contamination guidelines prescribed in "Guidelines for Decontamination of Facilities and Equipment Prior to Release for Unrestricted Use or Termination of Licenses for Byproduct, Source, or Special Nuclear Material" (NRC, 1993a).

The applicant will leak-test sources in accordance with the following NRC Branch Technical Positions:

1. "License Condition for Leak-Testing Sealed Byproduct Material Sources" (NRC, 1993b);

2. "License Condition for Leak-Testing Sealed Source Which Contains Alpha and/or Beta-Gamma Emitters" (NRC, 1993c); and

3. "License Condition for Leak-Testing Sealed Uranium Sources" (NRC, 1993d).

Sources containing plutonium will not be used at the facility.

The applicant has an access control program that ensures that: (a) signs, labels, and other access controls are properly posted and operative; (b) restricted areas are established to prevent the spread of contamination and are identified with appropriate signs; and (c) step-off pads, change facilities, protective clothing facilities, and personnel monitoring instruments are provided in sufficient quantities and locations.

The applicant has established action levels of 2.5 Bq/100cm^2 (150 dpm/100cm^2) alpha or beta/gamma contamination (corrected for background) at points of egress from Restricted Areas and any additional designated areas within the Restricted Area.

The applicant has established radiation survey and monitoring programs in accordance with the acceptance criteria.

4.3.8   Additional Program Requirements

The applicant has established a program to maintain records of the RP program, radiation survey results, and results of corrective action program referrals, RWPs, and planned special exposures.

The applicant will report, to NRC, any event that results in an occupational exposure to radiation exceeding the dose limits in Part 20, within the time specified in 10 CFR 20.2202, 10 CFR 30.50, 10 CFR 40.60, and 10 CFR 70.74. The applicant will prepare and submit, to NRC, an annual report of the results of individual monitoring, as required by 10 CFR 20.2206(b).

The applicant will refer to the facility's corrective action program any radiation incident that results in an occupational exposure that exceeds the dose limits in Part 20, Appendix B, or is required to be reported per 10 CFR 30.50, 10 CFR 40.60, and 10 CFR 70.74, and to reporting, to NRC, both the corrective actions taken (or planned) to protect against a recurrence, and the proposed schedule to achieve compliance.

The staff reviewed the ISA Summary (LES, 2005b) and performed an on-site review of the ISA, and agree with the applicant that, as stated in Section 3.6 of the ISA Summary (LES, 2005b), the hazards from radioactivity were evaluated in the ISA and found to be low consequence.

4.4 EVALUATION FINDINGS

The applicant has established and will maintain an acceptable RP program that includes:

1. An effective documented program to ensure that occupational radiological exposures are ALARA;

2. An organization with adequate qualification requirements for the RP personnel;

3. Approved written RP procedures and RWPs for RP activities;

4. RP training for all personnel who have access to restricted areas;

5. A program to control airborne concentrations of radioactive material with engineering controls and respiratory protection;

6. A radiation survey and monitoring program that includes requirements for controlling radiological contamination within the facility and monitoring of external and internal radiation exposures; and

7. Other programs to maintain records, report to NRC in accordance with Parts 20 and 70, and correct for upsets at the facility.

The applicant's RP program meets the requirements of Parts 19, 20, 30, 40, and 70. Conformance to the license application will ensure safe operation.

4.5 REFERENCES

(ANSI/HPS, 1999) American National Standards Institute/Health Physics Society (ANSI/HPS). ANSI/HPS 13.1, "Sampling and Monitoring Releases of Airborne Radioactive Substances From the Stacks and Ducts of Nuclear Facilities," 1999.

(ANSI/ANS, 1993) American National Standards Institute/American Nuclear Society (ANSI/ANS). ANSI/ANS 3.1, "Selection, Qualification, and Training of Personnel for Nuclear Power Plants," 1993.

(LES, 2005a) Louisiana Energy Services (LES). "National Enrichment Facility Safety Analysis Report," Revision 6, 2005.

(LES, 2005b) Louisiana Energy Services (LES). "National Enrichment Facility Integrated Safety Analysis Summary," Revision 4, 2005.

(NRC, 1973) U.S. Nuclear Regulatory Commission (NRC). Regulatory Guide 8.2, "Guide for Administrative Practice in Radiation Monitoring," 1973.

(NRC, 1977) U.S. Nuclear Regulatory Commission (NRC). Regulatory Guide 8.10, "Operating Philosophy for Maintaining Occupational Radiation Exposures As Low As Is Reasonably Achievable," 1977.

(NRC, 1979) U.S. Nuclear Regulatory Commission (NRC). Regulatory Guide 8.24, "Health Physics Surveys During Enriched Uranium-235 Processing and Fuel Fabrication," 1979.

(NRC, 1991) U.S. Nuclear Regulatory Commission (NRC). NUREG-1400, "Air Sampling in the Workplace," 1991.

(NRC, 1992a) U.S. Nuclear Regulatory Commission (NRC). Regulatory Guide 8.7, "Instructions for Recording and Reporting Occupational Radiation Exposure Data," 1992.

(NRC, 1992b) U.S. Nuclear Regulatory Commission (NRC). Regulatory Guide 8.25, "Air Sampling in the Workplace," 1992.

(NRC, 1992c) U.S. Nuclear Regulatory Commission (NRC). Regulatory Guide 8.34, "Monitoring Criteria and Methods to Calculate Occupational Radiation Doses," 1992.

(NRC, 1993a) U.S. Nuclear Regulatory Commission (NRC). Branch Technical Position, "Guidelines for Decontamination of Facilities and Equipment Prior to Release for Unrestricted Use or Termination of Licenses for Byproduct, Source, or Special Nuclear Material," 1993.

(NRC, 1993b) U.S. Nuclear Regulatory Commission (NRC). Branch Technical Position, "License Condition for Leak-Testing Sealed Byproduct Material Sources," 1993.

(NRC, 1993c) U.S. Nuclear Regulatory Commission (NRC). Branch Technical Position, "License Condition for Leak-Testing Sealed Source Which Contains Alpha and/or Beta-Gamma Emitters," 1993.

(NRC, 1993d) U.S. Nuclear Regulatory Commission (NRC). Branch Technical Position, "License Condition for Leak-Testing Sealed Uranium Sources," 1993.

(NRC, 2002) U.S. Nuclear Regulatory Commission (NRC). NUREG-1520, "Standard Review Plan for the Review of a License Application for a Fuel Cycle Facility," 2002.

5.0 NUCLEAR CRITICALITY SAFETY

The purpose of this review is to determine whether the applicant's nuclear criticality safety (NCS) program is adequate to support safe design, construction, and operation of the facility, as required by 10 CFR Part 70. In addition, the purpose of this review is to determine whether the Integrated Safety Analysis (ISA) and ISA Summary meet the regulatory requirements specified in 10 CFR Part 70, Subpart H, "Additional Requirements for Certain Licensees Authorized to Possess a Critical Mass of Special Nuclear Material," for NCS.

The NCS programmatic review will determine whether: (1) the applicant has provided for the appropriate management of the NCS program; (2) the applicant has identified, and committed to, the responsibilities and authorities of individuals for developing and implementing the NCS program; (3) the facility management measures described in 10 CFR 70.62 have been committed to and will support implementing and maintaining the NCS program; and (4) an adequate NCS program is described, which includes identifying and committing to the NCS methods and NCS technical practices used to ensure the safe operation of the facility, as required by Part 70.

The NCS ISA review was performed to determine whether: (1) the ISA program is acceptable for NCS; (2) the ISA has been acceptably performed and will be maintained for NCS; and (3) the ISA Summary contains necessary information, such that the NCS accident sequences are "highly unlikely" or else failures of NCS safe-by-design components meet "highly unlikely" with significant margin.

5.1 REGULATORY REQUIREMENTS

The NCS review of the applicant's NCS program should verify if the information the applicant provided (LES, 2004; LES, 2005a; LES, 2005b; and LES, 2005c) meets the requirements of 10 CFR 70.22 and 70.65, which, respectively, specify the general and additional content of a license application. In addition, the NCS review should verify compliance with the regulatory requirements in 10 CFR 70.24; 70.52; 70.61; 70.62; 70.64; 70.65; 70.72; and Appendix A to Part 70.

The NCS review of the applicant's ISA program and ISA Summary should verify if the information the applicant provided (LES, 2005a; LES, 2005c) meets the requirements of 10 CFR 70.62 and 70.65, which, respectively, specify: (1) the requirements for establishing and maintaining a safety program (10 CFR 70.62), including an ISA program that addresses NCS; (2) requirements for conducting and maintaining an ISA (10 CFR 70.62(c)) for NCS; and (3) requirements for the contents of an ISA Summary (10 CFR 70.65(b)) for NCS.

5.2   REGULATORY ACCEPTANCE CRITERIA

The acceptance criteria for the NCS review of the applicant's NCS program are outlined in Sections 5.4.3.1, 5.4.3.2, 5.4.3.3, and 5.4.3.4 of NUREG-1520 (NRC, 2002). This includes the commitment to use NRC NCS Regulatory Guide 3.71 (NRC, 1998), which modified the use of the American National Standards Institute/American Nuclear Society (ANSI/ANS) Series-8 NCS standards.

The acceptance criteria used for the NCS review of the applicant's ISA program and ISA Summary are outlined in Sections 3.4.3.1 and 3.4.3.2 of NUREG-1520 (NRC, 2002).

5.3   STAFF REVIEW AND ANALYSIS

5.3.1   Possession Limits

In Chapter 1 of the Safety Analysis Report (SAR) (LES, 2005a), the applicant described the NCS-important special nuclear material possession limits of the facility, as shown in Section 1.2.3.4 of this Safety Evaluation Report (SER), at an enrichment level of no greater than 5.0 weight (wt) percent, in the chemical form of uranium hexafluoride (UF_6), uranium tetrafluoride (UF_4), and uranyl fluoride (UO_2F_2) in the physical form of solid, liquid, and gas.

5.3.2   NRC NCS Regulatory Guide 3.71 and ANSI/ANS Series 8 NCS Standards

In Chapter 5 of the SAR (LES, 2005a), the applicant committed to following NRC NCS Regulatory Guide 3.71 (NRC, 1998) and to use the following ANSI/ANS Series-8 NCS standards:

- ANSI/ANS-8.1, "Nuclear Criticality Safety in Operations with Fissionable Materials Outside Reactors" (ANSI/ANS, 1988a);

- ANSI/ANS-8.1, "Nuclear Criticality Safety in Operations with Fissionable Materials Outside Reactors" (ANSI/ANS, 1998a);

- ANSI/ANS-8.3, "Criticality Accident Alarm System" (ANSI/ANS, 1997a);

- ANSI/ANS-8.5, "Use of Borosilicate-Glass Raschig Rings as a Neutron Absorber in Solutions of Fissile Material" (ANSI/ANS, 1996a);

- ANSI/ANS-8.7, "Guide for Nuclear Criticality Safety Criteria in the Storage of Fissile Materials" (ANSI/ANS, 1998b);

- ANSI/ANS-8.6, "Safety in Conducting Subcritical Neutron-Multiplication Measurements In Situ" (ANSI/ANS, 1995a);

- ANSI/ANS-8.10, "Criteria for Nuclear Criticality Safety Controls in Operations with Shielding and Confinement" (ANSI/ANS, 1988b);

- ANSI/ANS-8.12, "Nuclear Criticality Control and Safety of Plutonium-Uranium Fuel Mixtures Outside Reactors" (ANSI/ANS, 1993);

- ANSI/ANS-8.15, "Nuclear Criticality Control of Special Actinide Elements" (ANSI/ANS, 1995c);

- ANSI/ANS-8.17, "Criticality Safety Criteria for the Handling, Storage, and Transportation of LWR Fuel Outside Reactors" (ANSI/ANS, 1997b);

- ANSI/ANS-8.19, "Administrative Practices for Nuclear Criticality" (ANSI/ANS, 1996b);

- ANSI/ANS-8.20, "Nuclear Criticality Safety Training" (ANSI/ANS, 1991);

- ANSI/ANS-8.21, "Use of Fixed Neutron Absorbers in Nuclear Facilities Outside Reactors" (ANSI/ANS, 1995d);

- ANSI/ANS-8.22, "Nuclear Criticality Safety Based on Limiting and Controlling Moderators" (ANSI/ANS, 1997c); and

- ANSI/ANS-8.23, "Nuclear Criticality Accident Emergency Planning and Response" (ANSI/ANS, 1997d).

The applicant took exception to the ANSI/ANS-8.9 standard (ANSI/ANS, 1995b), in that piping configurations containing aqueous solutions of fissile material will be evaluated in accordance with the 1998 version of ANSI/ANS-8.1 (ANSI/ANS, 1998a), using validated methods to determine subcritical limits.

In addition, the applicant used a newer version of the ANSI/ANS-8.1 standard (ANSI/ANS, 1998a) than the version of the ANSI/ANS-8.1 standard (ANSI/ANS, 1988a) that the NRC endorsed, with exception, in NRC NCS Regulatory Guide 3.71 (NRC, 1998). NRC staff reviewed the differences between the two versions of ANSI/ANS-8.1, along with the NRC endorsement with exception. Since NRC's intent did not change, but the standard did change, the applicant also committed to the following, concerning validation using ANSI/ANS-8.1-1998 (ANSI/ANS, 1998a): "In addition, the details of validation should state computer codes used, operations, recipes for choosing code options (where applicable), cross-section sets, and any numerical parameters necessary to describe the input."

The applicant also used a newer version of ANSI/ANS-8.7 (ANSI/ANS, 1998b) than the version of the standard endorsed by NRC in Regulatory Guide 3.71 (NRC, 1998). NRC staff reviewed the differences between the two versions of ANSI/ANS-8.7 and determined that it was acceptable for the applicant to use the newer version without exception.

Based on the review of the information provided, the staff finds the applicant has identified appropriate ANSI/ANS Series-8 standards and NRC Regulatory Guides relating to NCS.

5.3.3 Management of the NCS Program

In Chapters 2, 3, 5, 8, and 11 of the SAR (LES, 2005a) as well as in the facility Emergency Plan (LES, 2005b), the applicant described how it meets the acceptance criteria for management of the NCS program.

Specific information about the management of the NCS program is in Sections 5.3.4 (NCS Organization and Administration) and 5.3.6 (NCS Program) of this SER.

Regarding the specific acceptance criteria in Section 5.4.3.1 of NUREG-1520 (NRC, 2002) for management of the NCS program, in Section 5.1 of the SAR (LES, 2005a), the applicant:

- Committed to develop, implement, and maintain an NCS program to meet the regulatory requirements of 10 CFR Part 70;

- Stated that the NCS program objectives are to:

    ▸ Prevent an inadvertent criticality;
    ▸ Protect against the occurrence of an identified accident sequence; in the ISA Summary that could lead to an inadvertent criticality;
    ▸ Comply with the NCS performance requirements of 10 CFR 70.61;
    ▸ Establish and maintain NCS safety parameters and procedures;
    ▸ Establish and maintain NCS limits;
    ▸ Conduct NCS Analyses and Evaluations to assure that under normal and credible abnormal conditions, all nuclear processes are subcritical and maintain an approved margin of subcriticality for safety;
    ▸ Establish and maintain NCS IROFS, based on current NCS Analyses and NCS Evaluations;
    ▸ Provide training in emergency procedures in response to an inadvertent criticality;
    ▸ Comply with NCS baseline design criteria in 10 CFR 70.64(a)(9);
    ▸ Comply with the NCS ISA Summary requirements in 10 CFR 70.65(b);

- Established NCS safety parameters and will establish NCS procedures;

- Outlined an NCS program structure and defined the responsibilities and authorities of key program personnel;

- Committed to keep NCS methodologies and NCS technical practices applicable to current configuration by means of the configuration management function;

- Committed to the preparation of NCS postings, NCS training, and NCS emergency procedure training;

- Committed to adhere to the NCS baseline design criteria requirements in 10 CFR 70.64(a)(9); and

- Committed to use the NCS program to evaluate modifications to operations, to recommend process parameter changes necessary to maintain the safe operation of the facility, and to select appropriate NCS IROFS and management measures to ensure the availability and reliability of the IROFS.

Based on its review of the information provided, the staff finds the applicant has committed to adequate management measures for the NCS program.

5.3.4 NCS Organization and Administration

In Chapters 2, 5, and 8 of the SAR (LES, 2005a) as well as in the facility Emergency Plan (LES, 2005b), the applicant described how it meets the NCS organization and administration acceptance criteria, including the technical qualifications, training, and experience of the applicant and members of its staff to engage in the proposed NCS activities.

5.3.4.1 NCS Organization

Specific information about the commitments to ANSI/ANS Series-8 NCS standards is in Section 5.3.2 of this SER. In Chapter 2 of the SAR (LES, 2005a), the applicant identified that during the design phase of the facility, the NCS function is performed within the design engineering organization, which is responsible for implementing the NCS program. During the operations phase of the facility, the NCS function will report to the Health, Safety, and Environment Manager, who is responsible for implementing the NCS program. Also, the applicant identified that the Chief Operating Officer; Plant Manager; Safety Review Committee; Health, Safety, and Environment Manager; Criticality Safety Engineer; Shift Crew; and NCS staff have and will have responsibilities related to NCS.

Regarding the specific acceptance criteria in Section 5.4.3.2 of NUREG-1520 (NRC, 2002) for NCS organization, the applicant:

- Identified and functionally described the organizational groups responsible for managing the design, construction, and operation of the facility for NCS;

- Committed to describe organizational positions and outline organizational relations among the individual positions for NCS; and

- Committed to ANSI/ANS-8.1, "Nuclear Criticality Safety in Operations with Fissionable Materials Outside Reactors" (ANSI/ANS, 1988a), ANSI/ANS-8.1, "Nuclear Criticality Safety in Operations with Fissionable Materials Outside Reactors" (ANSI/ANS, 1998a), and ANSI/ANS-8.19, "Administrative Practices for Nuclear Criticality" (ANSI/ANS, 1996b), as they relate to the NCS organization.

5.3.4.2 NCS Administration

In Chapters 2, 5, and 8 of the SAR (LES, 2005a) and in the facility Emergency Plan (LES, 2005b), the applicant identified that responsible managers have (for design) and will have (for operations) the authority to delegate tasks to other individuals; however, the responsible manager retains (for design) and will retain (for operations) the ultimate responsibility and accountability for implementing the applicable requirements. Additional information about NCS administration is in Sections 5.3.2 and 5.3.6 of this SER.

In Chapters 2, 5, and 8 of the SAR (LES, 2005a) and the facility Emergency Plan (LES, 2005b), the applicant identified that the Chief Operating Officer; Plant Manager; Safety Review Committee; Health, Safety, and Environment Manager; Criticality Safety Engineer; Shift Crew; and NCS staff have and will have the following responsibilities related to NCS:

The Chief Operating Officer will:

1.   Provide Criticality Safety Engineers in sufficient numbers to implement and support the operation of the NCS program; and

2.   Have the appropriate management and supervisory responsibilities identified in ANSI/ANS-8.19 (ANSI/ANS, 1996b).

The Plant Manager will:

1.   Provide Criticality Safety Engineers in sufficient numbers to implement and support the operation of the NCS program; and

2.   Have the appropriate management and supervisory responsibilities identified in ANSI/ANS-8.19 (ANSI/ANS, 1996b).

The Safety Review Committee will:

1.   Provide technical and administrative review and audit of operations that could impact facility worker, public safety, and environmental impacts, including NCS;

2.   Include experts on operations and all safety disciplines, including NCS; and

3.   Conduct at least one facility review and audit per year in areas, including NCS control.

The Health, Safety, and Environment Manager will:

1.   Assure safety at the facility through activities, including NCS;

2.   Work with the other facility managers to ensure consistent interpretations of Health, Safety, and Environment requirements, including NCS;

3.   Perform independent reviews, including NCS;

4.   Support facility and operations change control reviews, including NCS;

5.   Ensure objective Health, Safety, and Environment audit, review, and control activities by being independent from other management positions;

6.   Have the authority to shut down operations, if they appear to be unsafe, including NCS;

7.   Need to consult with the Plant Manager with respect to restart of shutdown operations, after the deficiency or unsatisfactory condition has been resolved, including NCS;

8.   Have the authority and responsibility to assign and direct activities for the NCS Staff;

9.   Provide Criticality Safety Engineers in sufficient numbers to implement and support the operation of the NCS program; and

10.   Have the appropriate management and supervisory responsibilities identified in ANSI/ANS-8.19 (ANSI/ANS, 1996b).

The Criticality Safety Engineer will:

1.   Report to the Health, Safety, & Environment Manager (via a designated supervisory position, if applicable);

2.   Prepare NCS Evaluations and NCS Analyses;

3.   Review NCS Evaluations and NCS Analyses;

4.   Conduct and report on periodic NCS Assessments;

5.   Implement the NCS program; and

6.   Have the appropriate NCS Staff responsibilities identified in ANSI/ANS-8.19 (ANSI/ANS,1996b).

The Shift Crew will have at least one Criticality Safety Engineer available to be contacted by the Shift Manager to respond to any routine request or emergency condition. This availability may be offsite, if adequate communication ability is provided to allow response as needed.

During the operations phase of the facility, the NCS Staff responsibilities includes:

1.   Establishing the NCS program, including design criteria, procedures, and training;

2.   Providing NCS support for the integrated safety analysis and configuration control;

3.   Assessing normal and credible abnormal conditions for NCS;

4.   Determining NCS limits for controlled parameters;

5.   Developing and validating methods to support NCS Evaluations (i.e., non-calculational engineering judgments regarding whether existing NCS Analyses bound the issue being evaluated or whether new or revised NCS Analyses are required);

6.   Performing NCS Analyses (i.e., calculations), writing NCS Evaluations, reviewing and approving proposed changes in process conditions on equipment involving fissionable material;

7.   Specifying NCS control requirements and functionality;

8.   Providing advice and counsel on NCS control measures, including reviewing and approving operating procedures;

9.   Supporting planning for emergency response and responding to events;

10.   Evaluating the effectiveness of the NCS program using audits and assessments; and

11.   Providing NCS postings that identify administrative controls for operators in applicable work areas.

In Chapters 2, 5, and 8 of the SAR (LES, 2005a) and the facility Emergency Plan (LES, 2005b), the applicant identified qualifications related to NCS for the Chief Operating Officer; Plant Manager; Safety Review Committee; Health, Safety, and Environment Manager; Criticality Safety Engineer; Shift Crew; and NCS staff.

There are no minimum requirements for the Chief Operating Officer, related to NCS.

The minimum requirements for the Plant Manager, related to NCS are:

1.   Knowledgeable of the enrichment process, enrichment process controls and ancillary processes, and other safety aspects that apply to the overall safety of the facility, including NCS control;

2.   Bachelor's degree (or equivalent) in an engineering or scientific field; and

3.   Ten years of responsible nuclear experience.

The minimum requirements for an NCS member of the Safety Review Committee related to NCS are:

1.   Academic degree in an engineering or physical science field;

2.   Five years technical experience; and

3.   Three years related directly to NCS, which is included in the 5 years of technical experience.

The minimum requirements for a Health, Safety, and Environment Manager related to NCS are:

1.   Bachelor's degree (or equivalent) in an engineering or scientific field;

2.   Five years of responsible nuclear experience in Health, Safety, and Environment or related disciplines; and

3.   One year of direct experience in the administration of NCS Evaluations and NCS Analyses.

The minimum requirements for a Criticality Safety Engineer related to NCS are:

1.   Bachelor's degree (or equivalent) in an engineering or scientific field;

2.   Two years experience in the implementation of an NCS program;

3.   Two years nuclear industry experience in NCS;

4.   Understanding and experience in the application and direction of NCS programs; and

5.   Successful completion of a training program, applicable to the scope of operations, in the physics of criticality and in associated safety practices.

In addition, should a change to the facility require an NCS Evaluation or an NCS Analysis, then the person who would do that NCS Evaluation or NCS Analysis must meet the following minimum requirements:

1.   Possess the equivalent qualifications of a Criticality Safety Engineer; and

2.   Have 2 years of experience performing NCS Analyses and implementing NCS programs.

The minimum requirement for the Shift Crew is for a Criticality Safety Engineer to be able to be contacted by the Shift Manager to respond to any routine request or emergency condition.

The minimum requirement for the NCS staff is the same as for a Criticality Safety Engineer.

The applicant will implement the intent of the administrative practices for NCS, as contained in Section 4.1.1 of ANSI/ANS-8.1 (ANSI/ANS, 1998a). Also, a policy will be established whereby, "Personnel must report defective NCS conditions and perform actions only in accordance with written, approved procedures. Unless a specific procedure deals with the situation, personnel must report defective NCS conditions and take no action until the situation has been evaluated and recovery procedures provided."

Regarding the specific acceptance criteria in Section 5.4.3.2 of NUREG-1520 (NRC, 2002) for NCS administration, the applicant:

•   Identified functional responsibilities, experience, and qualification of personnel responsible for NCS;

•   Committed to ANSI/ANS-8.1, "Nuclear Criticality Safety in Operations with Fissionable Materials Outside Reactors" (ANSI/ANS, 1988a), ANSI/ANS-8.1, "Nuclear Criticality Safety in Operations with Fissionable Materials Outside Reactors" (ANSI/ANS, 1998a), and ANSI/ANS-8.19, "Administrative Practices for Nuclear Criticality" (ANSI/ANS, 1996b), as they relate to NCS administration;

•   Committed to implement the intent of the administrative practices for NCS, as contained in Section 4.1.1 of ANSI/ANS-8.1 (ANSI/ANS, 1998a), which is to use personnel, skilled in the interpretation of data pertinent to NCS and familiar with the operation of the facility, as a resource in NCS management decisions and the specialists should be independent of operations supervision;

- Committed to provide NCS postings in areas, operations, work stations, and storage locations;

- Committed to the establish the following policy: "Personnel must report defective NCS conditions and perform actions only in accordance with written, approved procedures. Unless a specific procedure deals with the situation, personnel must report defective NCS conditions and take no action until the situation has been evaluated and recovery procedures provided;" and

- Committed to staff the NCS program with suitably trained personnel and to provide sufficient resources for its operation.

5.3.4.3   NCS Organization and Administration Conclusion

Based on its review of the information provided, the staff concludes that the applicant has described an adequate NCS organization and associated administration to meet the requirements of 10 CFR 70.22(a)(6).

5.3.5   NCS Management Measures

In Chapters 3 and 11 of the SAR (LES, 2005a), the applicant described how it meets the NCS management measures acceptance criteria.

Additional information about training and procedures is in Sections 5.3.2 and 5.3.4 of this SER. Specific information about training for the NCS criticality accident alarm system is in Section 5.3.6 of this SER.

In Section 3.0.3 of the SAR (LES, 2005a), the applicant identified that management measures are functions that will be applied to IROFS and will be applied to any items that may affect the function of IROFS. Management measures will be applied to particular structures, systems, equipment, components, and activities of personnel and may be graded commensurate with the reduction of the risk. Management measures will ensure that those structures, systems, equipment, components, and activities of personnel within the identified IROFS boundary are designed, implemented, and maintained, as necessary, to ensure that the IROFS are available and reliable to perform their intended function, when needed to comply with the performance requirements of 10 CFR 70.61.

In Section 11.1 of the SAR (LES, 2005a), the applicant identified that, for any change [i.e., new design or operation, or modification to the facility, or modification to activities of personnel (e.g., site structures, systems components, computer programs, processes, operating procedures, management measures)], that involve or could affect uranium on site, an NCS Evaluation and, if required, an NCS Analysis, will be prepared and approved. Before implementing the change, it will be determined that the entire process will be subcritical with an applicable margin for safety under both normal and credible abnormal conditions. Each modification will also be evaluated and documented for radiation exposure to minimize worker exposures. IROFS, any items that affect the function of an IROFS, and, in general, items required to satisfy regulatory requirements, will be designated as Quality Level 1. NCS Analyses and NCS Evaluations will be captured in the document control and records management procedures. During design, configuration control for NCS is accomplished, through the use of procedures for controlling design, including preparation, review (including interdisciplinary review and preparation of NCS Analyses and NCS Evaluations, as applicable), and design verification (where appropriate), approval, release, and distribution for use.

In Section 11.2 of the SAR (LES, 2005a), the applicant identified that maintenance procedures involving IROFS will have: (a) corrective and preventive maintenance; (b) functional testing after maintenance; and (c) surveillance/monitoring maintenance activities. For new procedures, or work activities that involve or could affect uranium on site, an NCS Evaluation (and, if required, an NCS Analysis) will be prepared and approved.

In Section 11.3 of the SAR (LES, 2005a), the applicant identified that all persons under the supervision of facility management, including contractors, must participate in the General Employee Training and that training encompasses nuclear safety training. Nuclear safety training programs will be established for the various types of job functions commensurate with the NCS responsibilities associated with each position. That training will be highlighted to stress the high level of importance placed on the NCS of protecting the worker. Those training sessions covering NCS will be conducted on a regular basis and will include principles of NCS. NCS training for personnel associated with fissile material operations outside reactors, where potential exists for an inadvertent criticality, will be in accordance with ANSI/ANS-8.20 (ANSI/ANS, 1991).

In Section 11.4 of the SAR (LES, 2005a), the applicant identified that all activities involving licensed materials or IROFS will be conducted in accordance with approved procedures. Procedures will be used to control activities to ensure that the activities are carried out in a safe manner and in accordance with regulatory requirements. The facility training program will train the required personnel in the use of the latest operating procedures. Administrative procedures will be used to perform activities that support processing operations, including NCS. Procedures will be established and implemented for NCS in accordance with ANSI/ANS-8.19 (ANSI/ANS, 1996b). The NCS procedures will be written such that no single, inadvertent departure from a procedure could cause an inadvertent criticality. NCS postings at the facility will be established that identify administrative controls applicable and appropriate to the activity or area in question. NCS procedures and postings will be controlled by procedure to ensure that they are maintained current. For proposed procedure changes having a potential impact on NCS, an NCS Evaluation, and, if required, an NCS Analysis, will be performed. Any necessary controlled safety parameters, limits, IROFS, management measures, or NCS Analyses that must be imposed or revised will be adequately reflected in appropriate procedures and/or design basis documents. A second Criticality Safety Engineer will independently review changes and then the Health, Safety, and Environment Manager will approve them (if appropriate) before they go into effect.

In Section 11.5 of the SAR (LES, 2005a), the applicant identified that there will be a two-tiered approach to verifying compliance to procedures and performance to regulatory requirements. Audits will be focused on verifying compliance with regulatory requirements, procedural requirements, and licensing commitments. Assessments will be focused on effectiveness of activities and ensuring that IROFS and any items that affect the function of IROFS are reliable and available to perform their intended safety functions. Audits and assessments will be conducted in the NCS functional area. NCS audits will be conducted and documented quarterly, such that all aspects of the NCS program will be audited at least every 2 years. The Operations Group will be assessed periodically to ensure that NCS procedures are being followed and that the process conditions have not been altered to adversely affect NCS. NCS assessments will be performed in accordance with ANSI/ANS-8.19 (ANSI/ANS, 1996b) and will ensure that operations conform to NCS requirements. The frequency of NCS assessments will be based on the controls identified in the NCS Analyses and NCS Evaluations and will be conducted at least semiannually. Weekly NCS walk-throughs of UF_6 process areas will be conducted and documented.

In Section 11.6 of the SAR (LES, 2005a), the applicant identified that the incident investigation process will be a mechanism available for use by anyone at the facility for reporting deficiencies, abnormal events, and potentially unsafe conditions or activities. NCS requirements will be addressed when performing an incident investigation.

In Section 11.7 of the SAR (LES, 2005a), the applicant identified that records management will be performed in a controlled and systematic manner to provide identifiable and retrievable documentation, including for NCS.

In Section 11.8 of the SAR (LES, 2005a), the applicant identified that the Quality Assurance Program and its supporting manuals, procedures, and instructions will be applicable to items and activities designated as Quality Level 1 and Quality Level 2. The Quality Level 1 program will be applied to those structures, systems, components, and administrative controls that have been determined to be IROFS, items that affect the function of the IROFS, and, in general, items required to satisfy regulatory requirements.

Regarding the specific acceptance criteria in Section 5.4.3.3 of NUREG-1520 (NRC, 2002) for NCS management measures, the applicant:

- Committed to ANSI/ANS-8.19 (ANSI/ANS, 1996b) and ANSI/ANS-8.20, "Nuclear Criticality Safety Training" (ANSI/ANS, 1991), as they relate to training;

- Committed to provide training to all personnel to recognize the criticality accident alarm system signal and to evacuate promptly to a safe area;

- Committed to provide training regarding the policy that: "Personnel must report defective NCS conditions and perform actions only in accordance with written, approved procedures. Unless a specific procedure deals with the situation, personnel must report defective NCS conditions and take no action until the situation has been evaluated and recovery procedures provided.";

- Committed to ANSI/ANS-8.19 (ANSI/ANS, 1996b), as it relates to procedures and to the policy that no inadvertent departure from a procedure could cause an inadvertent criticality;

- Committed to ANSI/ANS-8.19 (ANSI/ANS, 1996b), as it relates to audits and assessments;

- Committed to conduct and document weekly NCS walk-throughs of UF_6 process areas; and

- Committed to NCS audits to be conducted and documented quarterly, such that all aspects of the NCS program, including management measures, will be audited at least every 2 years.

Based on its review of the information provided, the staff concludes that the applicant has described adequate NCS management measures to meet the requirements of 10 CFR 70.62(d).

5.3.6 NCS Program

The NCS program includes the commitments and descriptions of how to meet those commitments to prevent an inadvertent criticality and to respond to an inadvertent criticality.

In Chapters 3 and 5 of the SAR (LES, 2005a) and the ISA Summary (LES, 2005c), the applicant described the NCS program, which included descriptions in the following areas: (1) NCS methodologies and technical practices, (2) NCS criticality accident alarm system, (3) NCS subcriticality of operations and margin of subcriticality for safety, (4) NCS baseline design criteria, (5) NCS in the integrated safety analysis (ISA) program and the ISA Summary, and (6) additional NCS program commitments.

5.3.6.1 NCS Methodologies and Technical Practices

In Section 5.1 of the SAR (LES, 2005a), the applicant identified that the facility will be designed, constructed, and operated, such that an inadvertent criticality will be prevented. NCS at the facility will be assured by designing the facility, systems, and components with safety margins, such that safe conditions are maintained under normal and credible abnormal conditions.

In Section 5.1.1 of the SAR (LES, 2005a), the applicant committed to the double contingency principle, as stated in ANSI/ANS-8.1 (ANSI/ANS, 1998a), which is that "Process designs should incorporate sufficient factors of safety to require at least two unlikely, independent, and concurrent changes in process conditions before a criticality accident is possible." Each process that has accident sequences that could result in an inadvertent criticality at the facility will meet the double contingency principle.

In Section 5.1.3 of the SAR (LES, 2005a), the applicant identified that process operations will require the establishment of NCS limits. The facility UF_6 systems will involve mostly gaseous operations. These operations will be carried out under reduced atmospheric conditions (i.e., vacuum) or at slightly elevated pressures, not exceeding three atmospheres. Within the Separations Building, significant accumulations of enriched UF_6 could reside only in the Product Low-Temperature Take-off Stations, Product Liquid Sampling Autoclaves, Product Blending System, and the UF_6 Cold Traps. Except for the UF_6 Cold Traps, all of these will contain the UF_6 in Type 30B or Type 48Y cylinders. All these significant accumulations will be within enclosures protecting them from water ingress. The facility design will minimize the possibility of accidental moderation (e.g., water or hydrogen) by eliminating direct water contact with these cylinders of accumulated UF_6. In addition, the facility's procedural controls for enriching the UF_6 will assure that it does not become unacceptably hydrogen-moderated while in process. The facility's UF_6 systems' operating procedures will contain measures against the loss of moderation control, according to ANSI/ANS-8.22 (ANSI/ANS, 1997c).

In Section 5.1.4 of the SAR (LES, 2005a), the applicant identified that each portion of the facility, system, or component that may possibly contain enriched uranium will be designed with NCS as an objective. Where there will be significant in-process accumulations of enriched uranium as UF_6, the facility design will include multiple features to minimize the possibilities for breakdown of the moderation control limits. These features will eliminate direct ingress of water to product cylinders while in-process.

Table 5.3-1 provides the safety criteria (i.e., parameter, critical value, safe value, and safety factor) for uniform aqueous solutions of enriched UO_2F_2 for the facility, assuming a single component and 6.0 wt percent U-235 enrichment. In the table, the term safety factor means the ratio between the safe value and the critical value and the term k_{eff} means the effective neutron multiplication factor. k_{eff} represents the neutron production rate relative to the neutron loss rate. Therefore, the critical value - when k_{eff} is one - represents the point at which an inadvertent criticality (i.e., nuclear chain reaction) occurs. When k_{eff} is less than one, the system is subcritical. NRC reviewed the 'Critical Value' in the Table 5.3-1 for 6.0 wt percent against the values in ANSI/ANS-8.1 (ANSI/ANS, 1998a) at 5.0 and 10.0 wt percent. By using knowledge, experience, and best estimate for interpolation, NRC determined that the applicant's values in Table 5.3-1 are consistent with the values in ANSI/ANS-8.1 (ANSI/ANS, 1998a).

Table 5.3-1
Safety Criteria

Parameter	Critical Value (k_{eff} = 1.0)	Safe Value (k_{eff} = 0.95)	Safety Factor
Volume	24 L (6.3 gal.)	18 L (4.8 gal.)	0.75
Cylinder Diameter	24.4 cm (9.6 in.)	21.9 cm (8.6 in.)	0.90
Slab Thickness	11.5 cm (4.5 in.)	9.9 cm (3.9 in.)	0.86
Water Mass	15.4 kg (34.0 lb)	11.5 kg (25.4 lb)	0.75
Areal Density	9.5 g/cm^2 (19.5 lb/ft^2)	7.5 g/cm^2 (15.4 lb/ft^2)	0.79
Uranium Mass with no double batching	27 kg (59.5 lb)	19.5 kg (43.0 lb)	0.72
Uranium Mass with double batching	27 kg (59.5 lb)	12.2 kg (26.9 lb)	0.45

Table 5.3-2 shows how the above safety criteria for single components will be specifically applied at the facility (i.e., building/system/component, control mechanism, safety criteria) to prevent an inadvertent criticality, assuming 6.0 wt percent U-235 enrichment, which is conservative, because the facility will be limited to 5.0 wt percent U-235 enrichment, except for the Contingency Dump Trap System that will be limited (based on physics) to 1.5 wt percent U-235.

Table 5.3-2
Application of Safety Criteria

Building, System, or Component	Control Mechanism	Safety Criteria
Enrichment	Enrichment	6.0 wt percent U-235
Centrifuges	Diameter	< 21.9 cm (8.6 in.)
Product Cylinders (30B)	Moderation	Hydrogen < 0.95 kg (2.09 lb)
Product Cylinders (48Y)	Moderation	Hydrogen < 1.05 kg (2.31 lb)
UF_6 Piping	Diameter	< 21.9 cm (8.6 in.)
Chemical Traps	Diameter	< 21.9 cm (8.6 in.)
Product Cold Trap	Diameter	< 21.9 cm (8.6 in.)
Contingency Dump System Traps	Enrichment	< 1.5 wt percent U-235
Tanks	Mass	< 12.2 kg U (26.9 lb U)
Feed Cylinders	Enrichment	< 0.72 wt percent U-235
Uranium Byproduct Cylinders	Enrichment	< 0.72 wt percent U-235
UF_6 Pumps (first stage)	Not Applicable	Safe, by explicit calculation
UF_6 Pumps (second stage)	Volume	< 18.0 L (4.8 gal.)
Individual Uranic Liquid Containers (e.g., Fomblin Oil Bottle, Laboratory Flask, Mop Bucket)	Volume	< 18.0 L (4.8 gal.)
Vacuum Cleaners Oil Containers	Volume	< 18.0 L (4.8 gal.)

In Section 5.2.1.2 of the SAR (LES, 2005a), the applicant described the process that was used to identify the margin of subcriticality for safety when doing calculations to demonstrate safety (i.e., not scoping calculations) for normal and credible abnormal conditions. The k_{eff} equation that the applicant used for those calculations is: $k_{eff} = k_{calc} + 3\sigma_{calc} < 0.95$. The applicant used the MONK8 Monte Carlo computer program (AEAT, 1998) to calculate neutron multiplication to ascertain the k_{eff} values.

In Section 5.1.2 of the SAR (LES, 2005a), the applicant identified that the major controlling safety parameters to be used in the facility will be enrichment control, geometry control, moderation control, or limitations on the mass as a function of enrichment. In addition, reflection, interaction, and heterogeneous effects are important parameters that will be considered and applied. NCS Evaluations and NCS Analyses will be used to identify the significant safety parameters affected within a particular system. All assumptions relating to process, equipment, material function, and operation, including credible abnormal conditions, will be justified, documented, and independently reviewed. Where possible, passive engineered controls will be used to ensure NCS.

In Section 5.1.1 of the SAR (LES, 2005a), the applicant committed to the following elements related to NCS methodologies and technical practices in the NCS program:

- Adherence to the double contingency principle, as stated in ANSI/ANS-8.1 (ANSI/ANS, 1998a);

- Safety parameters will be established and procedures will be established;

- The NCS program structure will be provided, including definition of the responsibilities and authorities of key program personnel;

- The NCS methods and technical practices will be kept applicable to current configuration by means of the configuration management function and the NCS program will be upgraded, as necessary, to reflect changes in the ISA or NCS methodologies and to modify operating and maintenance procedures in ways that could reduce the likelihood of occurrence of an inadvertent criticality;

- The NCS program will be used to establish and maintain NCS safety limits and NCS operating limits for IROFS in nuclear processes and there is a commitment to maintain adequate management measures, to ensure the availability and reliability of the IROFS;

- NCS postings will be provided and maintained current;

- NCS emergency procedure training will be provided;

- The NCS baseline design criteria requirements in 10 CFR 70.64(a)(9) will be adhered to (i.e., see Table 5.3-2, "Application of Safety Criteria");

- The NCS program will be used to evaluate modifications to operations, to recommend process parameter changes necessary to maintain the safe operation of the facility, and to select appropriate IROFS and management measures;

- The NCS program will be used to promptly detect NCS deficiencies by means of operational inspections, audits, and investigations and deficiencies will be entered into the corrective action program so as to prevent recurrence of unacceptable performance deficiencies in IROFS, NCS functions, or management measures;

- NCS program records will be retained as described in Section 11.7 of the SAR (LES, 2005a); and

- NCS training will be provided to individuals who handle nuclear material at the facility based on the training program described in ANSI/ANS-8.20 (ANSI/ANS, 1991) and the training program will be developed and implemented with input from the NCS staff, training staff, and management. The training will focus on the following:

- Appreciation of the physics of NCS;
- Analysis of jobs and tasks to determine what a worker must know to perform NCS tasks efficiently; and
- Design and development of learning objectives, based on the analysis of jobs and tasks, that reflect the knowledge, skills, and abilities needed by the worker.

In Section 5.1.1 of the SAR (LES, 2005a), the applicant committed to the following general NCS philosophy: (1) to prevent accidental uranium enrichment excesses; (2) to provide geometrical safety, when practical; (3) to provide for moderation control within the UF_6 processes; and (4) to impose mass limits on containers of aqueous, solvent-based, or acid solutions containing uranium. Interaction control will provide for safe movement and storage of components. In addition, facility and equipment features will assure prevention of excessive enrichment.

In Section 5.1.1 of the SAR (LES, 2005a), the applicant described that the facility will be divided into six distinctly separate Assay Units (i.e., Cascade Halls) with no common UF_6 piping. UF_6 blending will be done in a physically separate portion of the facility. Process piping, individual centrifuges, and chemical traps, other than the Contingency Dump Chemical Traps, will be safe by limits placed on their diameters. Product cylinders will rely on uranium enrichment, moderation control, and mass limits, to protect against the possibility of an inadvertent criticality. Liquid effluent collection tanks that hold uranium in solution will be mass-controlled because none will be geometrically safe.

In Section 5.1.2 of the SAR (LES, 2005a), the applicant identified that the major controlling safety parameters to be used in the facility will be enrichment control, geometry control, moderation control, or limitations on the mass as a function of enrichment. In addition, reflection, interaction, and heterogeneous effects will be important parameters that will be considered and applied where appropriate in NCS Analyses. NCS Evaluations and NCS Analyses will be used to identify the significant safety parameters affected within a particular system. All assumptions relating to process, equipment, material function, and operation, including credible abnormal conditions, will be justified, documented, and independently reviewed. Where possible, passive engineered controls will be used to ensure NCS.

Other than the Type 30B Cylinders, Type 48Y Cylinders, first-stage UF_6 Pumps, and Contingency Dump Chemical Traps, all Separation Plant components that handle enriched UF_6 will be safe by favorable geometry, to preclude an inadvertent criticality. A centrifuge array will use a probability argument with multiple operational procedure barriers to demonstrate NCS. A product cylinder will be safe by total moderator or hydrogen-to-uranium (H/U) ratio control, as appropriate, to preclude an inadvertent criticality. In the Technical Services Building, NCS for uranium loaded liquids will be ensured by limiting the mass of uranium in any single tank to less than or equal to 12.2 kg U (26.9 lb U) to preclude an inadvertent criticality. Individual liquid storage bottles will be safe by volume to preclude an inadvertent criticality. Interaction in storage arrays will be accounted for to preclude an inadvertent criticality.

The control safety parameters to be applied to the facility are as follows:

Enrichment

Enrichment will be controlled to limit the wt percent U-235 within any process, vessel, or container. The Contingency Dump System will have a maximum enrichment of 1.5 wt percent (based on physics), but all other locations in the facility will have a maximum enrichment of 5.0 wt percent. As an added non-quantified conservatism, even though those parts of the facility will be limited to a maximum enrichment of 5.0 wt percent, NCS will be analyzed at 6.0 wt percent.

Geometry/Volume

Geometry/volume control may be used to ensure NCS within specific process operations or vessels and within storage containers. When performing calculations to determine safety, the geometry/volume limits will be chosen to ensure that: $k_{eff} = k_{calc} + 3\sigma_{calc} < 0.95$. The favorable geometry/favorable volume safe values for control define the characteristic dimension of importance for a single unit such that NCS will not dependent on any other parameter, assuming an enrichment of 6.0 wt percent.

Moderation

Water and oil will be the moderators considered at the facility. The only system where moderation will be used as a control parameter is in the Product Cylinders. Moderation control will be established consistent with ANSI/ANS-8.22 (ANSI/ANS, 1997c) and will incorporate the criteria below:

- Controls will be established to limit the amount of moderation entering the cylinders;

- When moderation is not considered a control parameter, either optimum moderation or worst-case H/U ratio will be assumed when performing the NCS Analysis; and

- When moderation is the only parameter used for NCS control, the following additional criteria will be applied that will assure that at least two independent controls would have to fail before an inadvertent criticality is possible:

    ▸ Two independent controls will be established and used to monitor and limit uncontrolled moderator before returning a cylinder to production, thereby limiting the amount of uncontrolled moderator from entering a system to an acceptable limit; and
    ▸ The NCS Evaluation of the cylinders under moderation control will include the establishment of limits for the ratio of maximum moderator-to-fissile material for both normal and credible abnormal conditions and the NCS Analysis will be supported by parametric studies.

Mass

Mass control may be used to limit the quantity of uranium within specific process operations, vessels, or storage containers. Mass control may be used on its own or in combination with other control methods. Analysis or sampling (using instrumentation) will be employed to verify the mass of the material. Conservative administrative limits for each operation will be specified in operating procedures.

Whenever mass control is established for a container, records will be maintained for mass transfers into and out of the container. Establishment of mass limits for a container will involve consideration of potential moderation, reflection, geometry, spacing, and enrichment. The NCS Evaluation will consider normal and credible abnormal conditions for determination of the operating mass limit for the container and the definition of subsequent controls necessary to prevent reaching the safety limits. When only administrative controls are used for mass-controlled systems, double batching will be assumed in the NCS Analysis, which sets a more conservative limit.

Reflection

Reflection will be considered when performing NCS Evaluations and NCS Analyses. The possibility of full-water reflection was considered, but the layout of the facility will be a very open design and so it is not credible that those vessels and facility components requiring NCS control could become flooded from a source of water within the facility. In addition, neither automatic sprinkler nor standpipe and hose systems will be provided in the Technical Support Building, Separation Buildings, Blending and Liquid Sampling Areas, Cylinder Receipt and Dispatch Building, Centrifuge Assembly Building, and Centrifuge Post Mortem Areas. Therefore, full-water reflection of vessels will be discounted. For conservatism, some NCS Analyses will be performed using full-water reflection. Partial reflection of 2.5 cm (0.984 in.) of water will be assumed where limited moderating materials may be present (e.g., human beings). Since concrete can be a more efficient reflector than water; it will be modeled in NCS Analyses where it is present. When moderation control is identified in the ISA Summary, it will be established within the guidelines of ANSI/ANS-8.22 (ANSI/ANS, 1997c).

Interaction

NCS Evaluations and NCS Analyses will consider the potential effects of interaction. A non-interacting unit is defined as a unit that is spaced an approved distance from other units such that the multiplication of the subject unit is not increased. Units may be considered non-interacting when they are separated by more than 60 cm (23.6 in.). If units are considered interacting, then an NCS Analysis will be performed to determine individual unit multiplication and array interaction using the Monte Carlo computer code MONK8A to ensure that:
$k_{eff} = k_{calc} + 3\sigma_{calc} < 0.95$.

In Section 5.2.1.4 of the SAR (LES, 2005a), the applicant indicated that NCS will be evaluated for the design features of the facility system or component and for the operating practices that relate to maintaining NCS. The NCS Analysis of individual systems or components and their interaction with other systems or components containing enriched uranium will be performed to assure that the NCS criteria are met. The NCS Analyses and Table 5.3-1, "Safety Criteria," provide a basis for the facility design and for NCS hazards identification performed as part of the ISA process. Each portion of the facility, system, or component that may possibly contain enriched uranium will be designed with NCS as an objective.

The facility will be designed and operated in accordance with the parameters in Table 5.3-2, "Application of Safety Criteria." The applicant provided a general description of the elements of an NCS Analysis. During the design phase of the facility, a qualified Criticality Safety Engineer will perform the NCS Analysis and a second qualified Criticality Safety Engineer will independently review it. During the operation of the facility, a qualified Criticality Safety Engineer will perform the NCS Analysis, a second qualified Criticality Safety Engineer will independently review it, and the Health, Safety, and Environment Manager will approve it (if appropriate) before it goes into effect.

In Section 5.2.1.5 of the SAR (LES, 2005a), the applicant indicated that the facility NCS Analyses will be performed using the methods and assumptions in Section 5.2.1.4 of the SAR (LES, 2005a) and that NCS Analyses will be performed in order to demonstrate compliance with the following criteria:

- Methods will be validated and used only within demonstrated acceptable ranges;

- NCS Analyses will adhere to ANSI/ANS-8.1 (ANSI/ANS, 1998a), as it relates to analysis methods;

- Compliance with the intent of the validation report statement in NRC Regulatory Guide 3.71 (NRC, 1998), including that it applies to ANSI/ANS-8.1 (ANSI/ANS, 1998a), which states that the applicant should demonstrate: (1) the adequacy of the margin of safety for subcriticality, by assuring that the margin is large compared to the uncertainty in the calculated value of k_{eff}; (2) the calculation of k_{eff} is based on a set of variables whose values lie in a range for which the method used to determine k_{eff} has been validated; and (3) trends in the bias support the extension of the method to areas outside the area or areas of applicability;

- A specific reference to (including date and revision number), and summary description of, either a manual or a documented, reviewed, and approved validation report for each methodology, will be included. Any change in the reference manual or validation report will be reported to NRC by letter;

- The reference manual and documented reviewed validation report will be kept at the facility;

- The reference manual and validation report will be incorporated into the configuration management program;

- NCS Analyses will be performed in accordance with the methods specified and incorporated into the configuration management program;

- The NCS methods and technical practices in Section 5.4.3.4 of NUREG-1520 (NRC, 2002) will be used to evaluate NCS accident sequences in operations and processes;

- The acceptance criteria in Section 3.4 of NUREG-1520 (NRC, 2002), as they relate to the identification of NCS accident sequences; consequences of NCS accident sequences; likelihoods of NCS accident sequences; and descriptions of IROFS for NCS accident sequences, will be utilized;

- The applicant will use NCS controls and controlled safety parameters to assure that, under normal and credible abnormal conditions, all nuclear processes are subcritical – including use of an approved margin of subcriticality for safety;

- As stated in ANSI/ANS-8.1 (ANSI/ANS, 1998a), process specifications will incorporate margins to protect against uncertainties in process variables and against a limit being accidentally exceeded;

- ANSI/ANS-8.7 (ANSI/ANS, 1998b), as it relates to the requirements for subcriticality of operations, margin of subcriticality for safety, and selection of controls requirements in 10 CFR 70.61(d), will be used;

- ANSI/ANS-8.10 (ANSI/ANS, 1988b), as it relates to the determination of consequences of NCS accidents, as modified by NRC Regulatory Guide 3.71 (NRC, 1998), will be used;

- If administrative k_{eff} margins for normal and credible abnormal conditions are used, then NRC pre-approval of the administrative margins will be sought;

- Subcritical limits for k_{eff} calculations such that: k_{eff} subcritical = 1.0 - bias - margin, where the margin includes adequate allowance for uncertainty in the method, data, and bias, to assure subcriticality, will be used;

- Studies to correlate the change in a value of a controlled parameter and its k_{eff} value will be performed. The studies will include changing the value of one controlled parameter and determining its effect on another controlled parameter and k_{eff};

- The double contingency principle will be used in determining NCS controls and NCS IROFS; and

- Acceptance criteria in Section 3.4 of NUREG-1520 (NRC, 2002), as they relate to subcriticality of operations and margin of subcriticality for safety, will be utilized. That is:

  "For nuclear criticality accident sequences, the method evaluates compliance with 10 CFR 70.61(d). That is, even in a facility with engineered features to limit the consequences of nuclear criticalities, *preventive* control(s) must be in place that are sufficient to ensure that the likelihood of criticality is controlled to be 'highly unlikely.' A moderately higher standard of likelihood may be permitted in preventing such events, consistent with ANSI/ANS Standard 8.10 [(ANSI/ANS, 1988b)]. In particular, criticality cannot result from the failure of any single IROFS. In addition, potential criticality accidents must meet an [NRC] approved margin of subcriticality for safety. Acceptance criteria for such margins are reviewed as programmatic commitments, but the ISA methods must consider and the ISA Summary must document the actual magnitude of those margins when they are part of the reason why the postulated accident sequence resulting in criticality is highly unlikely."

In Section 5.2.1.6 of the SAR (LES, 2005a), the applicant indicated that the NCS Evaluation process will be in accordance with ANSI/ANS-8.19 (ANSI/ANS, 1996b). For any change [i.e., new design or operation, or modification to the facility, or modification to activities of personnel (e.g., site structures, systems, components, computer programs, processes, operating procedures, management measures)] that involves or could affect uranium, an NCS Evaluation will be prepared and approved. Before implementing the change, it will be determined that the entire process will be subcritical, with an approved margin of safety under both normal and credible abnormal conditions. If this condition cannot be demonstrated with the NCS Evaluation, then either a new or revised NCS Analysis will be generated that meets the criteria, or the changes will not be made. The NCS Evaluation will determine and explicitly identify the controlled safety parameters and associated limits on which NCS depends, assuring that no single inadvertent departure from a procedure could cause an inadvertent criticality and that the safety basis of the facility will be maintained during the lifetime of the facility. The NCS Evaluation will ensure that all potentially affected uranic processes will be evaluated to determine the effect of the change on the safety basis of the process, including the effect on bounding process assumptions, the effect on the reliability and availability of NCS controls, and the effect on the NCS of connected processes. The NCS Evaluation process will involve a review of the proposed change, discussions with the subject matter experts to determine the processes that need to be considered, development of the controls necessary to meet the double contingency principle, and identification of the assumptions and equipment (e.g., physical controls and/or management measures) needed to ensure NCS.

The applicant provided a general description of the elements of an NCS Evaluation. The NCS Evaluation will be performed and documented by a qualified Criticality Safety Engineer. Engineering judgment of the qualified Criticality Safety Engineer will be used to ascertain the NCS impact of the proposed change. The basis for this judgment will be documented with sufficient detail in the NCS Evaluation to allow the independent review by a second qualified Criticality Safety Engineer to confirm the conclusions of the judgment of results. After the NCS Evaluation is completed and documented and the independent review by the second Criticality Safety Engineer is performed and documented, the Health, Safety, and Environment Manager will approve it (as appropriate) before it goes into effect.

In Section 5.2.1.7 of the SAR (LES, 2005a), the applicant identified that the facility NCS Evaluations will be performed according to the following criteria:

- NCS Evaluations will be performed in accordance with the procedures specified and incorporated in the configuration management program;

- NCS methods and technical practices in Sections 5.4.3.4.1(10)(a), (b), (d), and (e) of NUREG-1520 (NRC, 2002), will be used to evaluate NCS accident sequences in operations and processes;

- Acceptance criteria in Section 3.4 of NUREG-1520 (NRC, 2002), as they relate to: identification of NCS accident sequences, consequence of NCS accident sequences, likelihood of NCS accident sequences, and descriptions of IROFS for NCS accident sequences, will be utilized;

- NCS controls and controlled parameters to assure that, under normal and credible abnormal conditions, all nuclear processes are subcritical, including use of an approved margin of subcriticality for safety, will be used;

- The double contingency principle will be used in determining NCS controls and NCS IROFS; and

- Acceptance criteria in Section 3.4 of NUREG-1520 (NRC, 2002), as they relate to subcriticality of operations and margin of subcriticality for safety, will be utilized. That is:

  "For nuclear criticality accident sequences, the method evaluates compliance with 10 CFR 70.61(d). That is, even in a facility with engineered features to limit the consequences of nuclear criticalities, *preventive* control(s) must be in place that are sufficient to ensure that the likelihood of criticality is controlled to be 'highly unlikely.' A moderately higher standard of likelihood may be permitted in preventing such events, consistent with ANSI/ANS Standard 8.10 [(ANSI/ANS, 1988b)]. In particular, criticality cannot result from the failure of any single IROFS. In addition, potential criticality accidents must meet an [NRC] approved margin of subcriticality for safety. Acceptance criteria for such margins are reviewed as programmatic commitments, but the ISA methods must consider and the ISA Summary must document the actual magnitude of those margins when they are part of the reason why the postulated accident sequence resulting in criticality is highly unlikely."

Regarding the specific acceptance criteria in Sections 5.4.3.4.1 and 5.4.3.4.2 of NUREG-1520 (NRC, 2002) for NCS methodologies and technical practices, the applicant:

For NCS Methodologies:

- Committed to appropriately apply NCS controlled parameters;

- Committed to appropriately determine NCS limits on IROFS;

- Committed to use acceptable methodologies to perform NCS Analyses and NCS Evaluations;

- Committed to establish NCS limits on controls and controlled parameters to ensure an adequate margin of subcriticality for safety;

- Committed to use validated methods to develop NCS limits to ensure that they are within acceptable ranges, by utilizing both appropriate assumptions and acceptable computer codes;

- Committed to promptly detect an inadvertent criticality to ensure that radiation exposures to workers are minimized;

- Committed to ANSI/ANS-8.1 (ANSI/ANS, 1988a) and ANSI/ANS-8.1 (ANSI/ANS, 1998a) as they relate to NCS methodologies;

- Committed to the intent of the validation report statement in NRC NCS Regulatory Guide 3.71 (NRC, 1998);

- Committed to include a reference to and summary description of the reference manual or validation report for each methodology that will be used in each NCS Analysis and NCS Evaluation;

- Committed to have, at the facility, the reference manual or documented, reviewed, and approved validation report for each methodology that will be used in each NCS Analysis and NCS Evaluation;

- Committed to incorporate each reference manual or documented, reviewed, and approved validation report for each methodology, including the assumptions used, into the configuration management program; and

- Committed to perform NCS Analyses and NCS Evaluations in accordance with the following principles:

    ▸ NCS safety limits, NCS operating limits, and limits on NCS controlled parameters will be established assuming credible optimum conditions, unless specified controls are implemented to control the limit to a certain range of values;
    ▸ NCS safety limits, NCS operating limits, and limits on NCS controlled parameters will be derived from the NCS Analyses and NCS Evaluations;
    ▸ NCS safety limits, NCS operating limits, and limits on NCS controlled parameters will be based on the proper application of the NCS methodology to the process;
    ▸ NCS operating limits will be derived from NCS safety limits by taking into consideration changes in operating parameters to ensure processes will remain subcritical under both normal and credible abnormal conditions;
    ▸ NCS operating limits will establish sufficient margins of safety for processes and take into consideration the variability and uncertainty in processes and the NCS subcritical limits;
    ▸ NCS safety limits will establish sufficient margins of safety for processes and take into consideration the variability and uncertainty in processes and the NCS operating limits;
    ▸ The margin of subcriticality for safety for a process should be relative compared to the calculated value of k_{eff}; and
    ▸ The k_{eff} will be calculated from a set of variables whose values lie in a range for which the validity of the NCS methodology will have been demonstrated.

For NCS Technical Practices:

- Committed to the policy that no single credible event or failure can result in an inadvertent criticality;

- Committed to the following general statements about controls: (1) the preferred use of passive engineered controls to ensure NCS, and (2) the order of preference for NCS controls will be passive engineered, active engineered, enhanced administrative, and administrative;

- Committed to consider heterogeneous effects when evaluating a controlled parameter;

- Committed to perform either an NCS Analysis or NCS Evaluation for all controlled parameters that demonstrates that during both normal and credible abnormal conditions, the controlled parameter will be maintained;

- Committed to mass as a controlled parameter:

  ▸ When the mass is measured, instrumentation will be used.
  ▸ If double batching is physically possible, then the limit will be no more than 45 percent of the minimum critical mass; and
  ▸ If double batching is not physically possible, then the limit will be no more than 72 percent of the minimum critical mass;

- Committed to geometry as a controlled parameter:

  ▸ As part of the Quality Assurance Program and that geometry controls will be Quality Level 1, before beginning operations, all dimensions and nuclear properties that use geometry control will be verified and the facility configuration management program will be used to maintain the dimensions and nuclear properties;
  ▸ The margin of safety for a single unit will be 90 percent of the minimum critical cylinder diameter;
  ▸ The margin of safety for a single unit will be 86 percent of the minimum critical slab thickness. The NUREG-1520 (NRC, 2002) acceptance criteria is 85 percent, however, the applicant's commitment to 86 percent is acceptable because the overall conservatism of limiting the enrichment to 5.0 wt percent, but calculating limits based on 6.0 wt percent; and
  ▸ The margin of safety for a single unit will be 75 percent of the minimum critical volume;

- Committed to enrichment as a controlled parameter: A method of segregating enrichments will be used to ensure differing enrichments will not be interchanged (i.e, Contingency Dump System will be physically limited to 1.5 wt percent and all other parts of the facility will be limited to 5.0 wt percent).

- Committed to reflection as a controlled parameter;

- Committed to moderation as a controlled parameter to use ANSI/ANS-8.22 (ANSI/ANS, 1997c);

- Committed to interaction as a controlled parameter when maintaining a physical separation between units, engineered controls to ensure a minimum spacing or enhanced administrative controls will be used. In addition, the structural integrity of the spacers or racks should be sufficient for normal and credible abnormal conditions.

Based on its review of the information provided, the staff finds the applicant's NCS methodologies and technical practices to be acceptable.

5.3.6.2   NCS Criticality Accident Alarm System

In Section 5.3 of the SAR (LES, 2005a), the applicant indicated that the facility will be provided with a CAAS, as required by 10 CFR 70.24. Areas where special nuclear material will be handled, used, or stored in amounts at or above the 10 CFR 70.24 mass limits will be provided with CAAS coverage. Emergency management measures are covered in the facility Emergency Plan (LES, 2005b).

In Chapter 8 of the SAR (LES, 2005a), the applicant committed to the facility Emergency Plan (LES, 2005b). The facility Emergency Plan (LES, 2005b) included a description of the intended activities related to NCS, the CAAS, and what the response will be if an inadvertent criticality occurs. No inadvertent criticality has ever occurred at a facility processing low enriched uranium (LEU). An inadvertent criticality with low enriched uranium requires a precise combination of conditions. The facility will include multiple design, administrative, and engineered controls to prevent an inadvertent criticality. The Technical Services Building, three Cascade Halls, Cylinder Receipt and Dispatch Building, Blending and Liquid Sampling Area, and UF_6 Handling Area are the only areas in the facility where an inadvertent criticality could occur. A CAAS will be provided to detect and alarm if an inadvertent criticality occurs in an area where uranium at or above the 10 CFR 70.24 mass limits will be handled, used, or stored. If an inadvertent criticality were detected, then: (1) the CAAS would interface with the Plant Control System to provide a visual and audible alarm of the CAAS and CAAS status in the Control Room; (2) facility personnel would be alerted to evacuate the specific area, by a specific CAAS visual and audible alarm; (3) such evacuated facility personnel would assemble immediately in the Assembly Areas as required by the facility Emergency Plan Implementing Procedures; and (4) emergency response and protective measures for a Site Area Emergency would be performed because the applicant classified an inadvertent criticality as a Site Area Emergency. Also, criticality dosimeters will be located within the facility buildings and, when recovered and evaluated, the dosimeters will provide spectrum information to assist in reconstruction of the inadvertent criticality event.

In Section 3.1.5 of the ISA Summary (LES, 2005c), the applicant indicated that the facility will be provided with a CAAS, as required by 10 CFR 70.24. Areas where special nuclear material will be handled, used, or stored in amounts at or above the 10 CFR 70.24 mass limits will be provided with CAAS coverage. The CAAS is designed, will be installed, and will be maintained in accordance with ANSI/ANS-8.3 (ANSI/ANS, 1997a), as modified by Regulatory Guide 3.71 (NRC, 1998). CAAS coverage will consist of an overlapping detection layout, where all required covered areas will be monitored by a minimum of a pair of gamma detectors. Detectors will trip based on both steady radiation rate and time integrated total radiation dose levels. The detectors will have a stated trigger response of 1 milligray per hour (mGy/hr) (0.1 rad/hr) as a gamma-radiation rate meter detector. Based on this design and the guidance provided in Appendix B of ANSI/ANS-8.3 (ANSI/ANS, 1997a), the radius of detection must be less than 106 m (348 ft). Because of building-steel spacing and equipment arrangement, as well as a desire to maintain a factor of two safety margin, a radius of detection of 40 m (131 ft) is used in the design of the facility. This ensures that the CAAS is capable of detecting a criticality that produces an absorbed dose, in soft tissue, of 0.2 Gy (20 rads) of combined neutron and gamma radiation at an unshielded distance of 2 m (6.6 ft) from the reacting material, within 1 minute. The CAAS will be uniform throughout the facility for the type of radiation detected, the mode of detection, the alarm signal, and the system dependability. When tripped, the CAAS will automatically initiate a clearly audible signal in areas that must be evacuated. The CAAS will be provided with emergency power and is designed to remain operational during credible events or conditions, including fire, explosion, corrosive atmosphere, or seismic shock (i.e., equivalent to the site-specific design-basis earthquake or the equivalent value specified by the uniform building code). After the CAAS is installed, whenever the CAAS is not functional, compensatory measures will be implemented (e.g., limiting access, restricting special nuclear material movement). Should the CAAS coverage be lost and not restored within a specified number of hours, the operations will be rendered safe by shutdown and quarantine, if necessary. Onsite guidance in this respect will be provided based on process specific considerations that consider applicable risk trade-off of the duration of reliance on compensatory measures versus the risk associated with process upset in shutdown. Also, the applicant provided a diagram of criticality alarm system detector locations.

Regarding the specific acceptance criteria in Sections 3.4.3.2(4)(c) and 5.4.3.4.3 of NUREG-1520 (NRC, 2002) for NCS criticality accident alarm system, the applicant:

- Described the method for evaluating an acceptable response of at least two detectors to an inadvertent criticality at any location where special nuclear material may be handled, used, or stored;

- Provided a diagram giving the locations of all detectors relative to the potential locations of special nuclear material;

- Provided information supporting the determination of the gamma and neutron emission characteristics of the minimum credible accident of concern capable of producing the effects specified in 10 CFR 70.24;

- Provided information to demonstrate that the applicant's equipment and procedures will be adequate to ensure that specific emergency preparation requirements in 10 CFR 70.24 will be met;

- Documented that the facility criticality accident alarm system will meet the requirements of 10 CFR 70.24;

- Committed to the ANSI/ANS-8.3 (ANSI/ANS, 1997a) standard, as modified by NRC NCS Regulatory Guide 3.71 (NRC, 1998);

- Committed to have a criticality accident alarm system that is uniform throughout the facility for the type of radiation detected, the mode of detection, the alarm signal, and the system dependability;

- Committed to have a criticality accident alarm system that is designed to remain operational during credible events;

- Committed to have a criticality accident alarm system that is clearly audible in areas that must be evacuated or else provide alternate notification methods that are documented to be effective in notifying personnel that an evacuation is necessary;

- Committed to render operations safe, by shutdown and quarantine, if necessary, in any area where criticality accident alarm system coverage has been lost and not restored within a specified number of hours, the number of hours will be determined on a process-by-process basis, because shutting down certain processes, even to make them safe, may carry a larger risk than being without a criticality accident alarm system for a short time, and there will be compensatory measures when the criticality accident alarm system is not functional;

- Committed to ANSI/ANS-8.23 (ANSI/ANS, 1997d), as they relate to NCS;

- Provided a facility Emergency Plan (LES, 2005b);

- Committed to provide fixed and personnel accident dosimeters in areas that require a criticality accident alarm system. These dosimeters will be readily available to personnel responding to an emergency; and

- Committed to provide emergency power for the criticality accident alarm system.

Based on its review of the information provided, the staff concluded that the applicant has described an adequate NCS criticality accident alarm system to meet the requirements in 10 CFR 70.24 and 70.65(b)(4).

5.3.6.3   NCS Subcriticality of Operations and Margin of Subcriticality for Safety

Specific information about the commitments to ANSI/ANS Series-8 NCS standards is in Section 5.3.2 of this SER. Information that demonstrates that the applicant met the acceptance criteria in this part of the SER are also in other parts of this section in this SER.

In Section 5.2.1 of the SAR (LES, 2005a), the applicant indicated that the MONK8A Monte Carlo code (AEAT, 1998) was used to perform the NCS Analyses. MONK8A has accuracy over a wide range of applications and is distributed with a generic validation database comprising of critical experiments covering uranium, plutonium, and mixed systems over a wide range of moderation and reflection. However, NRC does not allow a generic vendor validation to be used as a demonstration of meeting the regulatory requirements for NCS validation. Since NRC staff did not accept the applicant's generic vendor validation report, the applicant provided a specific validation report, as discussed below and in Section 5.2.1.1 of the SAR (LES, 2005a). In addition, by December 30, 2005, the applicant will provide NRC with a revised validation report that meets LES' commitment to ANSI/ANS-8.1-1998 (ANSI/ANS, 1998a) and includes details of validation that state computer codes used, operations, recipes for choosing code options (where applicable), cross-section sets, and any numerical parameters necessary to describe the input.

In Section 5.2.1.1 of the SAR (LES, 2005a), the applicant described the validation process. The applicant validated the MONK8A code with the JEF2.2 cross-section library against experiments in the 2002 version of the *International Handbook of Evaluated Criticality Safety Benchmark Experiments* (ICSBEP) (NEA, 2002). This specific application validation was performed using the uranium solution experiments from ICSBEP. Using 36 LEU solution experiments, the calculated k_{eff} was 1.0007 +/- 0.0005. In the applicant's "MONK8A Validation and Verification Report" (LES, 2004), the applicant identified modifications to the descriptions of the ICSBEP experiments that were described in the SAR (LES, 2005a). The results of the validation were documented in the ISA. The MONK8A computer code and JEF2.2 cross-section library are within the scope of the facility Quality Assurance Program in Appendix A of the SAR (LES, 2005a).

In Section 5.2.1.2 of the SAR (LES, 2005a), the applicant provided the basis for the k_{eff} equation (i.e., $k_{eff} = k_{calc} + 3\sigma_{calc} < 0.95$) used at the facility. The validation process established a bias by comparing calculations to measured critical experiments. With the bias determined, an upper safety limit (USL) was determined by using the following equation from NUREG/CR-6698 (NRC, 2001):

$$USL = 1.0 + bias - \sigma_{bias} - \sigma_{SM} - \sigma_{AoA}.$$

The critical experiments were assumed to have a $k_{eff} = 1.0$. The calculated k_{eff} (from above) was 1.0007, which was > 1.0, and so the bias was positive. Since a positive bias may be non-conservative, the bias was set to zero. The σ_{bias} (from above) was 0.0005. The arbitrary subcritical margin, σ_{SM}, was assigned a value of 0.05. The σ_{AoA} term is an additional margin to account for being beyond the area of applicability. Since the experiments were representative of the specific application, the σ_{AoA} was set to zero.

Thus, the USL = 1 - 0.0005 - 0.05 = 0.9495. NUREG/CR-6698 (NRC, 2001) states that for normal and credible conditions, $k_{calc} + 2\sigma_{Calc} < USL$. Since σ_{Calc} is greater than σ_{bias}, the facility was designed using the more conservative equation: $k_{eff} = k_{calc} + 3\sigma_{calc} < 0.95$. Thus, margin of subcriticality for safety will be 0.05.

Regarding the specific acceptance criteria in Section 5.4.3.4.4 of NUREG-1520 (NRC, 2002) for NCS subcriticality of operations and margin of subcriticality for safety, the applicant:

- Committed to the use of NCS controls and controlled parameters to assure that under normal and credible abnormal conditions, all nuclear processes are subcritical, including use of an approved margin of subcriticality for safety;

- Committed to the following policy: "Process specifications shall incorporate margins to protect against uncertainties in process variables and against a limit being accidentally exceeded;"

- Committed to the following standards, as they relate to these requirements, ANSI/ANS-8.7 (ANSI/ANS, 1998b), ANSI/ANS-8.10 (ANSI/ANS, 1988b), ANSI/ANS-8.12 (ANSI/ANS, 1993), ANSI/ANS-8.15 (ANSI/ANS, 1995c), and ANSI/ANS-8.17 (1997b);

- Requested NRC pre-approval of administrative k_{eff} margins for normal and credible abnormal conditions;

- Committed to determine subcritical limits for k_{eff} calculations such that $k_{subcritical} = 1.0 - bias - margin$, where the margin includes adequate allowance for uncertainty in the methodology, data, and bias to assure subcriticality; and

- Committed to perform studies to correlate the change in a value of a controlled parameter and its k_{eff} value and the studies will include changing the value of one controlled parameter and determining its effect on another controlled parameter.

Based on its review of the information provided, the staff concludes that the applicant has adequately described how it assures subcriticality of operations under normal and credible abnormal conditions and has defined an adequate margin of subcriticality for safety to meet the requirements of 10 CFR 70.61(d).

5.3.6.4   NCS Baseline Design Criteria

In Section 3.1.7(l) of the ISA Summary (LES, 2005c), the applicant indicated that safety margins and control methods are used at the facility for criticality control to meet the NCS baseline design criteria in 10 CFR 70.64(a)(9). The design of process and storage systems will include demonstrable margins of safety for the NCS parameters that are commensurate with the uncertainties in the process and storage conditions, in the data and methods used in calculations, and in the nature of the immediate environment under accident conditions. All process and storage systems will be designed and maintained with sufficient factors of safety to require at least two unlikely, independent, and concurrent changes in process conditions before an inadvertent criticality is possible. The major NCS controlling parameters used in the facility are enrichment control, geometry control, moderation control, or limitations on the mass as a function of enrichment.

Regarding the specific acceptance criteria in Section 5.4.3.4.5 of NUREG-1520 (NRC, 2002) for NCS baseline design criteria, the applicant committed to the double contingency principle in determining NCS controls and IROFS in the design of new facilities or new processes that require a license amendment under 10 CFR 70.72.

Based on its review of the information provided, the staff concludes that the applicant has described criticality control, including adherence to the double contingency principle, to meet the NCS baseline design criteria of 10 CFR 70.64(a)(9).

5.3.6.5   NCS in the Integrated Safety Analysis (ISA) Program and the ISA Summary

In Chapter 3 of this SER, NRC reviewed the ISA program that was used to meet 10 CFR 70.62. NRC reviewed the ISA method that the applicant used for NCS when performing the ISA of the process accident sequences. In this ISA method, the applicant identified the hazard, developed a risk matrix with consequence and likelihoods, defined "highly unlikely" and developed IROFS and general management measures to make NCS accident sequences meet "highly unlikely." The consequences of an inadvertent criticality were conservatively assumed to be high for the workers. For accident sequences postulated to result in an inadvertent criticality, IROFS were specified to ensure subcriticality under all normal and credible abnormal conditions and general management measures were specified. NRC reviewed the ISA method that the applicant used for NCS when performing the ISA for safe-by-design components. In this ISA method, "highly unlikely" is achieved with a significant margin and other conditions (i.e., there is no credible way to change the applicable geometric parameters without effecting a design change), rather than with accident sequences, IROFS, and management measures.

In Chapter 3 of this SER, NRC reviewed the ISA Summary (LES, 2005c) related to whether the applicant met certain general contents of the ISA Summary in 10 CFR 70.65. However, the review to determine whether the applicant met other contents of the ISA Summary for NCS are reviewed in this section of the SER.

5.3.6.5.1   Demonstration of Meeting the Contents of an ISA Summary for NCS

NRC staff reviewed the ISA Summary (LES, 2005c) as well as classified documents pertaining to meeting the performance requirements when using both the ISA method for NCS and the safe-by-design ISA method for NCS. The focus of the NRC review was on the risk-significant NCS aspects in the implementation of both the ISA method for NCS and the safe-by-design ISA method for NCS. Specific information about the commitments to NRC NCS Regulatory Guide 3.71 and the ANSI/ANS Series-8 NCS standards is in Section 5.3.2 of this SER.

5.3.6.5.1.1   NCS ISA Method

In Sections 3.6 and 3.7 of the ISA Summary (LES, 2005c), the applicant identified potential hazards and accident sequences which could result in an inadvertent criticality. Thus, the sequences were assumed to be high consequence and no specific consequence analysis was performed. The hazards and accident sequences were identified using the HAZOPS method.

The applicant identified IROFS for each accident sequence and described its expected function. There are no IROFS that are frequently or continuously challenged. Management measures will ensure that IROFS are designed, implemented, and maintained, as necessary, to be available and reliable to perform their safety function when needed. IROFS will be designed, constructed, tested, and maintained to Quality Level 1. The four types of IROFS are: Passive Engineered Control (PEC), Active Engineered Control (AEC), Enhanced Administrative Control (EAC), and Administrative Control (AC). For the PEC, AEC, and AC IROFS, the applicant used a generally conservative index score compared to the guidance in NUREG-1520 (NRC, 2002). For the EAC IROFS, the applicant included a justification for the less conservative index score within Section 3.8 of the ISA Summary (LES, 2005c).

In Table 3.8-2 of the ISA Summary (LES, 2005c), the applicant provided a descriptive list of sole IROFS (summarized in the table below) to meet 10 CFR 70.65(b)(8) for NCS when using the ISA method. All six are EAC IROFS, and Table 5.3-3 below identifies the IROFS number, the NCS process accident sequences, and the description of the IROFS.

In Sections 3.7 and 3.8 of the ISA Summary (LES, 2005c), the applicant provided a demonstration of meeting "highly unlikely" for each accident sequence. The applicant provided 43 specific process NCS accident sequences. NRC sorted them into 12 general process NCS accident sequence groups that have similar characteristics (i.e., description, initiating event, IROFS, and index scores). The applicant used either AC IROFS or EAC IROFS along with independent verification for 11 of the 12 groups. The applicant used sampling as part of 4 of the 12 groups.

NRC reviewed the 12 groups of sequences and determined the following:

•   Group 1 (sequence PT2-2, uses enrichment control):

      For this sequence, a criticality could occur if a product cylinder were placed in a feed station. The sequence is well described with appropriate specific IROFS and appropriate index numbers for the initiating event and IROFS failures. The PEC IROFS is extremely important and plays a significant role in minimizing the overall risk. Also, the AC IROFS are appropriately incorporated such that they contribute to minimizing the risks.

Table 5.3-3
Identification of IROFS

IROFS	APPLICANT IDENTIFIED SEQUENCE(S)	DESCRIPTION OF IROFS
14a	FR1-1, FR2-1, DS1-1, DS2-1, DS3-1,SW1-1, LW1-2	Verify use of safe-by-design transfer frame prior to movement.
14b	FR1-2, FR2-2, DS1-2, DS2-2, DS3-2, SW1-2, LW1-3	Verify acceptable use of storage array prior to movement.
15	PT3-5	Prevent additional transfer of material if the container has material in it and it is a non-safe-by-design container.
16a	PB2-2, CP1-2	Allow no visible oil and limit cylinder vapor pressure.
45	PB1-3, RD1-1	Verify non-stacked condition and no other cylinder in movement prior to moving cylinder.
C6	EC3-1	Calculate and set cascade enrichment control device.

- Group 2 (sequences PT2-3 and PB2-5, uses moderator control):

  In this group, a criticality could occur if excessive moderator is introduced into a cylinder. The group is clearly described with general IROFS and appropriate index numbers for the initiating event and IROFS failures. The two AC IROFS are appropriately incorporated such that they contribute to minimizing the risks.

- Group 3 (sequences PT2-5 and PB2-6, uses moderator control):

  In this group, a criticality could occur if moderator is introduced into a cylinder. The group is clearly described with general IROFS and appropriate index numbers for the initiating event and IROFS failures. The two AC IROFS are appropriately incorporated such that they contribute to minimizing the risks.

- Group 4 (sequences PT3-1, PT3-3, PB2-3, PB3-1, PB3-2, PB4-5, VR1-1, VR1-2, CL3-1, CP1-1, EC4-2, uses mass control):

  In this group, a criticality could occur if product forms on a filter over time. The group is clearly described with general IROFS and appropriate index numbers for the initiating event and IROFS failures. The two AEC IROFS are important and play a significant role in minimizing the overall risk.

- Group 5 (sequence PT3-5, uses item control):

  For this sequence, a criticality could occur if enriched material is transferred many times into a non-safe-by-design container. The sequence is clearly described with a general IROFS and appropriate index numbers for the initiating event and IROFS failure. This sequence has a sole NCS EAC IROFS. The NRC finds this acceptable because the failure of the IROFS would have to occur many times before a criticality could occur. The multiple conditions that would have to exist for enriched material to be in the container to begin with also contribute to minimizing the risks.

- Group 6 (sequences PB1-3, RD1-1, uses interaction control):

In this group, a criticality could occur if more than one product cylinder is moved and stacked. The group is clearly described with a general IROFS and appropriate index numbers for the initiating event and IROFS failure. This group has a sole NCS EAC IROFS. The NRC finds this acceptable because of the limited number of cranes that can be used to move product cylinders. The conditions that would have to exist for the moving and stacking also contribute to minimizing the risk.

- Group 7 (sequences PB2-2, CP1-2, uses moderator control):

In this group, a criticality could occur if excessive moderator is introduced into a receiver cylinder. The group is clearly described with a general IROFS and appropriate index numbers for the initiating event and IROFS failure. This group has a sole NCS EAC IROFS. The NRC finds this acceptable because the IROFS does not allow the operator to make qualitative judgements about the amount of moderation present. The conditions that would have to exist for moderation to exist contribute to minimizing the risks.

- Group 8 (sequence VR2-7, uses item control):

For this sequence, a criticality could occur if sufficient mass accumulates in a container. The sequence is clearly described with general IROFS and appropriate index numbers for the initiating event and IROFS failures. The two AC IROFS are appropriately incorporated such that they contribute to minimizing the risks.

- Group 9 (sequences FR1-1, FR2-1, DS1-1, DS2-1, DS3-1, SW1-1, LW1-2, uses interaction control):

In this group, a criticality could occur if appropriate distance is not maintained when moving a container. The group is well described with general IROFS and appropriate index numbers for the initiating event and IROFS failure. This group has a sole NCS EAC IROFS. The NRC finds this acceptable because this group uses a piece of equipment that meets the alternate safe-by-design ISA method for NCS for moving the container. The equipment is extremely important and plays a significant role in minimizing the overall risk. Also, the conditions that would have to exist for enriched material to exist in the container contribute to minimizing the risks.

- Group 10 (sequences FR1-2, FR2-2, DS1-2, DS2-2, DS3-2, SW1-2, LW1-3, uses interaction control):

In this group, a criticality could occur if the appropriate distance is not maintained when storing a container. The group is well described with general IROFS and appropriate index numbers for the initiating event and IROFS failure. This group has a sole NCS

5-33

EAC IROFS. The NRC finds this acceptable because this group uses a piece of equipment that meets the alternate safe-by-design ISA method for NCS for storing the container. The equipment is extremely important and plays a significant role in minimizing the overall risk. Also, the conditions that would have to exist for enriched material to exist in the container contribute to minimizing the risk.

- Group 11 (sequences DS1-3, DS2-3, LW1-1, LW2-1, LW3-1, LW5-1, uses mass control):

  In this group, a criticality could occur if sufficient mass accumulates in a tank. The group is clearly described with general IROFS and appropriate index numbers for the initiating event and IROFS failures. The two AC IROFS are appropriately incorporated such that they contribute to minimizing the risks.

- Group 12 (sequence EC3-1, uses enrichment control):

  For this sequence, a criticality could occur if the enrichment setting for the centrifuge process is not correct. The sequence is well described with a general IROFS and appropriate index numbers for the initiating event and IROFS failure. This sequence has a sole NCS EAC IROFS. The NRC finds this acceptable because more than one operator would have to independently incorrectly set the controls. Also, the conditions that would have to exist in the process contribute to minimizing the risks.

Regarding the specific acceptance criteria in Section 5.4.3.4.6 of NUREG-1520 (NRC, 2002) for both the ISA method for NCS and the safe-by-design ISA method for NCS, the applicant:

- Identified NCS accident sequences, consequences of NCS accident sequences, likelihoods of NCS accident sequences, and descriptions of IROFS for NCS accident sequences or else used the safe-by-design ISA method for NCS appropriately;

- Committed to use Appendix A of ANSI/ANS-8.1 (ANSI/ANS, 1988a) and ANSI/ANS-8.1 (ANSI/ANS, 1998a) in determining NCS accident sequences; and

- Committed to ANSI/ANS-8.10 (ANSI/ANS, 1988b), as modified by NRC NCS Regulatory Guide 3.71 (NRC, 1998), in determining the consequences of NCS accident sequences.

Based on the above review, the staff has reasonable assurance that: (1) the applicant identified all hazards and accident sequences for NCS; and (2) the applicant's IROFS and management measures will ensure that it is highly unlikely for each accident sequence (i.e., inadvertent criticality) to occur.

The staff notes that the above analysis does not apply when using the safe-by-design ISA method because when using the safe-by-design ISA method, "highly unlikely" is achieved with a significant margin and other conditions (i.e., there is no credible way to change the applicable geometric parameters without effecting a design change), rather than with accident sequences, IROFS, and management measures (see below, in Section 5.3.6.5.1.2 of this SER).

5.3.6.5.1.2   NCS Safe-By-Design ISA Method

In Section 3.1.2 of the ISA Summary (LES, 2005c), the applicant provided a demonstration of meeting "highly unlikely" for NCS when using the safe-by-design ISA method to meet 10 CFR 70.65(b)(4). The demonstration of significant margin to meet "highly unlikely" was provided for each of the components listed in Tables 3.7-6 through 3.7-21 of the ISA Summary (LES, 2005c) in the following classified documents: ETC4009554 through ETC4009559, ETC40009561, ETC4009565 through ETC4009567, ETC4009609, ETC4009614, ETC4009677, ETC4009679, ETC4009723, and ETC4009730. These classified documents are incorporated by reference into the ISA Summary. Also, the configuration management system required by 10 CFR 70.72, which is implemented by the facility Configuration Management Program, will ensure the maintenance of the safety function of these components and will assure compliance with both the double contingency principle and the defense-in-depth criterion of 10 CFR 70.64(b).

NRC reviewed the classified information above for all the applicant-identified safe-by-design components. For each piece of favorable geometry equipment, NRC reviewed the dimensions provided, calculated the geometry percentages for the equipment, and compared that percentage versus the geometry percentage criteria. For each non-favorable geometry equipment, NRC reviewed the appropriateness of the conservative assumption(s) and compared the calculated k_{eff} value versus the k_{eff} limit criteria. Therefore, NRC determined that the classified information met the safe-by-design criteria. The applicant slightly revised the classified information and then confirmed that all the information in the new classified documents met the criteria for using the safe-by-design ISA method for those components.

Based on the above review, the staff has reasonable assurance that: (1) the applicant used the safe-by-design ISA method appropriately; and (2) it is highly unlikely for an inadvertent criticality to occur with those safe-by-design components.

5.3.6.6   Additional NCS Program Commitments

Specific information about the commitments to ANSI/ANS Series-8 NCS standards is in Section 5.3.2 of this SER. Information that demonstrates that the applicant met the acceptance criteria in this part of the SER are also in Section 5.3.5 of this SER as well as other parts of this section in this SER.

In Section 5.4 of the SAR (LES, 2005a), the applicant committed to provide a program for evaluating the NCS significance of events and for making the required notification to the NRC Operations Center. Qualified individuals will make the determination of significance of NCS events. The determination of loss or degradation of IROFS or double contingency principle compliance will be made against the Materials License and Appendix A to Part 70. The reporting criteria of Appendix A to Part 70 and the report content requirements of 10 CFR 70.50 will be incorporated into the facility emergency procedures. The necessary report, based on whether the IROFS credited were lost, irrespective of whether the safety limits of the associated parameter(s) were actually exceeded, will be issued. If it cannot be ascertained, within 1 hour, whether the criteria in Appendix A to Part 70, paragraph (a) or (b) apply, then the event will be treated as a 1-hour reportable event.

Regarding the specific acceptance criteria in Section 5.4.3.4.7 of NUREG-1520 (NRC, 2002) for additional NCS program commitments, the applicant:

- Committed to use the NCS program to promptly detect any NCS deficiencies by means of operational inspections, audits, or investigations, and to refer to the facility's corrective action function any unacceptable performance deficiencies in IROFS, NCS function, or management measures, so as to prevent recurrence;

- Committed to support the facility change mechanism process by performing NCS Analyses and NCS Evaluations to evaluate changes to processes, operating procedures, IROFS, and management measures;

- Committed to upgrade the NCS program to reflect changes in the ISA or new NCS methodologies and to modify operating and maintenance procedures in ways that could reduce the likelihood of occurrence of an inadvertent criticality;

- Committed to retain records of NCS programs and to document any corrective actions taken;

- Committed to use the NCS methodologies and NCS technical practices to evaluate NCS accident sequences in operations and processes;

- Committed to have a program for evaluating the NCS significance of NCS events, to have an apparatus in place for making the required notification to the NRC Operations Center, to have the determination of NCS significance made by qualified individuals, and to have the determination made against the Materials License and the Regulations;

- Committed to incorporate the reporting criteria of Appendix A and the report contents of 10 CFR 70.50 into the facility emergency procedures;

- Committed to issue the necessary report based on whether the IROFS credited were lost, irrespective of whether the safety limits of the associated parameters were actually exceeded; and

- Committed to the following: "If the applicant cannot ascertain within one hour of whether the criteria of 10 CFR 70 Appendix A Paragraph (a) or (b) apply, the event will be treated as a one-hour reportable event."

Based on its review of the information provided, the staff finds the applicant's information regarding additional NCS program commitments to be acceptable and the staff concludes that the applicant has described reporting criteria to meet the requirements of Appendix A to 10 CFR Part 70.

5.4   EVALUATION FINDINGS

NRC staff reviewed the NCS and other information for the facility according to Chapters 3 and 5 of NUREG-1520 (NRC, 2002). The staff determined that:

1. The applicant will have in place a staff of managers, supervisors, engineers, process operators, and other support personnel who are qualified to develop, implement, and maintain the NCS program in accordance with the facility organization and administration, as well as management measures, according to the requirements in Part 70, including those in 10 CFR 70.62 and 70.65;

2. The applicant's conduct of operations will be based on NCS methods and NCS technical practices, which will ensure that the fissile material will be possessed, handled, stored, and used safely, according to the requirements in Part 70, including those in 10 CFR 70.22, 70.62, 70.65, and 70.72;

3. The applicant will develop, implement, and maintain a criticality accident alarm system in accordance with the facility emergency management program and the requirements in Part 70, including those in 10 CFR 70.24 and 70.65(b)(4);

4. The applicant will have in place an NCS program, in accordance with the requirements in Part 70, including those subcriticality of operations, margin of subcriticality for safety, and selection-of-controls requirements in 10 CFR 70.61(d);

5. The applicant will have in place an NCS program, in accordance with the requirements in Part 70, including those baseline design criteria for NCS in 10 CFR 70.64(a)(9); and

6. The applicant will have in place an NCS program, in accordance with the reporting requirements in Part 70, including those in 10 CFR 70.52 and Appendix A to Part 70.

Based on this NCS review, the staff concludes that the applicant's NCS program meets the requirements of Part 70 and provides reasonable assurance for the protection of public health and safety, including workers and the environment.

NRC staff reviewed the NCS and other information for the facility according to Chapter 3 of NUREG-1520 (NRC, 2002) for NCS. The staff determined that:

1. The applicant has in place an ISA program, that is adequately established and will be maintained pursuant to 10 CFR 70.72, such that the performance requirements for NCS will be met in the future, if changes are made;

2. The applicant has adequately performed an ISA for NCS;

3. The applicant has adequately documented the ISA results in an ISA Summary for NCS;

4. The applicant has adequately identified and evaluated the hazards for NCS;

5. The applicant has adequately identified and evaluated the potential accident sequences for NCS;

6. The applicant has adequately demonstrated that credible NCS accidents sequences are highly unlikely;

7. The applicant has adequately demonstrated that the failure of safe-by-design components for NCS is highly unlikely, as required by the regulations.

Based on this NCS review, the staff concludes that the applicant's ISA program commitments for NCS and the ISA Summary for NCS meet the requirements of 10 CFR Part 70 and provide reasonable assurance for the protection of public health and safety, including workers and the environment.

5.5 REFERENCES

(AEAT, 1998) AEA Technology (AEAT). "MONK: A Monte Carlo Program for Nuclear Criticality Safety and Reactor Physics Analyses," Version 8, 1998.

(ANSI/ANS, 1988a) American National Standards Institute/American Nuclear Society (ANSI/ANS). ANSI/ANS-8.1, "Nuclear Criticality Safety in Operations with Fissionable Materials Outside Reactors," 1988.

(ANSI/ANS, 1988b) American National Standards Institute/American Nuclear Society (ANSI/ANS). ANSI/ANS-8.10, "Criteria for Nuclear Criticality Safety Controls in Operations with Shielding and Confinement," 1988.

(ANSI/ANS, 1991) American National Standards Institute/American Nuclear Society (ANSI/ANS). ANSI/ANS-8.20, "Nuclear Criticality Safety Training," 1991.

(ANSI/ANS, 1993) American National Standards Institute/American Nuclear Society (ANSI/ANS). ANSI/ANS-8.12, "Nuclear Criticality Control and Safety of Plutonium-Uranium Fuel Mixtures Outside Reactors," 1993.

(ANSI/ANS, 1995a) American National Standards Institute/American Nuclear Society (ANSI/ANS). ANSI/ANS-8.6, "Safety in Conducting Subcritical Neutron-Multiplication Measurements In Situ," 1995.

(ANSI/ANS, 1995b) American National Standards Institute/American Nuclear Society (ANSI/ANS). ANSI/ANS-8.9, "Nuclear Criticality Safety Criteria for Steel-Pipe Intersections Containing Aqueous Solutions of Fissile Materials," 1995.

(ANSI/ANS, 1995c) American National Standards Institute/American Nuclear Society (ANSI/ANS). ANSI/ANS-8.15, "Nuclear Criticality Control of Special Actinide Elements," 1995.

(ANSI/ANS, 1995d) American National Standards Institute/American Nuclear Society (ANSI/ANS). ANSI/ANS-8.21, "Use of Fixed Neutron Absorbers in Nuclear Facilities Outside Reactors," 1995.

(ANSI/ANS, 1996a) American National Standards Institute/American Nuclear Society (ANSI/ANS). ANSI/ANS-8.5, "Use of Borosilicate-Glass Raschig Rings as a Neutron Absorber in Solutions of Fissile Material," 1996.

(ANSI/ANS, 1996b) American National Standards Institute/American Nuclear Society (ANSI/ANS). ANSI/ANS-8.19, "Administrative Practices for Nuclear Criticality Safety," 1996.

(ANSI/ANS, 1997a) American National Standards Institute/American Nuclear Society (ANSI/ANS). ANSI/ANS-8.3, "Criticality Accident Alarm System," 1997.

(ANSI/ANS, 1997b) American National Standards Institute/American Nuclear Society (ANSI/ANS). ANSI/ANS-8.17, "Criticality Safety Criteria for the Handling, Storage, and Transportation of LWR Fuel Outside Reactors," 1997.

(ANSI/ANS, 1997c) American National Standards Institute/American Nuclear Society (ANSI/ANS). ANSI/ANS-8.22, "Nuclear Criticality Safety Based on Limiting and Controlling Moderators," 1997.

(ANSI/ANS, 1997d) American National Standards Institute/American Nuclear Society (ANSI/ANS). ANSI/ANS-8.23, "Nuclear Criticality Accident Emergency Planning and Response," 1997.

(ANSI/ANS, 1998a) American National Standards Institute/American Nuclear Society (ANSI/ANS). ANSI/ANS-8.1, "Nuclear Criticality Safety in Operations with Fissionable Materials Outside Reactors," 1998.

(ANSI/ANS, 1998b) American National Standards Institute/American Nuclear Society (ANSI/ANS). ANSI/ANS-8.7, "Guide for Nuclear Criticality Safety Criteria in the Storage of Fissile Materials," 1998.

(ASME, 1994) American Society of Mechanical Engineers (ASME). Quality Assurance (QA) Standard NQA-1-1994, "Quality Assurance Program Requirements for Nuclear Plants," 1994.

(LES, 2004) Louisiana Energy Services (LES). letter to U.S. Nuclear Regulatory Commission, "MONK 8A Validation and Verification," May 7, 2004.

(LES, 2005a) Louisiana Energy Services (LES). "National Enrichment Facility Safety Analysis Report," Revision 5, 2005.

(LES, 2005b) Louisiana Energy Services (LES). "National Enrichment Facility Emergency Plan," Revision 3, 2005.

(LES, 2005c) Louisiana Energy Services (LES). "National Enrichment Facility Integrated Safety Analysis Summary," Revision 4, 2005.

(NEA, 2002) Nuclear Energy Agency (NEA). NEA/NSC/DOC(95)03, "International Handbook of Evaluated Criticality Safety Benchmark Experiments," 2002.

(NRC, 1998) U.S. Nuclear Regulatory Commission (NRC). NRC Regulatory Guide 3.71, "Nuclear Criticality Safety Standards for Fuels and Material Facilities," 1998.

(NRC, 2001) U.S. Nuclear Regulatory Commission (NRC). NUREG/CR-6698, "Guide for Validation of Nuclear Criticality Safety Calculational Methodology," 2001.

(NRC, 2002) U.S. Nuclear Regulatory Commission (NRC). NUREG-1520, "Standard Review Plan for the Review of a License Application for a Fuel Cycle Facility," 2002.

6.0 CHEMICAL PROCESS SAFETY

The primary purpose of this review is to determine that the applicant has designed a facility that will adequately protect workers, public, and the environment during normal operations, and also against chemical hazards of licensed material and its byproducts. It must also protect against facility conditions or operator actions that can affect the safety of licensed materials and thus present an increased chemical risk.

6.1 REGULATORY REQUIREMENTS

The regulatory bases for the review are the general and additional contents of an application that addresses chemical-process safety, as required by 10 CFR 30.33, 10 CFR 40.32, 10 CFR 70.22, and 70.65. In addition, the chemical-process safety review should provide a determination of compliance with 10 CFR 70.61, 70.62, and 70.64.

6.2 REGULATORY ACCEPTANCE CRITERIA

The acceptance criteria for the U.S. Nuclear Regulatory Commission's (NRC's) review of chemical-process safety for the proposed facility are outlined in Section 6.4.3 of NUREG-1520 (NRC, 2002).

6.3 STAFF REVIEW AND ANALYSIS

The NRC staff reviewed the Safety Analysis Report (SAR) (LES, 2005a) and the Integrated Safety Analysis (ISA) (LES, 2005b) submitted by LES (the applicant) and considered the following areas:

1. Chemical Process Description;
2. Chemical Accident Sequences;
3. Chemical Accident Consequences;
4. Chemical-Process Items Relied on for Safety (IROFS);
5. Management Measures; and
6. Baseline Design Criteria (BDC).

The staff reviewed the applicant's responses to requests for additional information and ISA documents during an on-site visit, as necessary, to have a better understanding of the process and safety requirements. The staff evaluated the information to determine if the facility's design complied with the BDC and defense-in-depth requirements specified in 10 CFR 70.64(a) and 70.64(b), respectively. The evaluation is summarized in the following sections.

6.3.1 Process Description

The plant process is designed to enrich natural uranium hexafluoride (UF_6) by separating a feed stream containing the naturally occurring proportions of uranium isotopes into a product stream enriched in the uranium-235 (U^{235}) isotope and a tails stream depleted in the U^{235} isotope. The process, entirely physical in nature, is a mechanical separation of isotopes using a fast rotating cylinder (centrifuge) based on a difference in centrifugal forces from differences in molecular weight of the uranic isotopes. No chemical changes nor nuclear reactions take place. The feed, product, and tails streams are all in the form of UF_6.

The nominal capacity of the facility is 3 million separative work units (SWUs) per year. The maximum gross output of the facility is sightly greater than 3 million SWUs thus allowing for a production margin for centrifuge failures and occasional production losses during the operational lifetime of the facility.

6.3.1.1   Gas Centrifuge Process

The enrichment process proposed by the applicant, housed in the Separation Building, is comprised of four major systems: a UF_6 Feed System, a Cascade System, a Product Take-off System, and a Tails Take-off System. Other product-related functions include the Product Liquid Sampling System and Product Blending System. Supporting functions include sample analysis, equipment decontamination and equipment rebuild, liquid effluent treatment, and solid waste management.

The major equipment used in the UF_6 feed process are Solid Feed Stations. UF_6 is delivered to the plant in American Nuclear Standards Institute (ANSI) N14.1 (ANSI, 1995) standard Type 48X or 48Y international transit cylinders. Feed cylinders are loaded into Solid Feed Stations, vented for removal of light gases, primarily air and hydrogen fluoride (HF), and heated to sublime the UF_6.

The light gases and UF_6 gas generated during feed purification are routed to the Feed Purification Subsystem. The major pieces of equipment in the Feed Purification Subsystem are UF_6 Cold Traps, a Vacuum Pump/Chemical Trap Set, and a Low-Temperature Take-off Station (LTTS). The UF_6 is captured in UF_6 cold traps and ultimately recycled as feed, whereas HF is captured on chemical traps.

After purification, gaseous UF_6 is flow-controlled through a pressure control system for distribution to the cascade system at subatmospheric pressure. Individual centrifuges are not able to produce the desired product and tails concentration in a single step. They are grouped together in series and in parallel to form arrays known as cascades. A typical cascade is comprised of many centrifuges. Each centrifuge has a thin-walled, vertical, cylindrically shaped rotor that spins around a central post within an outer casing. Feed enters, and product and tails streams leave the centrifuge through the central post. Control valves, restriction orifices, and controllers provide uniform flow of product and tails.

Depleted UF_6 from the cascades is desublimed in a Tails Take-off System comprised of vacuum pumps, Tails LTTS, and Vacuum Pumps/Chemical Trap Sets. Depleted UF_6 exiting the cascade is transported into Uranium Byproduct Cylinders (UBCs) for desublimation at subatmospheric pressure. The primary components of the Tails Take-off System are the vacuum pumps and the Tails LTTS. Chilled air flows over the cylinder in the Tails LTTS to effect the desublimation. Filling of the cylinders is monitored with a load cell system, and filled cylinders (solid UF_6) are transferred to an outdoor storage area (UBC Storage Pad).

Enriched UF_6 from the cascades is desublimed in a Product Take-off System comprised of vacuum pumps, Product LTTS, UF_6 cold traps, and Vacuum Pumps/Chemical Trap Sets. The vacuum pumps transport the UF_6 from the cascades to the Product LTTS at subatmospheric pressure. The heat of desublimation of the UF_6 is removed by cooling air routed through the LTTS. The product stream normally contains small amounts of light gases that have passed through the centrifuges. Therefore, a UF_6 Cold Trap and Vacuum Pump/Trap Set are provided to vent these gases from the product cylinder. Any UF_6 captured in the cold trap is periodically transferred to another product cylinder for use as product or blending stock. Filling of the product cylinders is monitored with a load cell system, and filled cylinders (solid UF_6) are transferred to the Product Liquid Sampling System for sampling.

Sampling is performed to verify product assay level (weight percent U^{235}). The Product Liquid Sampling Autoclave is an electrically heated, closed-pressure vessel used to liquify the UF_6 and allow collection of a sample. The autoclave is fitted with a hydraulic tilting mechanism that elevates one end of the autoclave so that liquid UF_6 pours into a sampling manifold connected to the cylinder valve. After sampling, the autoclave is brought back to the horizontal position and the cylinder is indirectly cooled by water flowing through coils located on the outer shell of the autoclave.

With the exception of liquid-sampling operations, the entire enrichment process operates at subatmospheric pressure. This safety feature helps ensure that releases of UF_6 or HF are minimized, because leakage would typically be inward to the system. During sampling operations, UF_6 is liquified within an autoclave that provides the heating required to homogenize the material for sampling. The autoclave is an American Society of Mechanical Engineers (ASME), Section VIII, Division 1, "Boiler and Pressure Vessel Code"-rated pressure vessel that serves as a secondary containment for the UF_6 product cylinder while the UF_6 is in liquid state.

6.3.1.2   Chemical Screening and Classification

The applicant classifies all site chemicals based on their potential for harm by placing each chemical or related waste into one of three categories: Chemicals of Concern (Class 1); Interaction Chemicals (Class 2); or Incidental Chemicals (Class 3).

Chemicals of Concern (Class 1)

The applicant determined the Chemicals of Concern (Class 1) based on one or more characteristics of the chemical or the quantity in storage/use at the facility. For licenced materials or hazardous chemicals produced from licensed materials, chemicals of concern are those that, in case of release, have the potential to exceed any of the concentrations defined in 10 CFR 70.61(b) and 70.61(c).

Chemicals of concern that are not related to licensed materials are those that are listed and above the specified threshold quantities of either 29 CFR 1910.119, "OSHA Process Safety Management," or 40 CFR Part 68, "EPA Risk Management Program."

These chemicals represent, based on their inherent toxic, reactive, or flammable properties, a potential for a severe chemical release or acute chemical exposure to an individual that:

1. Could endanger the life of a worker; or

2. Could lead to irreversible or other serious, long-lasting health effects to any individual located outside the controlled area.

UF_6 is the only licensed material-related chemical of concern (Class 1) that the applicant will be using at the facility. There are no non-licensed chemicals of concern at the facility.

Interaction Chemicals (Class 2)

Interaction chemicals (Class 2) are those chemicals/chemical systems that require evaluation for their potential to precipitate or propagate accidents in chemicals of concern (Class 1) systems, but by themselves are not chemicals of concerns. Class 2 chemicals are listed in Table 6.1-1 of the SAR (LES, 2005a).

Incidental Chemicals (Class 3)

The facility will use other chemicals that are neither chemicals of concern nor interaction chemicals. Some of these incidental chemicals (Class 3) include those that have the potential to result in injurious occupational or environmental exposure, but represent no potential for acute exposure to the public and that, via their nature, quantity, or use, have no potential for affecting chemicals of concern (Class 1). Class 3 chemicals are listed in Table 6.1-1 of the SAR (LES, 2005a), but do not impact the performance requirements of 10 CFR 70.61.

6.3.1.3   Hazardous Chemicals and Chemical Interactions

The only chemical present in significant quantities in the plant is UF_6, and it constitutes the main hazard in this facility. Any UF_6 that is released to the environment will react exothermically with water vapor, present in air, producing uranyl fluoride (UO_2F_2) and HF. The chemical form and rate of reaction of the products of this reaction will depend on the temperature and the relative humidity of the air at the time of the release. UO_2F_2 is a solid whereas HF is a gas (see Eq. 1).

$$UF_{6(gas)} + 2H_2O_{(vapor)} \rightarrow UO_2F_{2(solid)} + 4HF_{(gas)} + Heat \qquad (1)$$

At room temperature, hydrated forms of UO_2F_2 and HF can be produced depending on the relative humidity in the air (see Eq. 2).

$$UF_{6(gas)} + (2+4x)H_2O_{(vapor)} \rightarrow UO_2F_2 \bullet H_2O_{(solid)} + 4HF_{(gas)} + Heat \qquad (2)$$

In both cases, UO_2F_2 compounds are deposited or precipitated close to the point of the release.

6.3.1.4 UF_6 and Interaction Chemicals

Interaction chemicals represent potential hazards, depending on the quantities present in the facility.

Perfluorinated Polyether (Fomblin) Oil

Beside the interaction with moisture, UF_6 can react exothermically with hydrocarbons. Gaseous UF_6 reacts with hydrocarbons to form a black residue of uranium-carbon compounds. Hydrocarbons can be explosively oxidized if they are mixed with UF_6 in the liquid phase or at elevated temperatures. The applicant will not be using non-fluorinated hydrocarbon lubricants in any UF_6 system at the proposed facility.

The applicant's UF_6 vacuum pumps are lubricated using perfluorinated polyether (PFPE) oil that is commonly referred to by a manufacturer's trade name - Fomblin oil. Fomblin oil is inert, fully fluorinated, and does not react with UF_6 under any operating conditions.

Small quantities of uranium compounds and traces of hydrocarbons may be contained in the Fomblin oil. The UF_6 degrades in the oil or reacts with trace hydrocarbons to form crystalline compounds - primarily UO_2F_2 and uranium tetrafluoride (UF_4) particles - that gradually thicken the oil and reduce pump capacity.

Recovery of Flombin oil for reuse in the system is conducted remotely from the UF_6 process systems. The dissolved uranium compounds are removed through a process of precipitation, centrifugation, and filtration. Anhydrous sodium carbonate (Na_2CO_3) is added to contaminated Fomblin oil. Uranium compounds react to form sodium uranyl carbonate, which precipitates out. A filter removes the precipitate during subsequent centrifugation of the oil.

Trace amounts of hydrocarbons are then removed by adding activated carbon to the Fomblin oil and heating, causing absorption of the hydrocarbons. The carbon is in turn removed through a bed of celite.

Chemical Traps - Activated Carbon, Aluminum Oxide, and Sodium Fluoride

The applicant uses chemisorption for the removal of UF_6 and HF from gaseous effluent streams. The applicant also uses it to remove oil mist from vacuum pumps operating upstream of gaseous effluents ventilation systems. The applicant places absorbent materials on stationary beds in chemical traps downstream of the various cold traps. These materials capture HF and the trace amounts of UF_6 that escape desublimation during feed purification or during venting of residual UF_6 contained in hoses or piping that is bled down before disconnection.

The applicant will be using two different types of traps. The first type of trap contains a charge of activated carbon to capture the small amounts of UF_6 that escape desublimation. A second type of trap is used to absorb HF, since HF is not fully absorbed on carbon at low pressure. The second type of trap contains a charge of aluminum oxide (Al_2O_3).

Activated carbon cannot be used in the Contingency Dump System because the relatively high UF_6 flow rates during this non-routine operation could lead to severe overheating. A chemical trap containing sodium fluoride (NaF) is installed in the contingency dump flow path to trap UF_6. NaF is used because the heat of UF_6 chemisorption on NaF is significantly lower than the heat of UF_6 chemisorption on activated carbon.

Decontamination - Citric Acid

The applicant uses citric acid to decontaminate components (e.g., pumps, valves, piping) from the residuals of UO_2F_2 compound layers that are present on the surfaces, once the components are removed from the process area.

The reaction of the uranium compounds with the citric acid solution produces various uranyl citrate complexes. The applicant is using personnel protective features for the safe handling of decontamination chemicals and byproducts.

Nitrogen

The applicant uses gaseous nitrogen in the UF_6 systems for purging and filling lines that have been exposed to the atmosphere for any of several reasons including: connection and disconnection of cylinders; preparing lines/components for maintenance; providing an air-excluding gaseous inventory for system vacuum pumps; and filling the interstitial space of the liquid sampling autoclave (secondary containment) before cylinder liquefaction. Nitrogen is not reactive with UF_6 in any plant operational condition.

Silicone Oil

The applicant uses silicone oil as a heat-exchange medium for heating/chilling of various cold traps. This oil is external to the UF_6 process streams in all cases and is not expected to interact with UF_6.

Halocarbon Refrigerants

The applicant uses halocarbon refrigerants (including R23 trifluoromethane, R404A fluoromethane blend, and R507 penta/trifluoromethane) in individual package chillers that will provide cooling of UF_6 cylinders or silicon oil heat-exchanger media for take-off stations and cold traps. The applicant is using these halocarbon refrigerants because of their good heat-transfer properties; they satisfy environmental restrictions regarding ozone depletion; and are non-flammable. All halocarbon refrigerants are external to the UF_6 process streams in all cases and are not expected to interact with UF_6.

Plant Chilled Water

The applicant uses chilled water in coils as a heat-exchanger medium for cooling of the liquid-autoclave, after liquid samples have been drawn. Chilled water is external to the autoclave, which is a secondary containment for the product cylinder and sampling piping, representing three physical barriers between the water and the UF_6, so no interaction is anticipated.

Centrifuge Cooling Water

The applicant provides centrifuge cooling water from the Centrifuge Cooling Water Distribution System. The applicant uses this system to provide a supply of deionized cooling water to the cooling coils of the centrifuge. This system provides stringent control over the operating temperature of the centrifuge to enable its efficient operation. Centrifuge cooling water is external to the UF_6 process streams in all cases and is not expected to interact with UF_6.

6.3.2 Chemical Accident Sequences

ISA Summary Section 3.7, Table 3.7-1 (LES, 2005b), identifies the chemical accident sequences, and Table 3.7.2 provides a narrative description. The chemical accident sequences covered the Tails Take-Off, UF_6 Feed, Product Take-Off, Product Blending and Liquid Sampling, Ventilated Room, Chemical Laboratory, Cylinder Preparation Room, Contingency Dump, Cascade, and the Centrifuge Test/Centrifuge Post-Mortem areas. A total of 36 different chemical accident sequences were identified by the applicant. The accident sequences covered the range of events that could result in a loss of containment of UF_6 and the hazardous chemicals produced from UF_6 (i.e., hydrogen fluoride, uranyl fluoride, and interactions with organic materials such as hydrocarbons). The accident sequences addressed both intermediate- and high-consequence events. Staff review of the process and hazards involved did not identify any chemical accident sequences overlooked by the applicant. The staff concludes that the applicant has identified appropriate chemical accident sequences based on the applicant's use of an approved process hazards analysis method (HAZOP) to identify those sequences and the results of the above staff review.

6.3.3 Chemical Accident Consequences

The applicant proposed chemical exposure limits (LES, 2004a), based on the soluble uranium values referenced in NUREG-1391 (NRC, 1991), and proposed HF chemical exposure limits (LES, 2004a) based on Acute Exposure Guideline Level (AEGL) values, in a manner consistent with NUREG-1520, Table A-5, "Consequence Severity Categories Based on 10 CFR 70.61" (NRC, 2002). The applicant also proposed a worker-exposure strategy that incorporates 10-minute AEGL values for HF, as used in NUREG-1391 (NRC, 1991). The staff finds this approach to be acceptable for the determination of compliance with the performance criteria of 10 CFR 70.61.

The applicant used the methods prescribed in NUREG/CR-6410 (NRC, 1998) to determine the source terms. Many source term values are in the classified portion of the ISA. Staff review of the ISA and supporting documentation found the source term values to be reasonable. Site boundary atmospheric dispersion factors were generated based on Regulatory Guide 1.145 (NRC, 1982). Meteorological data collected at the Midland/Odessa, Texas, National Weather Service Station, during 1987-1991, are used. The applicant also used modeling methods for source-term determination, release fractions, dispersion factors, and meteorological conditions, as prescribed in Regulatory Guide 1.145 (NRC, 1982). The staff finds the applicant's proposed methodology for source-term determination to be acceptable.

6.3.4 Items Relied on for Safety (IROFS) and Management Measures

6.3.4.1 Chemical-Process IROFS

ISA Summary Section 3.7, Table 3.7-2 (LES, 2005b), describes each of the applicant identified accident sequences and the specific IROFS that are applied to prevent or mitigate the consequences of those accident sequences. ISA Summary Table 3.8-1 (LES, 2005b) describes the safety functions of all identified IROFS and the specific accident sequences that take credit for each IROFS. The staff reviewed the listed IROFS and the process descriptions and process flow diagrams provided in ISA Summary Sections 3.4 and 3.5 to identify where each IROFS would be used and how the IROFS would function to prevent or mitigate the consequences of the identified accident sequences. The identified IROFS provide protection to prevent a loss of confinement of licensed material during operation of the facility. Based on this system level review and the staff's on-site visit to a similar gas centrifuge uranium enrichment facility in Almelo, The Netherlands, which uses the same technology and process systems, the staff concludes that the applicant has identified chemical-process IROFS to prevent the consequences of accident sequences involving the chemical hazards of licensed material and hazardous chemicals produced from licensed material.

6.3.4.2   Management Measures

The applicant has identified management measures to ensure the availability and reliability of chemical-safety IROFS in SAR Section 6.4 (LES, 2005b). The applicant's Quality Assurance (QA) Program is described in Appendix A of the SAR. The applicant states that it will meet the requirements of 10 CFR Part 50, Appendix B, "Quality Assurance Criteria for Nuclear Power Plants and Fuel Reprocessing Plants," and the guidelines of ASME QA Standard NQA-1-1994, "Quality Assurance Program Requirements for Nuclear Facilities" (ASME, 1994). The applicant further states that it will apply all sections of the QA Program to the chemical-safety IROFS. The following provides a brief description of the management measures applied to chemical safety. Chapter 11 of this SER offers more detail of the management measures in the facility.

Configuration Management

The Configuration Management proposed for the facility includes those controls that ensure that the facility design basis is throughly documented and maintained, and that changes to the design basis are controlled. This includes the following:

–   That management commitment and staffing is appropriate to ensure configuration management is maintained;

–   That the proper QA is in place for design control, document control, and records management; and

–   That all structures, systems, and components (SSCs), including IROFS, are under appropriate configuration management.

Maintenance

The applicant proposes to help maintain chemical process safety through the implementation of administrative controls that ensure that process system integrity is maintained and that IROFS and other engineered controls are available and operate reliably. These controls include planned and scheduled maintenance of equipment and controls so that design features will function when required. Appropriate plant management is responsible for ensuring the operational readiness of IROFS under this control. For this reason, the maintenance function is closely coupled to operation. The maintenance function plans, schedules, tracks, and maintains records for maintenance activities.

Maintenance activities usually fall into the following categories:

1.   Surveillance/monitoring;

2.   Corrective maintenance;

3.   Preventive maintenance; and

4.   Functional testing.

Training

The applicant proposes to provide training to individuals who handle licensed materials and other chemicals at the facility. The training program is developed and implemented with input from the chemical safety staff, training staff, and management. The program includes the following:

1.   Analysis of jobs and tasks to determine what a worker must know to perform tasks efficiently;

2.   Design and development of learning objectives, based on the analysis of jobs and tasks that reflect the knowledge, skills, and abilities needed by the worker;

3.   Design and development of qualification requirements for positions where a level of technical capability must be achieved and demonstrated for safe and reliable performance of the job function;

4.   Development and implementation of standard and temporary operating procedures;

5.   Development and implementation of proper inspection, test, and maintenance programs and procedures;

6.   Development of chemical safety awareness throughout the facility so that all individuals know what their roles and responsibilities are in coordinating chemical-release mitigation activities, in support of the Emergency Plan, in case of a severe chemical release; and

7.   Coordination of the chemical-process safety training curriculum with that of other areas, including radiological safety, criticality safety, facility operations, emergency response, and related areas.

Procedures

The applicant proposes to use four types of plant procedures to control activities: (1) operating procedures; (2) administrative procedures; (3) maintenance procedures; and (4) emergency procedures.

Operating procedures, developed for workstation and control room operators, are used to directly control process operations. Operating procedures include:

1.	Direction for normal operations, including start-up and some testing, operation, and shutdown, as well as off-normal conditions of operation, including alarm response;

2.	Required actions to ensure radiological and nuclear criticality safety, chemical safety, fire protection, emergency planning, and environmental protection;

3.	Operating limits, controls, and specific direction regarding administrative controls to ensure operational safety; and

4.	Safety checkpoints such as hold points for radiological or criticality safety checks, QA verifications, or operator independent verification.

The applicant uses administrative procedures to perform activities that support the process operations, including, but not limited to, management measures such as:

1.	Configuration management;

2.	Nuclear criticality, radiation, chemical, and fire safety;

3.	QA;

4.	Design control;

5.	Plant-personnel training and qualification;

6.	Audits and assessments;

7.	Incident investigations;

8.	Record-keeping and document control; and

9.	Reporting.

Administrative procedures are also used for:

1.	Implementing the Fundamental Nuclear Material Control Plan (FNMCP);

2.	Implementing the Emergency Plan;

3.	Implementing the Physical Security Plan; and

4.	Implementing the Standard Practice Procedures Plan for the Protection of Classified Matter.

Maintenance procedures address:

1.	Preventive and corrective maintenance of IROFS;

2.	Surveillance (including calibration, inspection, and other surveillance testing);

3. Functional testing of IROFS; and

4. Requirements for pre-maintenance activity involving review of the work to be preformed and review of procedures.

Emergency procedures address the preplanned actions of operators and other plant personnel in case of an emergency.

<u>Audits and Assessments</u>

The applicant proposes to conduct audits to determine that plant operations are performed in compliance with regulatory requirements, license conditions, and written procedures. The applicant assesses activities related to radiation protection, criticality safety control, hazardous chemical safety, fire protection, and environmental protection.

The applicant performs the audits in accordance with a written plan, which identifies and schedules audits to be performed. Audit team members have no direct responsibility for the function and area being audited. Team members have technical expertise or experience in the area being audited and are indoctrinated in audit techniques. The applicant conducts audits annually, on selected functions and areas, as defined above. The chemical process safety functions and areas will be audited at least once every 3 years.

Personnel, qualified by the applicant, who are not directly responsible for production activities, are used to perform routine surveillance/assessments. Deficiencies noted during the inspection, requiring corrective action, are forwarded to the manager of the applicable area or function, for action. Future surveillance/assessments include a review to evaluate if corrective actions have been effective.

<u>Emergency Planning</u>

The facility has an emergency plan and program that includes responses to mitigate the potential impact of any process chemical release, including requirements for notification and reporting of accidental chemical releases. The LES fire brigade/emergency response team is outfitted, equipped, and trained for hazardous material response and local agencies can supplement LES with additional response teams.

<u>Incident Investigation and Corrective Actions</u>

The applicant has a facility-wide incident investigation process that includes chemical process-related incidents. This process is available for use by any person at the facility for reporting abnormal events and potentially unsafe conditions or activities. Events that potentially threaten or lessen the effectiveness of health, safety, or environmental protection will be identified and reported to and investigated by the Health Safety and Environment Manager. Each event is considered in terms of its requirements for reporting in accordance with regulations and is evaluated to determine the level of investigation required. These evaluations and investigations are conducted in accordance with approved procedures. The depth of the investigation depends on the severity of the incident in terms of the levels of uranium/chemical released or the degree of potential for exposure of workers, the public, or environment.

6.3.5 Baseline Design Criteria

In the SAR (LES, 2005a) and ISA Summary (LES, 2005b), the applicant provides design basis information for chemical-process-safety IROFS identified for the proposed facility. For chemical protection, 10 CFR 70.64(a)(5) states:

> "Chemical protection. The design must provide for adequate protection against chemical risks produced from licensed material, facility conditions which affect the safety of licensed material, and hazardous chemicals produced from licensed materials."

The only chemical of concern (Class 1) is UF_6. Details of design and safety features of all chemical process systems are found in Chapter 3, ISA Summary (LES, 2005b). The applicant's design of the chemical process systems includes numerous controls, in addition to the IROFS, for maintaining safe conditions during operation. The applicant accomplishes this through several means, including managing the arrangement and size of material containers and processes; selection and use of material compatible with process chemicals; providing inherently safer operating conditions (e.g., vacuum handling); and providing process interlocks, controls, and alarms within the process.

The staff reviewed the applicant's proposed design of the gas centrifuge uranium enrichment facility contained in ISA Summary Sections 3.3-3.5 and the process hazards description in Section 3.6 (LES, 2005b). The staff notes that the uranium enrichment process is basically a physical process that separates the U^{235} isotope from the U^{238} isotope based on their mass difference. The uranium is in the chemical form of UF_6. The entire process, with the exception of the product sampling step, is conducted under a significant vacuum. Furthermore, the process design involves very limited inventories of UF_6 throughout the entire process. As a result, any process leak would result in air in-leakage into the system. Any uranium that could escape through a system breach would be limited by the available inventory and molecular diffusion. The chemical behavior of UF_6 and its reaction products (upon contact with the moisture in air, discussed in Section 6.3.1.3 of this SER) are such that most of the uranium bearing material is likely to accumulate near any process breach. Maintenance of the vacuum is necessary to operate the centrifuge machines. Any significant leakage would be quickly detected, as operation with even relatively small amounts of air would result in damage to the machines. Based on the need to operate at, and maintain, a significant vacuum throughout the gaseous portion of the process, and the limited inventories of licensed material contained in any portion of the gaseous process, the staff concludes that the design basis provides for adequate protection against chemical risks.

The staff notes that the applicant's design of product liquid sampling system utilizes an ANSI N14.1 qualified cylinder as the primary confinement vessel and an ASME Code pressure vessel as a secondary confinement system. The staff concludes that this design approach for the liquid portion of the process is acceptable because it uses internationally recognized nuclear fuel cycle industry codes and standards.

The staff reviewed the results of the applicant's HAZOP analysis as discussed in SER Chapter 3. This method is widely used in the chemical industry during the design phase to identify operability and safety issues and is identified as an acceptable method in NUREG-1513 (NRC, 2001). As applied to the gas centrifuge uranium enrichment process, the HAZOP considered a variety of internal process, facility and external hazards that could breach the process and release licensed material and hazards chemicals produced from licensed materials. The results of the applicant's ISA are presented in ISA Summary Table 3.7-1 (LES, 2005b). The table contains information concerning the accident sequences identified as a result of the HAZOP, the unmitigated risk of each applicant identified accident sequence, and the IROFS applied to prevent or mitigate the accident sequence. The staff also reviewed selected high-consequence and intermediate consequence accident scenarios to confirm that chemical events that could exceed the performance requirements of 10 CFR 70.61 were addressed.

Based on the above, the staff concludes that the applicant's design provides for adequate protection against chemical risks produced from licensed material, facility conditions which affect the safety of licensed material, and hazardous chemicals produced from licensed material, and meets the requirements of 10 CFR 70.64(a)(5).

6.4   EVALUATION FINDINGS

The staff evaluated the application using the criteria previously listed. Based on the review of the license application, the NRC staff has concluded that the applicant has described and assessed accident consequences that can result from the handling, storage, or processing of licensed materials and that can potentially have significant chemical consequences and effects. The applicant has prepared a hazard analysis that identifies and evaluates those chemical process hazards and potential accidents and established safety controls providing reasonable assurance of safe facility operation. To ensure that the performance requirements in 10 CFR Part 70 are met, the applicant has stated that controls are maintained, available, and reliable to perform their safety-related functions when needed. The staff has reviewed these safety controls and the applicant's plan for managing chemical-process safety and finds them acceptable.

The staff concludes that the applicant's plan for managing chemical-process safety and chemical-process-safety controls meets the requirements of Parts 30, 40, and 70, and provides reasonable assurance that the public health and safety, and the environment, will be protected.

6.5   REFERENCES

(ANSI, 1995) American National Standards Institute (ANSI). ANSI N14.1, "Standard for Nuclear Materials - Uranium Hexafluoride - Packaging for Transportation," 1995.

(ASME, 1994) American Society of Mechanical Engineers (ASME). NQA-1, "Quality Assurance Program Requirements for Nuclear Facilities," 1994.

(LES, 2004a) Louisiana Energy Services (LES), letter to U.S. Nuclear Regulatory Commission, "License Application and Integrated Safety Analysis Summary Update Related too Chemical Exposure Limits," December 22, 2004.

(LES, 2005a) Louisiana Energy Services (LES). "National Enrichment Facility Safety Analysis Report," Revision 6, 2005.

(LES, 2005b) Louisiana Energy Services (LES). "National Enrichment Facility Integrated Safety Analysis," Revision 4, 2005.

(NRC, 1982) U.S. Nuclear Regulatory Commission (NRC). Regulatory Guide 1.145, "Atmospheric Dispersion Models for Potential Accident Consequence Assessments at Nuclear Power Plants," 1982.

(NRC, 1991) U.S. Nuclear Regulatory Commission (NRC). NUREG-1391, "Chemical Toxicity of Uranium Hexafluoride Compare to Acute Affects of Radiation," 1991.

(NRC, 1998) U.S. Nuclear Regulatory Commission (NRC). NUREG/CR-6410, "Nuclear Fuel Cycle Facility Accident Analysis Handbook," 1998.

(NRC, 2002) U.S. Nuclear Regulatory Commission (NRC). NUREG-1520, "Standard Review Plan for the Review of a License Application for a Fuel Cycle Facility," 2002.

7.0 FIRE SAFETY

The purpose of this review is to determine, with reasonable assurance, whether the applicant has designed a facility that provides adequate protection against fires and explosions that could affect the safety of licensed materials and thus present an increased radiological risk. The review should also establish that the applicant has considered the radiological consequences of fires and will institute suitable safety controls to protect workers, the public, and the environment.

7.1   REGULATORY REQUIREMENTS

The regulatory basis for the fire safety review should be the general and additional contents of application, as required by 10 CFR 30.33, 10 CFR 40.32, 10 CFR 70.22, and 10 CFR 70.65. In addition, the fire safety review should focus on providing reasonable assurance of compliance with 10 CFR 70.61, 70.62, and 70.64.

7.2   REGULATORY ACCEPTANCE CRITERIA

The acceptance criteria the U.S. Nuclear Regulatory Commission (NRC) uses for reviews of fire safety are outlined in Sections 7.4.3.1 through 7.4.3.5 of NUREG-1520 (NRC, 2002).

7.3   STAFF REVIEW AND ANALYSIS

7.3.1   Process Fire Hazards and Special Hazards

The applicant plans to use gas centrifuge machines to enrich uranium up to 5 weight (wt.) percent uranium-235 (U^{235}). The feed material will be uranium hexafluoride (UF_6), which will be enriched by using 48 enrichment cascades with hundreds of gas centrifuge machines per cascade. The UF_6 normal feed is 0.711 wt. percent U^{235}; the expected product is a maximum of 5.0 wt. percent U^{235}; and the depleted tails are typically 0.2 wt. percent U^{235}. Enriched and depleted UF_6 streams are withdrawn from the cascade by pumps and returned to a solid phase in product and tails low-temperature take-off stations, respectively. The remainder of this section describes the NRC staff review of key fire hazards and risks associated with the proposed facility and gas centrifuge enrichment process.

UF$_6$: UF_6 is not flammable and does not disassociate to flammable constituents under conditions at which it will be handled at the facility. UF_6 does not react with oxygen, nitrogen, carbon dioxide, or dry air, but does react with water or water vapor. Hydrocarbons can be explosively oxidized if they are mixed with UF_6 in the liquid state or at elevated temperatures. For this reason, non-fluorinated hydrocarbon lubricants are not utilized in the UF_6 processes at the facility. UF_6 pumps are lubricated using a perfluorinated polyether oil that is referred to by the manufacturer's trade name, Fomblin oil.

Hydrogen Fluoride (HF): HF is a byproduct of the chemical reaction of UF_6 with water vapor. HF is extremely reactive in both gaseous and aqueous form. HF by itself is not flammable nor combustible. It can, however, react exothermically with water to generate sufficient heat to ignite nearby combustibles.

Uranyl Fluoride (UO_2F_2): UO_2F_2 is also a byproduct of the chemical reaction of UF_6 with water vapor. UO_2F_2 is stable in air to 300° C (572° F). It is not flammable nor combustible and will not decompose to combustible constituents under conditions that will exist at the facility.

Centrifuge Machines and Components: The model TC-12 centrifuge contains a rotor assembly, under a vacuum, inside an aluminum outer casing. The casing also provides a vacuum enclosure outside the rotor to reduce drag. The rotor is driven by a electromagnetic motor. The only combustibles of any significance are the electrical cabling going to the drive motors. Therefore, any fire originating in one of the cascades will most likely result in limited damage to the centrifuge and its components, resulting in a small release.

Control Room: The control room will be provided with automatic smoke detection throughout. Additionally, the control room will house the fire alarm control panel and will be continuously staffed. Hand-portable fire extinguishers will be provided in accordance with National Fire Protection Association (NFPA) Standard 10 (NFPA, 1994). Items Relied on for Safety (IROFS) boundaries will include appropriate electrical separation from normal instrument and control functions to ensure that fire-induced spurious actuations do not occur. Based on the current design, all active engineered components that are IROFS will fail in the safe configuration.

Storage and Handling of UF_6: UF_6 cylinders are stored or handled in the Uranium Byproduct Cylinder (UBC) storage pad; the Cylinder Receipt and Dispatch Building (CRDB); the UF_6 handling areas; and the blending and liquid-sampling areas. On the UBC storage pad, fire concerns include the cylinder transport vehicle, a fire exposure from nearby vegetation, and fire exposure from a nearby vehicle accident. The licensee performed evaluations of these various fire scenarios and either concluded that they did not pose a threat to the stored cylinders or that with adequate controls, the threats could be adequately mitigated. In the CRDB, the primary fire concern was from a truck fire at the loading dock. The licensee also analyzed this and determined that the cylinders could be adequately protected by storing them at least 1 meter (m) [3.3 feet (ft)] from the edge of the loading dock. Combustible loadings in the UF_6 handling areas and the blending and liquid-sampling areas are limited and transient combustibles will be controlled. Therefore, any fire originating in these areas will be limited. In addition, smoke detection and an interlock with the ventilation system will limit releases to a low-consequence event in the Ventilated Room of the Technical Services Building (TSB).

Hydrogen Control: Hydrogen is used within the TSB Chemical Laboratory and may be generated at battery-charging stations in the facility. The laboratory will be protected by one or more of the following features:

- Hydrogen piping will be provided with excess flow control;

- Hydrogen supply will be isolated by emergency shutoff valves interlocked with hydrogen detection in the areas served by the hydrogen piping; and

- Natural or mechanical ventilation will be provided to ensure that hydrogen concentrations do not exceed 25 percent of the lower explosive limit (LEL). If mechanical ventilation is provided, it will be continuous or will be interlocked to start on detection of hydrogen in the area. Mechanical ventilation will also be provided with airflow sensors, to sound an alarm if the fan becomes inoperative.

Hydrogen control in battery-charging stations will be provided by measures identified in NFPA 70E (NFPA, 2004) and American National Standards Institute (ANSI) Standard C2 (ANSI, 1981). It is expected that natural or mechanical ventilation will be provided to ensure that hydrogen concentrations do not exceed 25 percent of the LEL. If mechanical ventilation is provided, it will be continuous or will be interlocked to start on detection of hydrogen in the area. Mechanical ventilation will also be provided with airflow sensors, to sound an alarm if the fan becomes inoperative.

Combustible Material Hazards: Materials of construction for the centrifuge process building, the supporting buildings, and centrifuge machines and components are predominantly noncombustible (e.g., steel, aluminum, concrete floors). A minimum of fixed combustibles is expected to be present in the operations areas, and the applicant plans to control transient combustibles to minimize potential fire hazards. The largest quantity of combustible material is two 19,000 L (5000 gal.) tanks of diesel fuel located outside. Other quantities of combustible materials are as follows:

- Silicone oil in the UF$_6$ handling area and the blending and liquid sampling area is contained within the heater and chiller units associated with the cold traps, with each unit containing approximately 72 L (19 gal.) of oil. Some units are paired, but each pair is located at least 9 m (30 ft) from any adjacent unit. The staff considers this distance to be sufficient to limit the potential involvement in a fire to one pair of cold trap heater or chiller units.

- Oxygen gas (oxidizer), acetylene gas, and propane gas in the Mechanical, Electrical, and Instrumentation Workshop;

- Acetone, toluene, petroleum ether, and peroxide in the Chemical Laboratory; and

- Primus gas, degreaser solvent, penetrating oil, and cutting oil in the vacuum pump rebuild workshop.

The NRC staff has reasonable assurance that the applicant has adequately identified all fire and explosion hazards. In accordance with guidance provided by NUREG-1520 (NRC, 2002), the applicant has reasonably identified and evaluated the facility and process hazards and risks associated with the proposed operations. In its review, NRC has taken into account the potential presence of these combustibles in the various accident scenarios. The identification of fire hazards and related analyses are documented in the applicant's "Integrated Safety Analysis (ISA) Summary" which provides the supporting safety basis for the license application.

7.3.2   Accident Scenarios

The applicant's ISA Summary describes, qualitatively, the potential credible fire-accident scenarios and associated risks for the facility. The applicant postulated and evaluated the following key fire-accident scenarios:

- Fire in the Centrifuge Test Facility;
- Fire in the CRDB;
- Fire involving Cylinder Transporters/Movers
- Fire inside the Cascade Halls;
- Fire in the Process-Services Area;
- Fire inside the UF_6 handling area/Blending and Liquid Sampling Area;
- Fire inside the TSB; and
- Fire affecting the UBC Storage Pad.

Exterior and interior building explosions as initiating events to accident sequences were evaluated and found to be highly unlikely, without the need for IROFS.

7.3.2.1   Fire in the Centrifuge Test Facility

The Centrifuge Test Facility is located in the Centrifuge Assembly Building with the Centrifuge Post-Mortem Facility. The centrifuge test facility has UF_6 contained in stainless steel components and piping. The fire loading in the centrifuge test area consists of cable insulation, activated carbon, and a low amount of transient combustibles. Transient combustibles will be controlled by administrative combustible loading controls (IROFS36a). With these controls and the estimated in-situ fire loading, the applicant determined that the UF_6 inventory would not be released from the stainless steel confinement. The applicant also evaluated the likelihood of a fire propagating into the area from an adjacent area. Fire propagation will be prevented by rated barriers (IROFS35) between adjacent fire areas and the Centrifuge Test Facility. Based on the above, the applicant has determined that the likelihood of a fire being initiated, and of the IROFS failing, so that a release exceeding the consequence threshold of 10 CFR 70.61(b) or (c) occurs, is highly unlikely.

The staff concludes with reasonable assurance that the applicant has demonstrated that the facility will be in compliance with the performance requirements of 10 CFR 70.61 in the event of a fire in the Centrifuge Test Facility.

7.3.2.2   Fire in the CRDB

All UF_6 feed cylinders and empty product and byproduct cylinders enter the facility through the CRDB. The applicant considered three separate fire scenarios in the CRDB. These fire scenarios were a fire at the CRDB loading dock, a fire in the CRDB general areas, and a fire involving a cylinder delivery vehicle.

At the CRDB loading dock, UF_6 is contained in 48Y, 48X, and 30B cylinders on the loading dock and scales adjacent to the dock. The most severe fire was postulated to be a vehicle fire at the loading dock. The applicant evaluated the effect of this fire by calculation and showed that there could be a potential fire threat to UF_6 cylinders, but that this threat could be eliminated by assuring that cylinders were stored with a 1 m (3.3 ft) setback from the edge of the loading dock. The 1 m (3.3 ft) setback is an administrative control (IROFS36b). Based on the above, the applicant has determined that the likelihood of a fire being initiated, and of the IROFS failing, so that a release exceeding the consequence threshold of 10 CFR 70.61(b) or (c) occurs, is highly unlikely. The staff reviewed the applicant's calculation (NRC, 2004) and agrees with this assessment.

The CRDB General Areas contain UF_6 in 48Y, 48X, and 30B cylinders. Combustible loading is expected to be very low and transient combustibles will be controlled with IROFS36a. No liquid combustibles are listed for the CRDB general areas. Hence, a fire with the intensity and duration to heat a large UF_6 cylinder to its critical temperature is not considered credible. Fire propagation will be prevented by rated barriers (IROFS35) between adjacent fire areas and the CRDB general areas. Based on the above, the applicant has determined that the likelihood of a fire being initiated, and of the IROFS failing, so that a release exceeding the consequence threshold of 10 CFR 70.61(b) or (c) occurs, is highly unlikely. The staff agree with this assessment.

At the staff's request, LES also evaluated a fire involving UF_6 cylinders present on a delivery vehicle. U.S. Department of Transportation regulations require thermal protection (e.g., overpack or other protective assembly), which will withstand the thermal test criteria, specified in 10 CFR 71.73(c)(4), without rupture of the containment system, for all off-site UF_6 shipments. Hence, both incoming cylinders and outgoing cylinders will be protected by approved thermal protection. The handling practice for incoming cylinders containing UF_6 will be to offload the integral cylinder in its protective assembly to the loading dock, before opening or removal of the protective assembly. Outgoing cylinders will be individually loaded into a protective assembly before placement on truck trailers. The applicant determined that the worst postulated truck fire involving diesel fuel and other combustibles associated with a truck fire would burn for no more than 22 minutes. Approved protective assemblies (IROFS36h) are designed to protect a cylinder for 30 minutes in an 800°C (1472°F) hydrocarbon fire. Because of the location of the cylinder in the assembly on the truck trailer and the duration of the fire, the UF_6 cylinders will be adequately protected from rupture from a truck fire. The staff also evaluated the applicant's calculation of fire duration and agrees with the result (NRC, 2005). Based on the above, the applicant has determined that the likelihood of a fire being initiated, and of the IROFS failing, so that a release exceeding the consequence threshold of 10 CFR 70.61(b) or (c) occurs, is highly unlikely. The staff agrees with this assessment.

The staff concludes with reasonable assurance that the applicant has demonstrated that the facility will be in compliance with the performance requirements of 10 CFR 70.61 in the event of a fire in the CRDB.

7.3.2.3 Fire Involving Cylinder Transporters/Movers

The assumed inventory for a fire involving a cylinder transporter/mover is the amount of UF_6 in a UF_6 cylinder (48X, 48Y, or 30B) in transit. Only electric drive cylinder transporters will be used for cylinder transport inside of the proposed facility's buildings. When filled 30B cylinders are transported outside of the buildings, they are protected by DOT approved overpacks (IROFS 36h). For other cylinders which may be transported outside of the buildings by diesel drive transporters, the fuel capacity of the transporters is limited to less than 280 L (74 gal.). This limit on fuel is IROFS 36c and is described in more detail in Section 7.3.2.8 of this SER. Based on the above, the applicant has determined that the likelihood of a fire being initiated, and of the IROFS failing, so that a release exceeding the consequence threshold of 10 CFR 70.61(b) or (c) occurs, is highly unlikely.

The staff concludes with reasonable assurance that the applicant has demonstrated that the facility will be in compliance with the performance requirements of 10 CFR 70.61 in the event of a fire on a cylinder transporter either inside or outside of the proposed facility's buildings.

7.3.2.4   Fire in the Cascade Halls

The six cascade halls contain the centrifuges and are located in the Separations Building. The fire scenario inside the cascade halls is assumed to take place inside a module that holds eight cascades, each cascade containing hundreds of centrifuges. If the entire module were engulfed in a fire and the total inventory released, a high-consequence event would result. However, a fire is prevented from propagating into the module by fire barriers (IROFS35). A fire originating inside the module is presumed to involve the cables feeding the centrifuge drive motors. If transient combustibles are controlled (IROFS36a), this fire, at worst, is expected to involve one and possibly an adjacent centrifuge and can be assumed to result in the release of no more than 1 percent of the inventory in the module, resulting in consequences below the threshold of the 10 CFR 70.61(c) intermediate-consequence limit. Based on the above, the applicant has determined that the likelihood of a fire being initiated, and of the IROFS failing, so that a release exceeding the consequence threshold of 10 CFR 70.61(b) or (c) occurs, is highly unlikely.

The staff concludes with reasonable assurance that the applicant has demonstrated that the facility will be in compliance with the performance requirements of 10 CFR 70.61 in the event of a fire in the Cascade Halls.

7.3.2.5   Fire in the Process Services Area

The Process Services Area contains the gas transport equipment (i.e, the piping to the Product Take-Off System and the piping to the Tails Take-Off System) and the Contingency Dump System. The Process Services Area is located in the Separations Building. The UF_6 inventory in the Process Services Area is UF_6 in feed, product, and tails piping manifolds. In addition, there could be significantly more in the 48 sodium fluoride traps that are part of the contingency dump system. If the entire Process Services Area were engulfed in a fire and the total inventory released, a high-consequence event would result. However, a fire is prevented from propagating into the Process Services Area by fire barriers (IROFS35). A fire originating in the Process Services Area involving expected in-situ combustibles is considered capable of causing failures in the aluminum piping and manifolds, but not in the aluminum traps. Combustible loading controls (IROFS36a) are intended to limit additional transient combustibles. With failure of the piping only, 50 percent of the UF_6 is assumed released. This would result in consequences that are below the 10 CFR 70.61(c) intermediate-consequence limit. Based on the above, the applicant has determined that the likelihood of a fire being initiated, and of the IROFS failing, so that a release exceeding the consequence threshold of 10 CFR 70.61(b) or (c) occurs, is highly unlikely.

The staff concludes with reasonable assurance that the applicant has demonstrated that the facility will be in compliance with the performance requirements of 10 CFR 70.61 in the event of a fire in the Process Services Area.

7.3.2.6   Fire in the UF$_6$ Handling Area/Blending and Liquid Sampling Area

The UF$_6$ Handling Area contains the UF$_6$ Feed System, Product Take-Off System, and Tails Take-Off System. The Blending and Liquid Sampling Area contains the Product Liquid Sampling System and the Product Blending System. The UF$_6$ inventory in the blending and liquid sampling area of the UF$_6$ Handling Area is contained in cylinders, piping, manifolds, and hoses. The applicant states that additional uranic material may be present on the carbon/alumina traps that capture residual traces of UF$_6$ from the various feed, product, and tails system cold traps. A fire is prevented from propagating into the UF$_6$ Handling Area, and Blending and Liquid Sampling Area, by fire barriers (IROFS35). A fire originating in these areas with improperly placed combustibles is considered capable of failing only a single-cylinder hose. A fire, involving expected in-situ and transient combustibles, could cause failure in the aluminum piping manifold and release 50 percent of the inventory feeding one module. This would result in consequences that are below the 70.61 (c) intermediate consequence limit. Severe fires will be prevented by IROFS36a, which will control the location and the amount of transient combustibles in the area. Based on the above, the applicant has determined that the likelihood of a fire being initiated, and of the IROFS failing, so that a release exceeding the consequence threshold of 10 CFR 70.61(b) or (c) occurs, is highly unlikely.

The staff concludes with reasonable assurance that the applicant has demonstrated that the facility will be in compliance with the performance requirements of 10 CFR 70.61 in the event of a fire in the UF$_6$ Handling Area, and the Blending and Liquid Sampling Area.

7.3.2.7   Fire in the TSB

The TSB contains support areas for the facility, such as the Solid Waste Collection Room; Vacuum Pump Rebuild Workshop; Decontamination Workshop; Ventilated Room; Cylinder Preparation Room; Mechanical, Electrical, and Instrumentation Workshop; TSB Gaseous Effluent Ventilation System Room; various laboratories; the Control Room; and monitoring centers. In the TSB, fires were postulated in all uranic material areas and IROFS were found to be needed in the Solid Waste Collection Room, the Decontamination Workshop, the Ventilated Room, and the Chemical Laboratory Sample Storage Room.

The uranium inventory in the Solid Waste Collection Room is contained in 12-L (3.2-gal.) metal containers and 210-L (55-gal.) metal drums. A fire is prevented from propagating into the Solid Waste Collection Room by rated fire barriers (IROFS35). For a fire originating in the area, and involving expected in-situ and transient combustibles, the applicant postulates that only a few kg of uranic materials would be present in open containers or drums, during transfer/packing operations, and driven off, in case of a fire. Preventive measures are to administratively limit transient combustible loading in areas containing uranic material, to ensure integrity of uranic components/containers and to limit the quantity of uranic material at risk, to ensure that consequences to the public are low (IROFS36d). Based on the above, the applicant has determined that the likelihood of a fire being initiated, and of the IROFS failing, so that a release exceeding the consequence threshold of 10 CFR 70.61(b) or (c) occurs, is highly unlikely.

The uranium inventory in the Decontamination Workshop is contained in up to three 12-L (3.2-gal.) metal containers and three steel tanks. A fire is prevented from propagating into the room by fire-rated barriers (IROFS35). For a fire originating in the area, and involving expected in-situ and transient combustibles, the applicant postulates that only a few kilograms of uranic materials would be present in open containers or drums during transfer/packing operations, and driven off, in case of a fire. Preventive measures are to administratively limit transient combustible loading in areas containing uranic material, to ensure integrity of uranic components/containers and to limit the quantity of uranic material at risk, to ensure that consequences to the public are low (IROFS36d). Based on the above, the applicant has determined that the likelihood of a fire being initiated, and of the IROFS failing, so that a release exceeding the consequence threshold of 10 CFR 70.61(b) or (c) occurs, is highly unlikely.

The uranium inventory in the ventilated room is contained in 12-L (3.2-gal.) metal and 210-L (55-gal.) drums. Additional inventory is present in single 48Y, 48X, or 30B cylinders, present in the room, for valve maintenance/change-out. Various fire scenarios were postulated for the ventilated room, each with different combinations of IROFS, consequences, or likelihood. A fire in improperly placed transient combustibles (but with IROFS36d in effect) could cause failure of the nitrogen hose or vent-line piping used to bleed gas from cylinders during valve servicing or subsequent nitrogen-pressure tests. This was determined to be a low-consequence event. Another fire, however, with expected in-situ and transient combustibles within the area, but with the uranic inventory in open containers, which could be driven off in a fire, is considered to be an intermediate-consequence event. This postulated accident sequence is reduced to a low-consequence sequence by adding a second IROFS (IROFS37), which is smoke detection interlocked to shut down the building ventilation system. The other sequences the applicant evaluated assumed various combinations of failure of the combustible loading controls and the smoke detector and evaluated the resulting consequence level and likelihood. Based on the above, the applicant has determined that the likelihood of a fire being initiated, and of the IROFS failing, so that a release exceeding the consequence threshold of 10 CFR 70.61(b) or (c) occurs, is highly unlikely or unlikely, respectively, in accordance with the performance requirements

The TSB Chemical Laboratory Sample Storage Room also contains a uranium inventory that could potentially result in a high consequence. A fire is prevented from propagating into the Chemical Laboratory Sample Storage room by rated fire barriers (IROFS35). For a fire originating in the Sample Storage Room with controls in place (IROFS36a), with expected in-situ and transient combustibles, the consequences would be low. The fire would not have sufficient combustibles to fail a sample cylinder. Based on the above, the applicant has determined that the likelihood of a fire being initiated, and of the IROFS failing, so that a release exceeding the consequence threshold of 10 CFR 70.61(b) or (c) occurs, is highly unlikely.

The staff concludes with reasonable assurance that the applicant has demonstrated that the facility will be in compliance with the performance requirements of 10 CFR 70.61 in the event of a fire in the TSB.

7.3.2.8   UBC Storage Pad

The UBC Storage Pad provides storage for UBCs and 6 months of empty feed cylinders. The UBC Storage Pad occupies approximately 9 hectares (23 acres) and is sized to accommodate enough cylinders for 30 years of operation.

For the UBC storage pad, fires were postulated on the transporter, on the UBC perimeter road, and in vegetation off the property. The fire on the transporter was assumed to be the same fire that NRC staff analyzed in Section 11.3.2.6, "UF_6 Storage Area Fire," of NUREG-1491 (NRC, 1994). In this event, the NRC staff assumed a hydrocarbon pool fire with a temperature of 800°C (1500°F). NRC estimated heat transfer to the cylinder from this fire and calculated that it would take 20 minutes for the fire to cause rupture of a UF_6 cylinder. The NRC staff further calculated that it would require 280 L (74 gal.) of diesel fuel for the fire to burn for 20 minutes. The applicant used the results of the NRC staff calculation as a basis for a control under IROFS36c, which limits fuel in a diesel-fueled UBC transporter to 280 L (74 gal). The applicant will also employ IROFS36e for other vehicles, which administratively limits transient combustible loading on the UBC storage pad, to ensure cylinder integrity. For the review, the staff re-evaluated the estimation of burning time, assuming 280 L (74 gal) of fuel and a 3.1-m (10-ft) long pool fire approximating a 48X cylinder. The staff calculated about 20 minutes or less of burn time for conservative assumptions of pool configuration (NRC, 2005). Based on the above, the applicant has determined that the likelihood of a fire being initiated, and of the IROFS failing, so that a release exceeding the consequence threshold of 10 CFR 70.61(b) or (c) occurs, is highly unlikely.

The applicant performed its own calculations to determine the possible effects of a service-vehicle fire on a perimeter road at the closest point to the UBC. This fire was based on a pool, confined by the access road, formed by 500 L (132 gal.) of diesel fuel contained in the truck along with other truck combustibles. The heat transfer from the equivalent pool fire was determined to be insufficient to heat the cylinder to the temperature where it might rupture. A control for fires of this type will be IROFS36f, which will administratively limit designated routes for bulk-fueling vehicles onsite, to ensure UBC cylinder integrity. Based on the above, the applicant has determined that the likelihood of a fire being initiated, and of the IROFS failing, so that a release exceeding the consequence threshold of 10 CFR 70.61(b) or (c) occurs, is highly unlikely.

The applicant also evaluated the fire in the off-site vegetation, by means of a heat-transfer calculation from an assumed large fire approximately 89 m (292 ft) from the UBC pad. The heat transfer from this fire was also found to be insufficient to heat the cylinder to a point where it could rupture. The applicant will control this scenario by IROFS36g, which will administratively limit on-site vegetation fire sources to ensure the integrity of important targets. Based on the above, the applicant has determined that the likelihood of a fire being initiated, and of the IROFS failing, so that a release exceeding the consequence threshold of 10 CFR 70.61(b) or (c) occurs, is highly unlikely.

The staff review of these analyses is documented in an in-office review report (NRC, 2004).

The staff concludes with reasonable assurance that the applicant has demonstrated that the facility will be in compliance with the performance requirements of 10 CFR 70.61 in the event of a fire in the UBC Storage Pad.

7.3.2.9   Conclusions

The NRC staff has reasonable assurance that the applicant has identified and evaluated all fire related accident scenarios credible for the proposed centrifuge process. The applicant has reasonably identified and evaluated possible fire initiators and consequences and has identified

IROFS for accident scenarios that could result in intermediate or large consequences leading to unacceptable performance, in accordance with 10 CFR Part 70, and as described in guidelines established in NUREG-1520 (NRC, 2002).

7.3.3 IROFS Related to Fire Safety

The NRC staff has reasonable assurance that the applicant has identified the required IROFS for preventing or mitigating fire accident scenarios that could lead to unacceptable performance in accordance with the requirements in 10 CFR 70.61. The applicant has identified a set of IROFS that would ensure that the likelihood of a fire causing high consequence events is highly unlikely and the likelihood of a fire causing intermediate consequence events is unlikely. These IROFS are listed in Table 7.3-1.

The NRC staff considers the failure probability indices assigned to these IROFS, with a failure probability index of -3, to be achievable, with their respective bases, as described in Section 3.8.3 of the ISA Summary. Section 3.8.3 of the ISA Summary provides proposed surveillance frequencies, safety margins, and other measures that will support the low-failure probabilities assumed for these measures. All of the above-listed IROFS will also be supported by the general management measures, as described in Section 3.1.8.3 of the ISA Summary. In conclusion, the staff finds the selection of accident sequences and the determination of IROFS to be acceptable to satisfy the performance requirements of 10 CFR 70.61.

The remaining features described in the license application are fire-protection measures that provide overall defense-in-depth protection of fire safety for operations. The applicant's ISA Summary has adequately addressed fire risks, in accordance with the regulation and the guidance established in NUREG-1520 (NRC, 2002).

7.3.4 Facility Fire Protection

7.3.4.1 Fire Safety Management Measures

The Health, Safety and Environmental manager is responsible for fire protection and is assisted by the Industrial Safety Manager, who is responsible for day-to-day safe operation of the facility, including fire safety. The Industrial Safety Manager is assisted by fire safety personnel who are trained in fire protection and have nuclear fire-safety experience. The fire protection staff is responsible for:

- Fire protection program and procedural requirements;
- Fire prevention activities (i.e., administrative controls and training);
- Maintenance, surveillance, and quality of the facility fire protection features;
- Control of design changes, as related to fire protection;
- Documentation and record-keeping, as related to fire protection;
- Organization and training of the fire brigade; and
- Pre-fire planning.

Fire prevention at the facility consists of administrative controls to: (a) govern the handling of transient combustibles; (b) control ignition sources; (c) ensure that open flames or combustion-generated smoke is not used for leak-testing; (d) conduct periodic fire prevention inspections;

(e) perform periodic house-keeping inspections; and (f) implement a system to control the disarming of fire-detection or fire-suppression systems. The inspection, testing, and maintenance of fire-protection systems will comply with industry standards. The applicant's

Table 7.3-1
Identification of IROFS

IROFS	Failure probability index	Description of IROFS
IROFS35	-3	Automatic closure of fire-rated-barrier opening protections (e.g., doors, dampers, penetration seals), to ensure that the integrity of area fire barriers prevents fires from propagating into areas containing uranic material.
IROFS36a	-3	Administratively limit transient combustible loading in areas containing uranic material, to ensure integrity of uranic material components/containers and limit the quantity of uranic material at risk, to ensure that consequences to the public are low.
IROFS36b	-3	Administratively limit storage of UF_6 cylinders in the CRDB, to ensure ≥ 1-m (3-ft) setback from the edge of the loading dock.
IROFS36c	-3	Administratively limit on-site UF_6 cylinder transporters/movers, to ensure only use of electric drive or diesel power with a fuel capacity of less than 280 L (74 gal.).
IROFS36d	-3	Administratively limit transient combustible loading in areas containing uranic material, to ensure integrity of uranic material components/containers and limit the quantity of uranic material at risk, to ensure that consequences to the public are low.
IROFS36e	-3	Administratively limit transient combustible loading on the UBC Storage Pad, to ensure cylinder integrity.
IROFS36f	-3	Administratively limit designated routes for bulk-fueling vehicles on-site, to ensure UBC cylinder integrity.
IROFS36g	-3	Administratively limit on-site vegetation fire sources, to ensure integrity of important targets.
IROFS36h	-3	Use of cylinder overpack/thermal protective assemblies
IROFS37	-2	Automatic trip of the Ventilated Room HVAC and isolation from TSB GEVS on smoke detection, and Ventilated-Room-design leakage limited to ensure offsite exposure from building out-flow maintains consequences to the public low.

ISA Summary has adequately addressed fire safety management measures, in accordance with the guidance established in NUREG-1520 (NRC, 2002). In addition, the applicant's fire safety management measures meet the requirements of 10 CFR 30.33, 10 CFR 40.32, 10 CFR 70.22, 10 CFR 70.64, and 10 CFR 70.65 as they pertain to the fire protection aspects of the facility.

7.3.4.2   Facility Passive-Engineered Fire-Protection Systems

Buildings containing UF_6 are the TSB, CRDB, Cascade Halls, Blending and Liquid Sampling Areas, and the Centrifuge Assembly Building (including the Centrifuge Test and Centrifuge Post Mortem Area), which are of pre-cast concrete-frame and concrete-panel construction. This construction is classified as Type I-443, in accordance with NFPA 220 (NFPA, 1999). This means that the exterior and bearing walls, interior-bearing walls supporting more than one floor, columns supporting more than one floor, and structural members supporting more than one floor, have a fire resistance of at least 4 hours. All interior-bearing walls, columns, and structural members supporting one floor only, or a roof only, shall have a fire resistance of at least 3 hours. Floors have a resistance of at least 3 hours and roofs have a fire resistance of at least 2 hours.

The Administration Building, Fire Water Pump Building, and the Central Utilities Building are unprotected steel-frame buildings, with insulated metal panel exterior walls, and with built-up roofing on a metal deck roof. The Visitor Center and the site security buildings are also unprotected steel frame with insulated metal-panel exterior walls and built-up roofing on a metal deck roof. The above buildings are classified as Type II-000, in accordance with NFPA 220 (NFPA, 1999). This means that all structural members are of non-combustible or limited combustible construction, but no fire rating is required. The applicant's ISA Summary has adequately addressed passive engineered fire protection systems, in accordance with the guidance established in NUREG-1520 (NRC, 2002). In addition, the applicant's passive-engineered fire protection systems meet the requirements of 10 CFR 30.33, 10 CFR 40.32, 10 CFR 70.22, 10 CFR 70.64, and 10 CFR 70.65 as they pertain to the fire protection aspects of the facility.

7.3.4.3   Facility Active-Engineered Fire-Protection Systems

Fire Alarm System

Each building of the facility is equipped with a listed modular, multi-zone fire alarm control panel installed in accordance with NFPA 72 (NFPA, 1996d). Each panel has a dual power supply, consisting of normal building power and backup power, by either 24-hour battery or the facility uninterruptible power supply (UPS). Sprinkler system and hose-station water-flow detection are connected to separate control-panel zone modules. Fire-detector and manual-pull station-alarm circuits are also on separate modules. Each zone module has separate alarm and trouble contacts for connection to the central alarm panel in the Control Room. Activation of a fire detector, manual-pull station, or water-flow detector results in an audible and visual alarm at the building control panel and the central alarm panel.

The central alarm panel, located in the control room, is a listed, microprocessor-based addressable console. The central alarm panel has dual power supplies, consisting of normal building power and backup power by either 24-hour battery or the facility UPS. The central alarm panel monitors all functions associated with the individual alarm panels and fire pump controllers. All alarm and trouble functions are audibly and visually annunciated by the central alarm panel and automatically recorded via printout. Failure of the central alarm panel will not result in failure of any building-alarm control-panel functions.

The following conditions are monitored by the central alarm console through the fire pump controllers:

- Pump running;
- Pump failure to start;
- Pump controller in off or manual position;
- Battery failure;
- Diesel overspeed;
- Diesel high engine-jacket coolant temperature;
- Diesel low oil pressure; and
- Battery charger failure.

Both pumps are maintained in the automatic-start position at all times, except during periods of maintenance and testing. Remote manual-start switches are provided in the Control Room adjacent to the alarm console. Fire pumps can only be shut off at the controllers.

Portable Fire Extinguishers

Portable fire extinguishers are installed throughout all buildings, in accordance with NFPA 10 (NFPA, 1994). Multipurpose fire extinguishers are provided generally for Class A (ordinary combustibles); Class B (flammable and combustible liquids); and Class C (electrical equipment) fires. Specialized extinguishers are located in areas requiring protection of particular hazards. Wheeled extinguishers are provided for use in water-exclusion areas. In areas with moderator-control issues, the extinguishers are filled with carbon dioxide or dry chemical.

Fire-Water Supply

The facility fire-water supply consists of two 946,000-L (250,000-gal.) water storage tanks designed and constructed in accordance with NFPA 22 (NFPA, 1996c). Within each tank, 473,000-L (125,000-gal.) are reserved for fire protection. Fill and makeup to the tanks are from the city water supply to the site, and the water supply is capable of filling either tank within an 8-hour period. The fire pumps consist of two diesel-driven 3785-Lpm (1000-gpm), at 10.35-bar (150-psia), pumps. Both pumps are horizontal centrifugal pumps designed and installed in accordance with NFPA 20 (NFPA, 1996b). The maximum anticipated fire demand is 3785 Lpm (1000 gpm), based on 1892 Lpm (500 gpm) from a building-sprinkler system, plus 3785 Lpm (500 gpm) for hose streams, for a duration of 2 hours. The combination of two water tanks and two fire pumps provides 100 percent redundancy for fire protection. In addition to fixed standpipes and fire hose stations, the facility will be provided with fire hose on mobile apparatus or at strategic locations throughout the facility. The amount of hose provided will be sufficient to ensure that all points within the facility will be able to be reached by at least two 38-mm (1½-in.) diameter hoses and one 64-mm (2 ½-in.)-diameter backup hose, consistent with NFPA 1410 (NFPA, 2000). These lines will have a minimum nozzle pressure of 4.5 bar (65 psia) for the 38-mm (1 ½-in.) hose, and 6.9 bar (100 psia) for the 64-mm (2 ½-in.) hose.

Engineered Automatic Fire-Suppression Systems

Wet pipe sprinkler systems are provided in the following buildings:

- Administration Building;
- Central Utilities Building; and
- Fire Pump House

These systems are designed and tested in accordance with NFPA 13 (NFPA, 1996a). Sprinkler-system control valves are monitored under a periodic inspection program, and their proper positioning is supervised in accordance with NFPA 801 (NFPA, 2003).

Fire-rated enclosures are provided for several chemical traps on the second floor of the Process Services Area, in each Separations Building Module. These enclosures will be protected with a gaseous-suppression system. The type of system used will be determined in the final design and will either be a carbon dioxide system designed in accordance with NFPA 12 (NFPA, 1993) or a clean-agent system designed in accordance with NFPA 2001 (NFPA, 1996f).

The NRC staff concludes that the applicant has reasonably determined the required fire-protection features for preventing or mitigating fire-accident scenarios that could lead to unacceptable performance, in accordance with the requirements in 10 CFR 70.61.

The applicant's ISA Summary has adequately addressed active engineered fire protection systems, in accordance with the guidance established in NUREG-1520 (NRC, 2002). In addition, the applicant's active-engineered fire protection systems meet the requirements of 10 CFR 30.33, 10 CFR 40.32, 10 CFR 70.22, 10 CFR 70.64, and 10 CFR 70.65 as they pertain to the fire protection aspects of the facility.

7.3.4.4   Fire Safety and Emergency Response

The facility will maintain a fire brigade made up of employees trained in fire fighting techniques, first aid procedures, and emergency response. The fire brigade is organized, operated, trained and equipped in accordance with NFPA 600 (NFPA, 1996e) for incipient fire fighting capability. The intent of the facility fire brigade is to be able to handle all minor fires and to provide a first-response effort designed to supplement the local fire department for major fires at the plant. The plant fire brigade, working with the plant's emergency operations center, will coordinate off-site fire department activities to ensure moderator control and criticality safety. The fire brigade is staffed so that there is a minimum of five brigade members available per shift.

Periodic training is provided to off-site assistance organization personnel in the facility emergency training procedures. Facility emergency response personnel meet at least annually with each off-site assistance group, to accomplish training and review items of mutual interest, including relevant changes to the program. A Memoranda of Understanding between Louisiana Energy Services and the City of Eunice, New Mexico Fire and Rescue Agency, and the City of Hobbs, New Mexico, Fire Department defines the fire-protection and emergency-response commitments between the organizations. The Eunice Fire and Rescue Agency is the initial response agency and can respond between 11 and 15 minutes. The department has three pumpers, three grass trucks, one tanker, and a roster of approximately 20 volunteers. The

Hobbs Fire Department can respond in about 25 to 30 minutes. The Hobbs Fire Department has five pumpers, one ladder truck, three grass trucks, one tanker, and a roster of about 70 paid personnel.

The NRC staff concludes with reasonable assurance that the on-site fire brigade, on-site water supply, on-site hose lines, and mutual aid from adequately equipped fire departments can provide defense-in-depth protection from releases from all identified credible fire scenarios and satisfy the requirements of 10 CFR 70.64(b), and are in accordance with the guidance in NUREG-1520 (NRC, 2002). In addition, the applicant's emergency response capability meet the requirements of 10 CFR 30.33, 10 CFR 40.32, 10 CFR 70.22, 10 CFR 70.64, and 10 CFR 70.65 as they pertain to the fire protection aspects of the facility.

7.4   EVALUATION FINDINGS

The dominant fire risk to safety and health of workers and the public for the proposed process is a fire that could lead to loss of confinement of UF_6. This includes a fire damaging the centrifuge machines and piping that provide UF_6 confinement, or UF_6 cylinders inside or on the outdoor storage pad. The applicant's submittals provide sufficient information, in accordance with requirements of 10 CFR 30.33, 10 CFR 40.32, 10 CFR 70.22, and 10 CFR 70.65, regarding potential fire hazards, consequences, and required controls for the proposed processes. The NRC staff determined that the applicant demonstrated compliance with the performance requirements of 10 CFR 70.61 for fire protection related to postulated accident scenarios. The applicant has identified a reasonable set of IROFS and defense-in-depth protection to ensure acceptable risks within the performance requirements of 10 CFR 70.61.

Based on the design of the facility, relative to fire protection and the designation of IROFS and measures that provide defense-in-depth, the staff concludes with reasonable assurance that the facility also meets the requirements of 10 CFR 30.33, 10 CFR 40.32, 10 CFR 70.22, and 10 CFR 70.64 (a)(3), regarding baseline design criteria for protection against fires and explosions, and 10 CFR 70.64 (b) defense-in-depth.

7.5   REFERENCES

(ANSI, 1981) American National Standards Institute (ANSI). ANSI C2, "National Electrical Safety Code," 1981.

(NFPA, 1993) National Fire Protection Association (NFPA). NFPA 12, "Standard on Carbon Dioxide Extinguishing Systems," 1993.

(NFPA, 1994) National Fire Protection Association (NFPA). NFPA 10, "Standard for Portable Fire Extinguishers," 1994.

(NFPA, 1996a) National Fire Protection Association (NFPA). NFPA 13, "Standard for the Installation of Sprinkler Systems," 1996.

(NFPA, 1996b) National Fire Protection Association (NFPA). NFPA 20, "Standard for Installation of Stationary Pumps for Fire Protection," 1996.

(NFPA, 1996c) National Fire Protection Association (NFPA). NFPA 22, "Standard for Water Tanks for Private Fire Protection," 1996.

(NFPA, 1996d) National Fire Protection Association (NFPA). NFPA 72, "National Fire Alarm Code," 1996.

(NFPA, 1996e) National Fire Protection Association (NFPA). NFPA 600, "Standard on Industrial Fire Brigades," 1996.

(NFPA, 1996f) National Fire Protection Association (NFPA). NFPA 2001, "Standard on Clean Agent Fire Extinguishing Systems," 1996.

(NFPA, 1999) National Fire Protection Association (NFPA). NFPA 220, "Standard on Types of Building Construction," 1999.

(NFPA, 2000) National Fire Protection Association (NFPA). NFPA 1410, "Standard on Training for Initial Emergency Scene Operations," 2000.

(NFPA, 2003) National Fire Protection Association (NFPA). NFPA 801, "Standard for Fire Protection for Facilities Handling Radioactive Materials," 2003.

(NFPA, 2004) National Fire Protection Association (NFPA). NFPA 70E, "Standard for Electrical Safety in the Workplace," 2004.

(NRC, 1994) U.S. Nuclear Regulatory Commission (NRC). NUREG-1491, "Safety Evaluation Report for the Claiborne Enrichment Center, Homer, Louisiana," January 1994.

(NRC, 2002) U.S. Nuclear Regulatory Commission (NRC). NUREG-1520, "Standard Review Plan for the Review of a License Application for a Fuel Cycle Facility," March 2002.

(NRC, 2004) Johnson, T.C., U.S. Nuclear Regulatory Commission (NRC). March 9-10, 2004, "Meeting Summary: Louisiana Energy Services' Integrated Safety Analysis In-Office Review." March 25, 2004.

(NRC, 2005) Wescott, R.G., U.S. Nuclear Regulatory Commission (NRC). "Note to File: Confirmatory Calculations for Fire Protection Review of National Enrichment Facility Integrated Safety Analysis Summary." March 22, 2005.

8.0 EMERGENCY MANAGEMENT

The purpose of this review of the applicant's emergency management plan is to determine if the applicant has established, before the start of operations, adequate emergency management facilities and procedures to protect workers, the public, and the environment.

Emergency capability is incorporated into the baseline design criteria of 10 CFR Part 70, and is intended to ensure control of licensed material, evacuation of personnel, and availability of emergency facilities.

8.1 REGULATORY REQUIREMENTS

10 CFR 30.32(i)(1), 10 CFR 40.31(j)(1), and 10 CFR 70.22(i)(1)(i) specifies when an applicant is not required to submit an Emergency Plan (EP) to the U.S. Nuclear Regulatory Commission (NRC). If an applicant is required to submit an EP, as described in 10 CFR 30.32(i)(1)(ii), 10 CFR 40.31(j)(1)(ii), and 10 CFR 70.22(i)(1)(ii), then 10 CFR 30.32(i)(3), 10 CFR 40.31(j)(3), and 10 CFR 70.22(i)(3) contains the information that must be included in the EP. In addition, 10 CFR 70.64(a)(6) requires applicants to address the control of licensed material, evacuation of personnel, and availability of emergency facilities for the design of new facilities.

8.2 REGULATORY ACCEPTANCE CRITERIA

The acceptance criteria for NRC's review of the applicant's emergency management program are outlined in Section 8.4.3.1 of "Standard Review Plan for the Review of a License Application for a Fuel Cycle Facility," NUREG-1520 (NRC, 2002).

8.3 STAFF REVIEW AND ANALYSIS

8.3.1 Facility Description

Section 1.0 of the EP (LES, 2005a) contains descriptions of the licensed activity, the facility and site, and the area near the site. The information provided includes:

- A description of the enrichment process;
- A discussion of chemicals of concern, including form, physical state, location, and quantity [uranium hexafluoride (UF_6) has been identified as the only direct chemical of concern to be used at the facility];
- A detailed description of the site location and layout;
- A description of the major structures to be located at the site;
- A description of the Gaseous Effluent Vent Systems, including stack heights, maximum flow rates, and filter efficiencies;
- A description of the area near the site, including airport, railroad, pipeline, topography, and area land-use information;

- Demography of the area, water use, and climate; and
- Detailed maps of the facility and surrounding area.

This fulfills the requirements of 10 CFR 30.32(i)(3)(i), 10 CFR 40.31(j)(3)(i), and 10 CFR 70.22(i)(3)(i).

The applicant will maintain compliance with the *Emergency Planning and Community Right-to-Know Act of 1986*, in accordance with 10 CFR 30.32(i)(3)(xiii), 10 CFR 40.31(j)(3)(xiii), and 10 CFR 70.22(i)(3)(xiii). This is accomplished by conducting annual inventories; compiling the inventory information, and providing it to the appropriate agencies; and advising or training construction and operating personnel regarding the marking, storage, and location of all hazardous chemicals, including reporting accidental releases of these substances.

8.3.2   Onsite and Offsite Emergency Facilities

Onsite emergency facilities are discussed in Sections 6.1 and 6.3 of the EP (LES, 2005a). The Control Room will be the primary Emergency Operations Center (EOC) and will have current as-built drawings, procedures, and operational engineering information. The designated alternate EOC is the Security Building, which will be used depending on the nature and location of the emergency situation, or if the Control Room becomes uninhabitable. The Security Building will have the same documentation that is available to emergency responders, as is available in the primary EOC. An offsite location will be available, if necessary, and is described in the Safeguards Contingency Plan.

Offsite emergency support and equipment is discussed in Section 4.3, including fire, emergency medical services, and local law enforcement. Support is available from a number of offsite agencies, if necessary, as discussed in Section 4.3.

Section 6.4 of the EP (LES, 2005a) describes the emergency monitoring equipment that is available for personnel and area monitoring, which includes criticality accident alarms, personnel monitoring equipment, liquid effluent monitors, air monitors, hydrogen fluoride monitors, and a meteorological measurement system for wind speed, direction, and temperature.

Emergency equipment will be inventoried and tested on a quarterly basis, as discussed in Section 7.6 of the EP (LES, 2005a). Deficiencies identified will be reported to the Emergency Preparedness Manager, who will ensure timely corrective action is taken.

Section 6.2 of the EP (LES, 2005a) describes the communication systems that will be used at the facility, which includes the facility telephone system with facsimile capability, the public address system, alarms, and two-way radios that are powered by diesel-backed alternating current sources. The bandwidths of the radio systems will allow communication with the hospital, ambulance services, fire department, and state, county, and local law enforcement agencies and the New Mexico Department of Public Safety, Office of Emergency Management. The telephone system will be the primary means of offsite emergency communications.

8.3.3 Types of Accidents

In Section 2.1 of the EP (LES, 2005a), both postulated high- and intermediate-consequence events are identified. Accident sequences, as well as mitigating and preventive measures, are described. Nuclear criticality and loss of containment leading to a very large release of UF_6 are the only postulated events identified for which protective actions may be necessary. A consequence analysis has been performed for a nuclear criticality scenario and for the Blending Donor Station Heater Controller Failure/Heater Run-Away scenario. The results of these analyses are provided in Section 3.7.3 of the "Integrated Safety Analysis Summary" (LES, 2005b). This meets the requirements of 10 CFR 30.32(i)(3)(ii), 10 CFR 40.31(j)(3)(ii), and 10 CFR 70.22(i)(3)(ii).

8.3.4 Classification of Accidents

Section 3.1 of the EP (LES, 2005a) explains the system used to classify an emergency as either an Alert or a Site-Area Emergency, and defines both types of incidents. The applicant has established that the threshold for escalating an event from an Alert to a Site-Area Emergency is based on indications of a release (e.g., indications such as the presence of a white vapor cloud issuing from buildings that house UF_6) that could require a response by an offsite response organization, to protect persons offsite from reaching the offsite exposure limits set forth in 10 CFR 30.32(i)(1)(i), 10 CFR 40.31(j)(1)(i), and 10 CFR 70.22(i)(1)(i). This system for classifying events is acceptable to the staff and meets the requirements of 10 CFR 30.32(i)(3)(iii), 10 CFR 40.31(j)(3)(iii), and 10 CFR 70.22(i)(3)(iii). The processes for making the appropriate classification are provided in the Emergency Plan Implementing Procedures (EPIPs).

Table 3.1-1 in the EP (LES, 2005a) provides examples of site-specific incidents and the emergency classification that will be declared for each event. The Shift Manager is responsible for accident classification and assumes the responsibility of the Emergency Director until relieved by the Plant Manager or his designee.

8.3.5 Detection of Accidents

Section 2.2 of the EP explains the methods and systems available to detect accidents at the facility, including:

- Automatic seismic trip of ventilation systems and associated alarms;
- Criticality accident alarm system;
- Facility-wide fire detection system;
- Visual detection of uranyl fluoride (UO_2F_2) and odor of hydrogen fluoride (HF);
- HF monitors;
- Radiation monitors; or
- Other alarms that detect abnormal process conditions, including temperature, pressure, weight, etc.

Actions in response to accidents will be outlined in EPIPs and directed by the Shift Manager. The methods and systems presented in Section 2.2 of the EP (LES, 2005a) fulfill the requirements of 10 CFR 30.32(i)(3)(iv), 10 CFR 40.31(j)(3)(iv), and 10 CFR 70.22(i)(3)(iv).

8.3.6   Mitigation of Consequences

Section 5.3 of the EP (LES, 2005a) describes actions and equipment that will be used to mitigate the consequences of accidents at the facility. The applicant states that since the facility will operate with only natural and low-enriched uranium in the form of UF_6, there will be no radiological hazards that could likely result in significant offsite radiation doses. The major hazard would be the chemical hazard caused by a release of UF_6.

The main features used at the facility to mitigate the consequences of accidents include automatic interruption or termination of specific operations, fire detection and suppression systems, operator response to abnormal conditions/alarms, and shutdown of the ventilation system, in case of a UF_6 release or a criticality. The actions, features, and means for maintaining them, which are adequately described in Sections 5.3 and 7.6 of the EP (LES, 2005a), fulfill the requirements of 10 CFR 30.32(i)(3)(v), 10 CFR 40.31(j)(3)(v), and 10 CFR 70.22(i)(3)(v).

8.3.7   Assessment of Releases

Section 5.2 of the EP (LES, 2005a) explains the actions that will be taken to assess the extent of an accident at the facility. In case of an Alert, dose projections of offsite radiation and hazardous material exposures will be made and provided to offsite response agencies. In addition, during a Site-Area Emergency, radiation or chemical surveys of the Assembly Area(s), the EOC, and the facility will be performed. Environmental air sampling will be performed off-site, if necessary. During a UF_6 release, water, air, and soil samples will be obtained in the release path, and urine and fecal samples will be obtained from any workers exposed to the release. Projections of offsite radiation and chemical exposures will be based on the estimated amount of UF_6 released, the point of release, and the meteorological conditions at the time of the release, and will be performed using the Radiological Assessment System for Consequence Analysis (RASCAL) software (NRC, 2001). These actions meet the requirements of 10 CFR 30.32(i)(3)(vi), 10 CFR 40.31(j)(3)(vi), and 10 CFR 70.22(i)(3)(vi).

8.3.8   Responsibilities

The responsibilities of facility personnel during normal operations and during emergency situations are described in Sections 4.1 and 4.2, respectively, of the EP (LES, 2005a). In case of an emergency, the Shift Manager assumes the duty of the Emergency Director until the Plant Manager or designee arrives. Responsibilities of the Emergency Director include:

- Coordinating the response;
- Deciding to provide protective action recommendations to authorities responsible for offsite emergency measures (shall not be delegated);
- Coordinating the staff and offsite personnel who augment the staff;
- Approving information before release to the press;
- Authority to request support from offsite agencies;
- Authority to delegate responsibilities;
- Directing prompt notification of local and state emergency organizations;
- Directing assessment of onsite and offsite actual or potential consequences;
- Implementing protective actions for onsite personnel; and
- Downgrading/terminating the emergency.

Section 4.2.2 of the EP (LES, 2005a) summarizes the responsibilities of the remaining onsite staff. This includes the Community Relations Coordinator, who will have a direct line of communication to the Emergency Director, to ensure current and factual news releases. Other information concerning public and media access to information is included in Section 5.8 of the EP (LES, 2005a). Memoranda of Understanding have been established with local agencies and will be reviewed annually and renewed at least every 4 years, or more frequently, if necessary.

Responsibilities for developing, maintaining, and updating the EP are discussed in Section 8.3.14 of this SER. The description of the responsibilities in the EP fulfills the requirements of 10 CFR 30.32(i)(3)(vii), 10 CFR 40.31(j)(3)(vii), and 10 CFR 70.22(i)(3)(vii).

8.3.9   Notification and Coordination

As discussed in Section 8.3.4 of this SER, classification of emergencies is outlined in Section 3.1 of the EP (LES, 2005a) and is the responsibility of the Shift Manager.

Section 3.2 of the EP (LES, 2005a) provides a clear commitment to promptly notify offsite response organizations of an emergency, including notification of the NRC Operations Center immediately after calling the offsite response organizations, but no later than 1 hour after declaring an emergency. Sections 4.3 and 4.4 of the EP (LES, 2005a) provide an adequate description of provisions for assistance from offsite response organizations. Sections 5.6 and 5.7 of the EP (LES, 2005a) provide an adequate description of provisions for medical treatment of contaminated workers. These sections adequately describe the agreements held between LES and local off-site organizations and agencies, procedures for these response organizations access to the site, and equipment/services available from these organizations.

Section 5.2.4 of the EP (LES, 2005a) describes how projections of offsite radiation and chemical exposures are made, and how this information is disseminated to appropriate personnel, including State authorities and NRC. Section 5.4.2 of the EP (LES, 2005a) discusses the use of offsite field monitoring data which will be compared with source-term data and included in the protective-action recommendation process.

As discussed previously, the Community Relations Manager will have a direct line of communication to the Emergency Director, to ensure current and factual news releases. Also, mutual aid agreements are in place to provide additional support if other services or equipment are not available.

As discussed previously, it is the responsibility of the Emergency Director (initially the Shift Manager) to: (1) declare an alert or site-area emergency; (2) activate the onsite emergency response organization; (3) notify offsite response authorities of an emergency and recommended initial protective actions; (4) notify the NRC Operations Center; (5) decide what onsite protective actions to initiate; (6) decide what offsite protective actions to recommend; (7) decide to request support from offsite organizations; and (8) decide to terminate the emergency or enter recovery mode.

This meets the requirements of 10 CFR 30.32(i)(3)(viii), 10 CFR 40.31(j)(3)(viii), and 10 CFR 70.22(i)(3)(viii).

8.3.10 Information To Be Communicated

Section 3.3 of the EP (LES, 2005a) provides an adequate description of the type of information to be given to offsite response organizations during an emergency. The applicant will use Table 3.3-1, "National Enrichment Facility Notification Form," as a script to be used for initial notification of an emergency at the facility, to appropriate offsite facilities. Information included in Table 3.3-1 is contact information for state and local authorities; what type of event has occurred (Alert or Site-Area Emergency); a description of the event, time and date of the event; radiological conditions; meteorological data; and protective-action recommendations, if any.

In the event of a Site Area Emergency, a standard recommendation will be provided to offsite assistance organizations with more specific data, if available, as discussed in Section 3.3 of the EP (LES, 2005a).

This meets the requirements of 10 CFR 30.32(i)(3)(ix), 10 CFR 40.31(j)(3)(ix), and 10 CFR 70.22(i)(3)(ix).

8.3.11 Training

Section 7.2 of the EP (LES, 2005a) provides a description of the training the licensee will provide to workers on how to respond to an emergency. All workers receive general employee training, which includes Quality Assurance; radiation protection (including the use of dosimetry and protective clothing); and safety, emergency, and administrative procedures. Training on criticality safety, radiation protection, and emergency procedures specific to each type of job function is also provided under the nuclear safety training program.

Emergency response personnel receive additional training annually, to provide specific information on how the emergency organization responds during emergency conditions, including staffing; determining and estimating potential offsite releases of radiation and chemicals; and interface with offsite assistance organizations. This training is required before being assigned to the emergency organization, and refresher training is provided at least once every year.

The nuclear safety training program includes: (a) information on instructions to workers; (b) as low as is reasonably achievable (ALARA) methods of controlling radiation exposures; (c) contamination control methods; (d) use of monitoring equipment; (e) emergency procedures and actions; (f) nature and sources of radiation; (g) safe use of chemicals; (h) biological effects of radiation; (i) use of personnel monitoring devices; (j) principles of nuclear criticality safety; (k) risk to pregnant females; (l) radiation-protection practices; (m) protective clothing; (n) respiratory protection; and (o) personnel surveys. Specific topics, performance objectives, content, training schedules, and number of training hours required for each position are contained in administrative procedures.

The Emergency Preparedness Manager is responsible for emergency preparedness training. Individuals requiring unescorted access to the Controlled Access Area receive annual retraining. The Health, Safety, and Engineering Manager, or designee, reviews and updates the content of the formal nuclear safety training programs, as required, at least every 2 years to ensure that the programs are current and adequate.

Facility tours and classroom training are also provided to offsite response organizations. The training includes: (a) information concerning facility access control; (b) potential accident scenarios; (c) emergency action levels; (d) notification procedures; (e) exposure guidelines; (f) personnel-monitoring devices; (g) communications; (h) contamination control; and (i) the role of the offsite assistance organization in responding to an emergency at the facility. Each group will meet at least annually, with facility personnel, to accomplish this training and review items of mutual interest, including relevant changes to the program.

This fulfills the requirements of 10 CFR 30.32(i)(3)(x), 10 CFR 40.31(j)(3)(x), and 10 CFR 70.22(i)(3)(x).

8.3.12   Safe Shutdown (Recovery and Facility Restoration)

Section 9.1 of the EP (LES, 2005a) states that during an emergency, immediate actions will be directed toward limiting the consequences of the incident to afford maximum protection to facility personnel and the general public. Once control of the facility has been reestablished, a systematic and planned approach to full facility recovery will be taken. Criteria to be considered before terminating an emergency and entering the recovery phase include:

- Conditions requiring emergency classification no longer exist;
- Radioactive and/or hazardous chemical releases are under control;
- Environmental assessment activities in progress are those necessary to determine the extent of environmental impact from the emergency, not the active tracking of material still being released;
- Facility radiation, contamination, and hazardous chemical levels are stable or decreasing and acceptable, given current facility conditions;
- Any fire, flood, earthquake or similar initiating emergency event no longer exists;
- All required notifications have been made;
- At an Alert or Site-Area Emergency, the facility Operations Center is staffed and activated; and
- Offsite conditions do not unreasonably limit access of outside support to the facility.

Sections 5.2, 5.3, and 5.4 of the EP (LES, 2005a) describe the methods to be used for assessing the extent of the event and the status of the facility, and the mitigative actions necessary to reduce or stop any ongoing releases of radioactive material or hazardous chemicals.

Section 9.2 of the EP (LES, 2005a) contains information regarding the staffing of the facility during the recovery phase of an event. It addresses minor events requiring only the normal on-shift personnel, as well as more significant events, which would require the Emergency Organization (EO), or portions of the EO, to perform recovery tasks. This fulfills the requirements of 10 CFR 30.32(i)(3)(xi), 10 CFR 40.31(j)(3)(xi), and 10 CFR 70.22(i)(3)(xi).

8.3.13   Exercises and Drills

Section 7.3 of the EP (LES, 2005a) provides adequate provisions for drills, exercises, and biennial exercises that are used to test the adequacy of EPIPs, emergency equipment, instrumentation, and to ensure all emergency response personnel are familiar and proficient with their duties. The following guidelines are used to control and evaluate drill and exercise conduct:

- Drill and exercise performance is assessed against specific scenario objectives, using postulated accidents that adequately test personnel, equipment, and resources, including any previously identified weaknesses;
- Participant, controller, evaluator, and observer pre-drill/exercise briefings are conducted;
- Controllers who maintain the timeline but do not interfere with the EO's response, except where safety considerations are concerned, provide scenario data and exercise messages;
- Trained evaluators and independent observers are used to identify and record participant performance, scenario strengths and deficiencies, equipment problems, and to provide recommendations for improvement;
- The pre-staging of equipment and personnel is minimized to realistically test the activation and staffing of emergency facilities; and
- Post-drill and post-exercise critiques will be conducted by those involved, and appropriate improvements will be implemented.

Each exercise is critiqued according to Section 7.4 of the EP (LES, 2005a). Areas evaluated include the adequacy of the EP, procedures, equipment, facilities, personnel training, and overall response effectiveness. The Emergency Preparedness Manager is responsible for tracking deficiencies and ensuring that corrective actions are implemented and effective.

The drill and exercise program is established to demonstrate:

- Effectively use available resources to control the site, and to obtain additional resources when necessary;
- Mitigate further damage to the facility;
- Control any radiological or hazardous material releases;
- Perform required onsite activities under simulated elevated radiation/hazardous material airborne conditions;
- Accurately assess the facility's status during emergency conditions;
- Initiate recovery;
- Demonstrate personnel protection measures both onsite and offsite;
- Control and minimize hazards to individuals during fires, medical emergencies, mitigation activities, and search and rescue operations;
- Effectively use communication systems to notify offsite agencies and to support emergency response activities; and
- Disseminate accurate, reliable, timely, and understandable information.

Offsite organizations are invited to participate in the biennial exercise and NRC is invited to participate or observe. Exercise objectives and scenarios will be submitted to NRC, for review and comment, at least 60 days before the exercise.

Section 7.3.2 of the EP (LES, 2005a) includes an adequate provision for quarterly communications checks, to verify the operability of initial notification points. The Emergency Preparedness Manager verifies and updates emergency telephone numbers contained in the facilities EPIPs.

The information provided in Sections 7.3 and 7.4 of the EP (LES, 2005a) meets the requirements of 10 CFR 30.32(i)(3)(xii), 10 CFR 40.31(j)(3)(xii), and 10 CFR 70.22(i)(3)(xii).

8.3.14 Responsibilities for Developing and Maintaining the Emergency Program and Its Procedures Current

Section 7.1 of the EP (LES, 2005a) explains there will be site procedures for maintaining the EP and procedures current. As outlined in Section 7.5 of the EP (LES, 2005a), an annual audit of the EPIPs will be conducted by an independent group, to ensure the continued effectiveness of the EP. Administrative procedures are established which assign responsibility for the development, review, approval, and update of the Emergency Plan and its supporting procedures.

Any proposed change in the EP that affects an offsite organization will be provided to that organization for review and comment at least 60 days before the change being implemented, unless mutually agreed otherwise. The applicant may incorporate changes to the EP without receiving prior NRC approval, provided those changes do not decrease the effectiveness of the EP, and NRC and affected offsite response organizations receive copies of the revised EP or procedures within 6 months, in accordance with 10 CFR 30.34(f), 10 CFR 40.35(f), and 10 CFR 70.32(i).

8.4 EVALUATION FINDINGS

The staff has evaluated the EP for the facility. In accordance with 10 CFR 30.32(i), 10 CFR 40.31(j), and 10 CFR 70.22(i), the licensee has established an EP for responding to the radiological hazards resulting from a release of radioactive material or hazardous chemicals incident to the processing of licensed material. The NRC staff reviewed the EP with respect to 10 CFR 30.32(i), 10 CFR 40.31(j), and 10 CFR 70.22(i) and the acceptance criteria in Section 8.4.3 of NUREG-1520 (NRC, 2002). It determined that the applicant's EP is adequate to demonstrate compliance with 10 CFR 30.32(i), 10 CFR 40.31(j), and 10 CFR 70.22(i), in that: (1) the facility is properly configured to limit releases of radioactive materials in case of an accident; (2) a capability exists for measuring and assessing the significance of accidental releases of radioactive materials; (3) appropriate emergency equipment and procedures are provided onsite, to protect workers against radiation and other chemical hazards that might be encountered after an accident; (4) a system has been established to notify Federal, State, and local Government agencies and to recommend appropriate protective actions to protect members of the public; and (5) necessary recovery actions are established to return the facility to a safe condition after an accident.

8.5 REFERENCES

(LES, 2005a) Louisiana Energy Services (LES). "National Enrichment Facility Emergency Plan," Revision 3, 2005.

(LES, 2005b) Louisiana Energy Services (LES). "National Enrichment Facility Integrated Safety Analysis Summary," Revision 4, 2005.

(NRC, 1992) U.S. Nuclear Regulatory Commission (NRC). Regulatory Guide 3.67, "Standard Format and Content for Emergency Plans for Fuel Cycle and Materials Facilities," 1992.

(NRC, 2001) U.S. Nuclear Regulatory Commission (NRC). NUREG-1741, "RASCAL 3.0: Description of Models and Methods," 2001.

(NRC, 2002) U.S. Nuclear Regulatory Commission (NRC). NUREG-1520, "Standard Review Plan for the Review of a License Application for a Fuel Cycle Facility," 2002.

9.0 ENVIRONMENTAL PROTECTION

The purpose of the U.S. Nuclear Regulatory Commission's (NRC's) review of the applicant's environmental protection plan is to determine whether the applicant's proposed environmental protection measures are adequate to protect the environment, and the health and safety of the public, as required by 10 CFR Parts 20, 30, 40, 51, and 70.

9.1   REGULATORY REQUIREMENTS

To be considered acceptable, the applicant must satisfy the following regulatory requirements regarding environmental protection:

1.   Part 20 specifies the effluent control and treatment measures necessary to meet the dose limits and dose constraints for members of the public specified in Subparts B, D, and F; the survey requirements of Subpart F; the waste disposal requirements of Subpart K; the records requirements of Subpart L; and the reporting requirements of Subpart M.

2.   10 CFR 30.33 specifies in part that an application for the possession and use of byproduct material will be granted provided that, among other things, the applicant's proposed equipment and facilities are adequate to protect health and minimize danger to life or property, and that the applicant is qualified by training and experience to use the byproduct material for the purpose requested in such a manner as to protect health and minimize danger to life and property.

3.   10 CFR 40.31(k) states that, "A license application for a uranium enrichment facility must be accompanied by an Environmental Report required under subpart A of Part 51 of this chapter."

4.   10 CFR 40.32(e) states that, "In the case of an application for a license for a uranium enrichment facility, or for a license to possess and use source and byproduct material for uranium milling, production of uranium hexafluoride, or for the conduct of any other activity which the Commission determines will significantly affect the quality of the environment, the Director of Nuclear Material Safety and Safeguards or his designee, before commencement of construction of the plant or facility in which the activity will be conducted, on the basis of information filed and evaluations made pursuant to subpart A of part 51 of this chapter, has concluded, after weighing the environmental, economic, technical and other benefits against environmental costs and considering available alternatives, that the action called for is the issuance of the proposed license, with any appropriate conditions to protect environmental values. Commencement of construction prior to this conclusion is grounds for denial of a license to possess and use source and byproduct material in the plant or facility. As used in this paragraph, the term 'commencement of construction' means any clearing of land, excavation, or other substantial action that would adversely affect the environment of a site. The term does not mean site exploration, roads necessary for site exploration, borings to determine

foundation conditions, or other preconstruction monitoring or testing to establish background information related to the suitability of the site or the protection of environmental values."

5. Part 51 specifies that the applicant must submit an environmental report for construction and operation of a uranium enrichment facility, as required by 10 CFR 51.60(b)(1)(vii).

6. 10 CFR 70.22(a)(7) specifies that the applicant must provide a description of the equipment and facilities which will be used by the applicant to protect health and minimize danger to life or property.

7. 10 CFR 70.59 outlines the radiological effluent monitoring reporting requirements for a Part 70 licensee.

8. 10 CFR 70.65(b) specifies that an applicant for a facility must provide an "Integrated Safety Analysis (ISA) Summary" that includes a list of the Items Relied on for Safety (IROFS) established by the applicant.

9.2   REGULATORY ACCEPTANCE CRITERIA

The acceptance criteria for NRC's review of the applicant's environmental protection program are outlined in Section 9.4.3.2 of NUREG-1520 (NRC, 2002).

9.3   STAFF REVIEW AND ANALYSIS

9.3.1   Radiation Safety

9.3.1.1   As Low As is Reasonably Achievable (ALARA) Goals for Air and Liquid Effluent Control

Air Effluent ALARA Goal

The applicant estimated the maximum individual committed effective dose equivalent (CEDE) for air effluents during normal operations at the proposed facility. The applicant estimated that the total air effluent releases would be less than 9 megabecquerel (MBq) [240 microcuries (µCi)] of uranium-234 (U-234), uranium-235 (U-235), and uranium-238 (U-238) isotopes – the principal constituents of natural, depleted, and enriched uranium that would be processed at the facility. Uranium-236 (U-236) is not considered to be a principal constituent in that it contributes significantly less than 1 percent to the total releases. Similar Urenco facilities in Europe typically maintain quantities of radioactivity in air effluent less than 10 grams (g) [0.35 ounces (oz)] (LES, 2005b). Even if all of the 10 g (0.35 oz) of uranium were conservatively assumed to be composed of 6 percent enriched uranium, then air effluent from Urenco facilities is generally limited to no more than 1.1 MBq (30 µCi) of uranium isotopes. Therefore, the staff finds that a value of 9 MBq (240 µCi) uranium represents a reasonably conservative upper bound on the annual quantity of air effluent from the facility.

As noted in Section 8.7 of the applicant's ER, the CEDE to the maximally exposed member of the public located at the south side of the controlled area boundary, resulting from the release to the atmosphere of 9 MBq (240 µCi) of uranium from gaseous release points, would be less than 0.17 microsieverts (µSv) (0.017 mrem) (LES, 2005b), or 0.017 percent of the 1 mSv (100-mrem) limit on dose to the public in Part 20. The estimated maximum public dose is also well below the 0.1 mSv (10-mrem) ALARA constraint on air emissions described in 10 CFR 20.1101 (e.g., between 1 and 2 percent).

NUREG-1520 provides that the applicant's radiological monitoring will be acceptable if, among other things, action levels of radionuclide concentrations are selected such that below those concentrations, doses to the public will be ALARA and will be below the limits specified in 10 CFR Part 20, Subpart B. In Section 6.1.1 of the ER (LES, 2005b), the applicant committed to develop a program of corrective actions that will be implemented based on established action levels that are sufficiently low to permit implementation of corrective actions before regulatory limits are exceeded. As proposed by the applicant, action levels are divided into three priorities:

1. The sample parameter lowest alarm point will be selected based on conditions of service, for example, 3 times the normal background level;

2. The sample parameter exceeds any of the existing administrative limits; and

3. The sample parameter exceeds any regulatory limit.

The applicant will implement corrective actions to ensure that: (a) the cause for the action-level exceedance can be identified and immediately corrected; (b) applicable regulatory agencies are notified, if required; (c) lessons-learned are communicated to appropriate personnel; and (d) applicable procedures are revised accordingly, if needed.

A reasonable initial ALARA goal for air effluents, described in NRC Regulatory Guides 8.34 and 8.37 (NRC, 1992; NRC, 1993) is 10-20 percent of the regulatory limit. The applicant will implement corrective actions, under Priority 1, above, that will likely result in doses that are a small fraction of the regulatory limits in Part 20 and 40 CFR Part 190. In addition, the calculated dose to the maximally exposed member of the public is a small fraction of the ALARA goals identified in the above-referenced Regulatory Guides. Therefore, staff finds this initial estimate of air effluent quantities to be a reasonable ALARA goal for air effluent.

Liquid Effluent ALARA Goal

The applicant estimated the maximum individual CEDE for liquid effluents during normal operations at the proposed facility. The applicant estimated maximum annual quantity of radiological material in liquid effluent would be 14.4 MBq (390 µCi) of uranium as discharge to the Treated Effluent Evaporative Basin (TEEB) (LES, 2005b). There are no offsite releases to any surface waters or publicly owned treatment works. Therefore, the release pathway the applicant assumed is airborne resuspension of particulate matter from the bottom of the basin, if waste water should evaporate. The applicant evaluated the dose for the 30th year of operations, which represents a realistically conservative upper bound on the amount of uranium available for resuspension and atmospheric dispersion.

As noted in Section 8.7 of the applicant's ER, potential radiological impacts from operation of the proposed facility would result from controlled releases of small quantities of uranium hexafluoride (UF_6) during normal operations. The CEDE to the maximally exposed member of the public located at the south side of the controlled area boundary, resulting from the annual release to the atmosphere of 2.1 MBq (56 µCi) of uranium from the TEEB, would be less than 0.017 µSv (0.0017 mrem) per year if the TEEB is dry only 10 percent of the year (LES, 2005b). If the TEEB is dry the entire year, the CEDE to the maximally exposed member of the public located at the south side of the controlled area boundary would be less than 0.17 µSv (0.017 mrem) per year. The estimated maximum public dose is also well below the 0.1 mSv (10-mrem) ALARA constraint on liquid emissions described in 10 CFR 20.1101 (e.g., less than 2 percent).

In Section 6.1.1 of the ER (LES, 2005b), the applicant committed to develop a program of corrective actions that will be implemented based on established action levels that are sufficiently low to permit implementation of corrective actions before regulatory limits are exceeded. As proposed by the applicant, action levels are divided into three priorities:

1. The sample parameter lowest alarm point will be selected based on conditions of service, for example, 3 times the normal background level;

2. The sample parameter exceeds any of the existing administrative limits; and

3. The sample parameter exceeds any regulatory limit.

The applicant will implement corrective actions to ensure that: (a) the cause for the action-level exceedance can be identified and immediately corrected; (b) applicable regulatory agencies are notified, if required; (c) lessons-learned are communicated to appropriate personnel; and (d) applicable procedures are revised accordingly, if needed.

A reasonable initial ALARA goal for liquid effluent described in NRC Regulatory Guide 8.34 (NRC, 1992) is 10-20 percent of the regulatory limit. The applicant committed to implement corrective actions under Priority 1, above, that will likely result in doses that are a small fraction of the regulatory limits in Part 20 and Part 190. In addition, the calculated dose to the maximally exposed member of the public is a small fraction of the ALARA goals identified in the above referenced Regulatory Guide. The applicant commits to implement corrective actions well below this level. Therefore, staff finds this initial estimate of liquid effluent to be a reasonable ALARA goal for liquid effluent.

9.3.1.2   Air Effluent Controls to Maintain Public Doses ALARA

Air effluent controls at the proposed facility include: (a) the Separations Building Gaseous Effluent Vent System (GEVS); (b) the Technical Services Building (TSB) GEVS; (c) the TSB Heating, Ventilating, and Air Conditioning (HVAC) system that services potentially contaminated areas; and (d) the Centrifuge Test and Post-Mortem Facilities Exhaust Filtration System.

GEVS

The function of each GEVS is to remove particulate matter containing uranium and hydrogen fluoride (HF) from potentially contaminated gas streams. Each GEVS includes ducts; prefilters;

high-efficiency particulate air (HEPA) filters; potassium carbonate-impregnated activated carbon filters; fans; monitors and controls; inlet and outlet isolation dampers; and a discharge stack. As described below, one GEVS unit includes electrostatic filters to remove oil vapor from gaseous effluent.

The TSB GEVS contains one filter station and one fan that accommodates 100 percent of the effluent. The filter train includes a 85-percent efficient prefilter; a 99.97-percent-efficient HEPA filter; and a 99-percent-efficient activated charcoal filter for removal of HF. Cleaned air effluent is discharged through a roof-top vent stack.

The Separation Building GEVS contains two 100-percent-capacity filter/fan trains, with one train in a standby configuration. Like the TSB GEVS, each Separation Building GEVS filter train is composed of an 85-percent-efficient prefilter and a 99.97-percent-efficient HEPA filter. The activated charcoal filter for removal of HF is rated to 99-percent efficiency. The exhaust of the vacuum pump/trap sets is also routed through an electrostatic filter with a removal efficiency of 97 percent. The operational and standby filter/fan trains share the same roof-top vent stack.

TSB HVAC - Potentially Contaminated Areas

A portion of the TSB HVAC services potentially contaminated rooms in the TSB. These rooms are the Decontamination Workshop, Cylinder Preparation Room, and Ventilated Room. The HVAC system for these rooms consists of two 50-percent-capacity Air-Handling Units. Airflow from the potentially contaminated rooms is exhausted through two 50-percent-capacity Bag-In/Bag-Out HEPA filters by one of two 100-percent-capacity filtration exhaust fans. The exhaust air is then discharged to the exhaust stack and monitored for alpha radiation and HF. Some potentially contaminated areas have fume hoods that are connected to the TSB GEVS.

Centrifuge Test and Post-Mortem Facilities Exhaust Filtration System

The Centrifuge Test and Post-Mortem Facilities Exhaust Filtration System consists of a duct network connected to one filter station and one of two 100-percent-capacity fans. One fan will normally be in standby. The Centrifuge Test and Post-Mortem Facilities Exhaust Filtration System filter train consists of: (a) an 85-percent-efficient prefilter; (b) a 99-percent-efficient activated charcoal filter for removal of HF; and (c) a 99.97-percent-efficient HEPA filter. Cleaned air effluent is discharged through a roof-top vent stack on the Centrifuge Assembly Building.

Staff Evaluation of Air Effluent Controls

The staff evaluated the air effluent controls, including the TSB and Separation Building GEVS, the TSB HVAC systems, and the Centrifuge Test and Post-Mortem Facilities Exhaust Filtration System, the placement of air effluent controls at the proposed facility, and the type of controls selected for the hazards that are present in various processes. The applicant's proposed use of HEPA filters is standard industry practice for control of air effluents that potentially contain airborne particle matter. Also, the use of activated charcoal for capture and control of HF is well-established. The applicant will also prepare and maintain operational procedures that limit activities with dispersible forms of uranium to areas with appropriate air effluent controls. In some cases, these designated areas will also be fitted with local control devices (flexible hoses attached to a GEVS, hoods, and glove boxes). Therefore, based on the above, the staff finds that the applicant has demonstrated that its air effluent controls will reduce releases to assure adequate protection of the environment and of health and safety of the public.

9.3.1.3   Liquid Effluent Controls to Maintain Public Doses ALARA

Liquid effluent controls at the proposed facility will be located in the Liquid Effluent Collection and Treatment System in the TSB. The applicant has identified seven major sources of liquid waste from processes in the TSB and Separations Building. The system is designed to use combinations of in-tank precipitation, centrifuge-based oil/water separation, a filter press, an evaporator/dryer, and mixed-bed demineralizers, to decontaminate liquid wastes before evaporation of decontaminated aqueous wastes in the outdoor TEEB.

The proposed liquid effluent control systems will reduce the quantity of uranium entering the TEEB by about two orders of magnitude, from 56.7 kilograms (kg) (125 lbs) to less than 0.6 kg (1.3 lbs) per year. The uranium removed by the liquid effluent control systems inside the TSB will be disposed as Class A low-level radioactive waste (LLW) at one or more licensed disposal facilities. As described in the staff's Safety Evaluation Report (SER) Section 9.3.1.1, the dose to the maximally exposed offsite member of the public from the resuspension and atmospheric dispersion of uranium-bearing dust from a dry TEEB is less than 0.17 µSv (0.017 mrem) per year. This dose is less than one percent of the regulatory limit. A reasonable initial ALARA goal for liquid effluent described in NRC Regulatory Guide 8.34 (NRC, 1992) is 10-20 percent of the regulatory limit. Therefore, based on the above, the staff finds that the applicant has demonstrated that it will reduce releases to adequately assure protection of the environment and of health and safety of the public.

9.3.1.4   ALARA Reviews and Reports to Management

In Section 4.2 of its Safety Analysis Report (SAR), the applicant describes an ALARA program for the proposed facility. The ALARA program would include annual review of the content and implementation of the radiation protection program, including the effluent control program. The Radiation Protection Manager is responsible for implementing the ALARA program and ensuring that adequate resources are committed to make the program effective. The Radiation Protection Manager prepares an annual ALARA program evaluation report. The report reviews: (1) radiological exposure and effluent release data for trends; (2) audits and inspections; (3) use, maintenance, and surveillance of equipment used for exposure and effluent control; and (4) other issues, as appropriate, that may influence the effectiveness of the radiation protection and ALARA programs. Copies of the report are submitted to the Plant Manager and the Safety Review Committee. As described more fully in chapter 4 of this SER, the ALARA program is acceptable to the staff.

9.3.1.5   Waste Minimization

In Section 4.13.4 of the ER (LES 2005b), the applicant described facility features and systems that will minimize the generation of radioactive waste. These features and systems are based on principles of control, conservation, reprocessing, and recovery. Specific examples include: (a) the Fomblin Oil Recovery System; (b) the Decontamination System; (c) a General Decontamination process; (d) Sample Bottle Decontamination System; (e) Flexible Hose Decontamination System; and (f) the Laundry System. These systems are consistent with the guidance provided in NRC Information Notice 94-23, "Guidance to Hazardous, Radioactive, and

Mixed-Waste Generators on the Elements of a Waste Minimization Program" (NRC, 1994b) and are, therefore, acceptable to the staff.

9.3.1.6   Safe Handling of Radioactive Wastes

The applicant has identified design features and procedures for safe handling of air and liquid effluent and solid wastes from both construction and operation of the proposed facility. The staff's evaluation of air and liquid effluent is described in SER Sections 9.3.1.2 and 9.3.1.3.

The applicant has described a solid waste management program at the proposed facility for industrial (nonhazardous), radioactive and mixed, and hazardous wastes. In addition, the applicant will segregate solid radioactive and mixed waste according to the quantity of liquid that is not readily separable from the solid material. The applicant does not propose to process and/or treat solid waste.

The applicant proposes to dispose of all solid radioactive wastes as Class A LLW. Industrial waste, including miscellaneous trash, vehicle air filters, empty cutting oil cans, miscellaneous scrap metal, and paper will be shipped offsite for minimization and then sent to a permitted waste landfill.

Radioactive waste will be collected in labeled containers in each Restricted Area and transferred to the Radioactive Waste Storage Area for inspection. Suitable waste will be volume-reduced and all radioactive waste disposed of at a licensed LLW disposal facility.

Hazardous wastes (e.g., spent blasting sand; empty spray-paint cans; empty propane-gas cylinders; solvents such as acetone and toluene; degreaser solvents; diatomaceous earth; hydrocarbon sludge; and chemicals such as methylene chloride and petroleum ether) and some mixed wastes will be generated at the facility. These wastes will also be collected at the point of generation, transferred to the Waste Storage Area, inspected, and classified. Any mixed waste that may be processed to meet land disposal requirements may be treated in its original collection container and shipped as LLW for disposal.

As noted in Section 3 of the applicant's ER, the operation of the facility would yield an annual production of 625 cylinders of depleted UF_6 per year, or approximately 7800 metric tons (8,600 tons). The Uranium Byproduct Cylinder (UBC) Storage Pad would have a capacity of 15,727 cylinders. The NRC has evaluated the environmental impacts of alternatives for disposition of depleted UF_6 in its "Environmental Impact Statement for the Proposed National Enrichment Facility in Lea County, New Mexico" (NRC, 2005). During temporary storage of this material on the UBC Storage Pad, the applicant has committed to a number of mitigation measures to minimize public and occupational health impacts. Among these measures is a cylinder management program to monitor storage conditions on the UBC Storage Pad; to monitor cylinder integrity by conducting routine inspections for breaches; and to perform cylinder maintenance, as needed. The UBCs will be stored on saddles made of materials that will not cause corrosion of the cylinders. The storage array of the UBCs will permit easy visual inspection of the cylinders. The cylinders will be surveyed for external contamination before being placed on the UBC Storage Pad and before being transported offsite. In addition, the UBCs will be re-inspected annually for damage or surface coating defects. These inspections are to insure that the cylinders are free from bulges; dents; gouges; cracks; or significant corrosion, and that the cylinder plugs are undamaged and not leaking. If significant deterioration is detected, then the contents of the cylinder will be transferred to another cylinder. This will be followed by a root cause assessment of any significant deterioration.

On the basis of this analysis, the staff find that the applicant's implementation of its program for management of solid radiological and nonradiological wastes related to facility operation will reduce unnecessary exposures to adequately assure protection of the environment and of health and safety of the public.

9.3.2 Effluent and Environmental Monitoring

9.3.2.1 Air Effluent Monitoring

As described in SER section 9.3.1.1, the staff finds that the expected concentrations of radioactive materials in airborne effluents would be well below the regulatory limits specified in 10 CFR 20.1302(c). The applicant has proposed to demonstrate compliance with air effluent limits by calculation of the total effective dose equivalent to the individual who is likely to receive the highest dose. Such a demonstration of regulatory compliance is in accordance with 10 CFR 20.1302(b)(1).

The staff reviewed the applicant's assumptions and conclusions used in its calculations (ER Sections 4.12 and 8.7) and determined that they are reasonable. The staff's evaluation of radiation exposures from normal operations is found in Chapter 4 (section 4.2.12.2) of the Final Environmental Impact Statement (EIS).

The applicant has identified all airborne effluent discharge locations, and will monitor discharges from potentially contaminated areas in accordance with NRC Regulatory Guide 4.16 (NRC, 1985). The applicant will perform continuous monitoring at the discharge stacks for alpha-emitting radioactivity and HF. The applicant will also conduct periodic grab sampling of areas that contain dispersible uranium, but which are not expected to have airborne contamination. Sampling of each contributing source will not be part of the effluent monitoring design. However, each potentially contaminated area will be monitored by health physics surveys for airborne contamination, should such surveys be required for effective process and effluent control.

A list of HVAC systems, GEVS, and exhaust filtration systems that discharge building ventilation exhaust is provided in Table 9.3-1.

Weekly samples of stack air effluent will be measured for gross alpha radioactivity. Weekly samples will be composited quarterly for isotope-specific analyses for U-234, U-235, U-236, and U-238.

The applicant has proposed measurement sensitivities of 10^{-15} µCi/mL for weekly gross alpha measurements and 10^{-17} µCi/mL for uranium isotopes.

As noted in section 9.3.1.1 above, the applicant committed to developing a program of corrective actions to be taken when established action levels of radiation are exceeded for any of the measured parameters. As proposed by the applicant, action levels are divided into three priorities:

Table 9.3-1
Building Ventilation Air Discharge Locations

Process Area	No. of Air-Handling Units (Capacity, each)	Estimated Exhaust Flow	Room Pressurization	Controls	Stack	Monitoring
Cascade Hall (Each of 3 modules)	None	None	Ambient	None	No	No
Process Service Corridor (Each of 3 modules)	8 (12.5%)	34,300 cfm	Neutral	None	No	Continuous α, HF
Link Corridor (Each of 3 modules)	2 (50%)	3,700 cfm	Neutral	None	No	Continuous α, HF
Above Cascade Hall (Each of 3 modules)	4 (100%)	123,000 cfm	Neutral	None	No	Continuous α, HF
UF₆ Handling Area (Each of 3 modules)	3 (33%)	90,600 cfm	Neutral	None	No	Continuous α, HF
Blending & Liquid Sampling Workshop	2 (50%)	34,000 cfm	Neutral	None	No	Continuous α, HF
TSB HVAC Contaminated Areas (Decon. Workshop, Cylinder Prep. and Vent. Room)	2 (50%)	38,100 cfm	Negative	Roughing filter Carbon filter HEPA filter	Yes	α, HF
TSB HVAC Non-contaminated Areas	2 (50%)	38,100 cfm	Neutral	None	No	No
Vacuum Pump Rebuild and ME&I Workshops	(1 exhaust fan) (100%)	38,100 cfm	Neutral	None	No	No

Process Area	No. of Air-Handling Units (Capacity, each)	Estimated Exhaust Flow	Room Pressurization	Controls	Stack	Monitoring
Chem./Mass Spect.& Environmental Monitoring Lab	1 (100%)	38,100 cfm	Pos. (EM Lab) Neg. (others)	None	No	Quarterly grab
Centrifuge Assembling Building Test and Post Mortem Facility	2 (50%)	16,100 cfm	Neutral / Neg.	Roughing filter Carbon filter HEPA filter	Yes	α, HF
Cylinder Receipt and Dispatch Building	10 (10%)	288,000 cfm	Neutral	None	No	Quarterly grab
TSB GEVS	N/A	11,000 cfm	Neg.	Roughing filter Carbon filter HEPA filter	Yes	α, HF
Separation Building GEVS	N/A	6,500 cfm	Neg.	Roughing filter Carbon filter HEPA filter	Yes	α, HF

1. The sample parameter lowest alarm point will be selected based on conditions of service, for example, 3 times the normal background level;

2. The sample parameter exceeds any of the existing administrative limits; and

3. The sample parameter exceeds any regulatory limit.

The program of corrective actions will be implemented to ensure that: (a) the cause for the action-level exceedance can be identified and immediately corrected; (b) applicable regulatory agencies are notified, if required; (c) lessons-learned are communicated to appropriate personnel; and (d) applicable procedures are revised accordingly, if needed.

As described in Section 1.3 of the applicant's SAR, in addition to meeting NRC requirements, the applicant will also obtain required Federal and State permits for hazardous air pollutants, as required.

As described in Section 6.1.2 of the applicant's ER, the applicant's reporting procedures comply with the requirements of 10 CFR 70.59 and the guidance specific in Regulatory Guide 4.16 (NRC, 1985).

On the basis of this analysis, including a review of specific applicant commitments noted above, the staff finds that the applicant's air effluent monitoring during operation of the facility will ensure that concentrations of radioactivity in air effluent are below the regulatory limits in Part 20.

9.3.2.2   Liquid Effluent Monitoring

As described in SER section 9.3.1.2, the expected concentrations of radioactive materials in liquid effluents are ALARA and are well below the limits specified in 10 CFR 20, Appendix B Table 2. The applicant will demonstrate compliance with liquid effluent limits by calculation of the total effective dose equivalent to the individual who is likely to receive the highest dose in accordance with 10 CFR 20.1302(b)(1).

The staff reviewed the applicant's assumptions and conclusions used in its calculations (ER Sections 4.12 and 8.7) and determined that they are reasonable. The staff's evaluation of radiation exposures from normal operations is found in section 4.2.12.2 of the Final EIS.

The applicant has identified one liquid effluent discharge location, the TEEB. Treated liquid effluent will be discharged in batches to the TEEB. The Liquid Effluent Treatment System Monitor Tank, which holds treated effluent before discharge, will be sampled and analyzed for uranium isotope concentrations before it is discharged to the TEEB. The sample from this tank is representative by virtue of continuous mixing with mechanical agitators or recirculation. The applicant has proposed measurement sensitivities of 3×10^{-9} µCi/mL for uranium isotopes in liquid discharges. The applicant has proposed action levels that are divided into three priorities (see SER 9.3.2.1).

The UBC Storage Pad stormwater retention pond for the site have the potential to contain small amounts of radioactivity as a result of runoff from the UBC Storage Pad. The applicant proposes to include sampling and analysis of water and sediment from each of the stormwater runoff retention-detention basins in the Radiological Effluent Monitoring Program.

Table 9.3-2 is a summary of the air and liquid effluent sampling points proposed by the applicant from the Radiological Effluent Monitoring Program.

Table 9.3-2
Summary of Air and Liquid Effluent Sampling
Points in the Radiological Effluent Monitoring Program

Effluent Type	Sample Location	Sample Type	Analysis	Frequency
Air	Separation Building GEVS Stack	Continuous, Particulate Matter	Gross Alpha/Beta	Weekly
			U-234, 235, 236, 238	Quarterly Composite
Air	TSB GEVS Stack	Continuous, Particulate Matter	Gross Alpha/Beta	Weekly
			U-234, 235, 236, 238	Quarterly Composite
Air	TSB HVAC Stack	Continuous, Particulate Matter	Gross Alpha/Beta	Weekly
			U-234, 235, 236, 238	Quarterly Composite
Air	Centrifuge Test and Post-Mortem Facilities Exhaust Filtration System Stack	Continuous, Particulate Matter	Gross Alpha/Beta	Weekly
			U-234, 235, 236, 238	Quarterly Composite
Air	Process Areas	Continuous, Particulate Matter	Gross Alpha/Beta	Weekly
			U-234, 235, 236, 238	Quarterly Composite
Air	Non-Process Areas	Continuous, Particulate Matter	Gross Alpha/Beta	Quarterly
Liquid	Monitor Tank	Representative Grab Sample	U-234, 235, 236, 238	Before Discharge

As described in Section 1.3 of the applicant's ER (LES, 2005b), in addition to meeting NRC regulatory requirements, the applicant has committed to obtaining required Federal and State permits relating to liquid discharges (primarily related to groundwater protection), as required. For example, the applicant will obtain a discharge permit for use of a septic system at the Proposed facility. The permit is required by New Mexico Water Quality Control Commission Regulations, under Section 20.6.2.3.3104 of the New Mexico Administrative Code. In addition to having direct permitting authority, the State of New Mexico also has oversight responsibility for EPA water permits granted through the New Mexico Water Quality Bureau.

As described in Section 6.1.2 of the applicant's ER, the applicant has developed reporting procedures that when implemented will assure compliance with the regulatory requirements of 10 CFR 70.59 and the guidance provided in Regulatory Guide 4.16 (NRC, 1985).

On the basis of this analysis, the staff finds that the applicant's liquid effluent monitoring during operation of the facility ensures that concentrations of radioactivity in liquid effluent will be below the limits in Part 20.

9.3.2.3 Laboratory Quality Control

The applicant has a quality assurance program that includes controlled written procedures for sample collection, lab analysis, chain of custody, reporting of results, and corrective actions. An on-site laboratory and any contractor laboratory used to analyze samples will participate in third-party laboratory inter-comparison programs appropriate to the media and analytes being measured. Examples the applicant cites are the Mixed Analyte Performance Evaluation Program and the U.S. Department of Energy (DOE) Quality Assurance Program that DOE administers. All radiological and non-radiological laboratory vendors will be certified by the National Environmental Laboratory Accreditation Conference, or an equivalent State laboratory accreditation agency, for the analytes being tested.

The staff finds that these procedures are adequate to validate the analytical results produced by the Radiological Environmental Monitoring Program (REMP), which is described in the applicant's ER (Section 6.1.2).

9.3.2.4 Environmental Monitoring

The applicant has establish its REMP for the facility. The REMP is a major part of the applicant's effluent compliance program. The effectiveness of the applicant's effluent controls will be confirmed through implementation of the REMP. The purpose of the REMP is to verify confinement integrity at the facility and to support the primary means of demonstrating compliance with applicable radiation protection standards, which is effluent monitoring (see SER Sections 9.3.2.1 and 9.3.2.2).

As part of the REMP, the applicant will establish background and baseline concentrations of radionuclides in environmental media through sampling and analysis. The REMP will be initiated at least 2 years before plant operations in order to develop a sufficient environmental baseline. The types of samples to be collected under the REMP are listed in Table 9.3-3. Staff evaluation of proposed corrective action levels, which apply to both effluent monitoring and the REMP, is described in SER Section 9.3.2.1. The scope of the applicant's REMP meets the environmental monitoring criteria found in NUREG-1520, Section 9.4.3.3.2(2). As noted in the applicant's ER (Section 6.1.2), the REMP sampling locations are based on NRC guidance found in NUREG-1302 (NRC, 1991).

Table 9.3-3
Types of Samples Collected under the REMP

Sample Type	No. of Sample Locations	Quantity and Collection Frequency	Type of Analysis
Continuous Airborne Particulate Matter	7	Biweekly	Gross alpha/beta
		Quarterly	U-234, 235, 236, 238
Vegetation	8	1- to 2-kg samples, semiannually	U-234, 235, 236, 238
Groundwater	5	4 liters, semiannually	U-234, 235, 236, 238
Basins	1 from each of 3 basins	4 liters water and 1- to 2-kg sediment, quarterly	U-234, 235, 236, 238
Soil	8	1- to 2-kg samples, semiannually	U-234, 235, 236, 238
Septic Tank(s)	1 from each tank	1- to 2-kg sludge sample from the affected tank(s) before pumping	U-234, 235, 236, 238
TLD	16	Quarterly	Gamma / neutron

The applicant has an adequate and timely program to collect information to determine baseline concentrations of radionuclides, which information will be used to demonstrate compliance with applicable radiation protection standards. The staff finds that the applicant's environmental monitoring program adequately addresses applicable regulatory requirements and is acceptable.

9.3.3   ISA Summary

In SER Chapter 3, the staff provides its evaluation of the Integrated Safety Assessment (ISA) Summary, and documents its conclusion that the ISA Summary is complete, provides reasonable estimates of the likelihood and consequences of each accident sequence, and provides sufficient information to determine whether adequate engineering or administrative controls are identified for each accident sequence. SER Chapter 11 contains the staff's evaluation of management measures used to ensure that IROFS will satisfactorily perform their intended safety functions. In this section, the staff evaluates whether the ISA uses acceptable methods to estimate environmental effects that may result from accident sequences.

Under 10 CFR Part 70, Subpart H, an applicant is to assure, among other things, compliance with various performance requirements. Section 70.61(c)(3) identifies the environmental performance requirement that the applicant apply controls such that a credible intermediate-consequence event is unlikely to occur or that the consequence of such an event will not exceed a 24-hour averaged release of radioactive material outside the restricted area in concentrations 5000 times the values in Table 2 of Appendix B to Part 20. The restricted areas are defined by the applicant in Figure 4.7-2 of the SAR (LES, 2005a). The restricted area is one contiguous area encompassing the main processing buildings and the UBC Storage Pad. Specifically, main processing buildings included within the restricted area are: (1) the Centrifuge Assembly Building; (2) the Cylinder Receipt and Dispatch Building; (3) the six Cascade Halls; (4) the six UF_6 Handling Areas; (5) the TSB; and (6) the Blending and Liquid Sampling Area.

The applicant derived enhanced definitions of consequence severity levels based on the following rationale. For releases to the atmosphere of soluble uranium compounds (i.e., UF_6), the applicable value of Table 2 of Appendix B to Part 20 is 3×10^{-12} µCi/mL. This is the value for uranium isotopes U-234, -235, and -238. The performance requirement is, therefore, a 24-hour averaged concentration of any mixture of these isotopes that will not exceed $3 \times 10^{-12} \times 5000$, or 1.5×10^{-8} µCi/mL total radioactivity. The staff assumed a bounding enrichment of 6 percent. Using footnote 3 of Part 20, Appendix B, the specific activity is approximately 2.8 µCi per gram U. Therefore, the environmental performance requirement, after conversion to a mass concentration, is a 24-hour averaged concentration of 5.4 mg U/m³. The environmental performance requirement, expressed as the enhanced severity level of 5.4 mg U/m³, was evaluated at the restricted-area boundary. In sections 3.7 and 3.8 of the ISA Summary, the applicant has shown that it has adequately reduced the risks to the environment from accidents for which the consequences could otherwise exceed the environmental-consequence severity level.

In its ISA Summary, the applicant identified various sequences for radiological and non-radiological accidents which were evaluated to assure adequate protection of worker health and safety. By assuring that all credible high-consequence events are rendered highly unlikely and that all intermediate-consequence events are unlikely, the applicant also assured that the environmental performance requirements of section 70.61(c)(3) will be met. In addition, the staff independently evaluated the accident sequences to identify whether a credible accident sequence could occur in which an environmental hazard could be created without also creating a worker-related hazard. Even when postulating conservative, multiple independent equipment failures combined with human error, the staff did not identify any accident sequence that would fail to meet the environmental performance requirements of section 70.61(c)(3).

The applicant's approach to risk reduction is to be accomplished through a combination of preventive and mitigative measures, with an emphasis on preventive measures. A more complete discussion is found in Chapter 3 of the SER, which addresses accident sequences for high and intermediate consequences. It also addresses preventive and mitigative measures.

9.4 EVALUATION FINDINGS

The applicant has developed a program to implement adequate environmental protection measures during operation, which measures include: (1) environmental and effluent monitoring; and (2) effluent controls to maintain public doses ALARA as part of the radiation protection program. The NRC staff concludes that the applicant's program, as described in its application, is adequate to protect the environment and the health and safety of the public and complies with regulatory requirements imposed by the Commission in Parts 20, 30, 40, 51, and 70.

The NRC staff issued a Final EIS in June 2005, for this licensing action, as required by 10 CFR 51.20. After weighing the environmental impacts of the proposed construction, operation, and decommissioning of the proposed facility and comparing alternatives, the NRC staff recommends in its Final EIS that, unless safety issues mandate otherwise, the proposed license be issued to LES.

9.5   REFERENCES

(LES, 2005a) Louisiana Energy Services (LES). "National Enrichment Facility Safety Analysis Report," Revision 6, 2005.

(LES, 2005b) Louisiana Energy Services (LES). "National Enrichment Facility Environmental Report," Revision 5, 2005.

(NRC, 1985) U.S. Nuclear Regulatory Commission (NRC). Regulatory Guide 4.16, "Monitoring and Reporting Radioactivity in Releases of Radioactive Materials in Liquid and Gaseous Effluents from Nuclear Fuel Processing and Fabrication Plants and Uranium Hexafluoride Production Plants," 1985.

(NRC, 1991) U.S. Nuclear Regulatory Commission (NRC). NUREG-1302, "Offsite Dose Calculation Manual Guidance: Standard Radiological Effluent Controls for Boiling Water Reactors," 1991.

(NRC, 1992) U.S. Nuclear Regulatory Commission (NRC). Regulatory Guide 8.34, "Monitoring Criteria and Methods to Calculate Occupational Radiation Doses," 1992.

(NRC, 1993) U.S. Nuclear Regulatory Commission (NRC). Regulatory Guide 8.37, ALARA Levels for Effluents From Materials Facilities," 1993.

(NRC, 1994a) U.S. Nuclear Regulatory Commission (NRC). NUREG-1491, "Safety Evaluation Report for the Claiborne Enrichment Center, Homer, Louisiana," 1994.

(NRC, 1994b) U.S. Nuclear Regulatory Commission (NRC). Information Notice 94-23, "Guidance to Hazardous, Radioactive, and Mixed-Waste Generators on the Elements of a Waste Minimization Program," 1994.

(NRC, 2002) U.S. Nuclear Regulatory Commission (NRC). NUREG-1520, "Standard Review Plan for the Review of a License Application for a Fuel Cycle Facility," 2002.

(NRC, 2005) U.S. Nuclear Regulatory Commission (NRC). NUREG-1790, "Environmental Impact Statement for the Proposed National Enrichment Facility in Lea County, New Mexico," June 2005.

10.0 DECOMMISSIONING

The purpose of this review of the applicant's decommissioning plan is to determine that the applicant will be able to decommission the facility safely and in accordance with U.S. Nuclear Regulatory Commission (NRC) requirements.

At the time of the initial license application for a uranium enrichment facility, the applicant is required to submit a decommissioning funding plan (DFP). The purpose of NRC's review of the DFP is to determine whether the applicant has considered decommissioning activities that may be needed in the future, has performed a credible site-specific cost estimate for those activities, and has presented NRC with financial assurance to cover the cost of those activities in the future. The DFP, therefore, should contain an overview of the proposed decommissioning activities, the methods used to determine the cost estimate, and the financial assurance mechanism. This overview must contain sufficient details to enable the reviewer to determine whether the decommissioning cost estimate is reasonably accurate.

10.1 REGULATORY REQUIREMENTS

The following NRC regulations require planning, financial assurance, and record-keeping for decommissioning, as well as procedures and activities to minimize waste and contamination:

10 CFR 20.1401-1406	"Radiological Criterial for License Termination" (Subpart E)
10 CFR 30.35	"Financial Assurance and Recordkeeping for Decommissioning"
10 CFR 30.36	"Expiration and Termination of Licenses and Decommissioning of Sites and Separate Buildings or Outdoor Areas"
10 CFR 40.36(d)	"Decommissioning Funding Plan"
10 CFR 40.36	"Financial Assurance and Recordkeeping for Decommissioning"
10 CFR 40.42	"Expiration and Termination of Licenses and Decommissioning of Sites and Separate Buildings or Outdoor Areas"
10 CFR 70.22(a)(9)	"Decommissioning Funding Plan"
10 CFR 70.25	"Financial Assurance and Recordkeeping for Decommissioning"
10 CFR 70.38	"Expiration and Termination of Licenses and Decommissioning of Sites and Separate Buildings or Outdoor Areas"

10.2 REGULATORY ACCEPTANCE CRITERIA

The "Standard Review Plan for the Review of a License Application for a Fuel Cycle Facility," NUREG-1520 (NRC, 2002a) and "Consolidated NMSS Decommissioning Guidance," NUREG-1757 (NRC, 2003), define relevant regulatory guidance and appropriate acceptance criteria for decommissioning and DFPs contained in license applications.

10.3 STAFF REVIEW AND ANALYSIS

10.3.1   Conceptual Decontamination and Decommissioning Plan

10.3.1.1   Decommissioning Program

10.3.1.1.1   Radioactive Contamination Control

The applicant states, in Section 10.1.5.2 of the safety analysis report (SAR) (LES, 2005) that the following features will primarily serve to minimize the spread of radioactive contamination during operation and, therefore, simplify eventual plant decommissioning. As a result, worker exposure to radiation and radioactive waste volumes are minimized as well.

- Certain activities during normal operation are expected to result in surface and airborne radioactive contamination. Specially designed rooms are provided for these activities to preclude contamination spread. These rooms are isolated from other areas and are provided with ventilation and filtration. The Solid Waste Collection Room, Ventilated Room, and the Decontamination Workshop meet these specific design requirements.

- All areas of the plant are sectioned off into Unrestricted and Restricted Areas. Restricted Areas limit access for the purpose of protecting individuals against undue risks from exposure to radiation and radioactive materials. Radiation Areas and Airborne Contamination Areas have additional controls to inform workers of the potential hazard in the area and to help prevent the spread of contamination. All procedures for these areas fall under the Radiation Protection Program and serve to minimize the spread of contamination and simplify the eventual decommissioning.

- Non-radioactive process equipment and systems are minimized in locations subject to potential contamination. This limits the size of the Restricted Areas and limits the activities occurring inside these areas.

- Local air filtration is provided for areas with potential airborne contamination, to preclude its spread. Fume hoods filter contaminated air in these areas.

- Curbing, pits, or other barriers are provided around tanks and components that contain liquid radioactive wastes. These serve to control the spread of contamination in case of a spill.

10.3.1.1.2   Worker Exposure and Waste Volume Control

The applicant states, in Section 10.1.5.3 of the SAR (LES, 2005), that the following features will primarily serve to minimize worker exposure to radiation and minimize radioactive waste volumes during decontamination activities. As a result, the spread of contamination is minimized as well.

- During construction, a washable epoxy coating is applied to floors and walls that might be radioactively contaminated during operation. The coating will serve to lower waste volumes during decontamination and simplify the decontamination process. The coating is applied to floors and walls that might be radioactively contaminated during operation that are located in the Restricted Areas.

- Sealed, nonporous pipe insulation is used in areas likely to be contaminated. This will reduce waste volume during decommissioning.

- Ample access is provided for efficient equipment dismantling and removal of equipment that may be contaminated. This minimizes the time of worker exposure.

- Tanks are provided with accesses for entry and decontamination. Design provisions are also made to allow complete draining of the wastes contained in the tanks.

- Connections in the process systems provided for required operation and maintenance allow for thorough purging at plant shutdown. This will remove a significant portion of radioactive contamination before disassembly.

- Design drawings, produced for all areas of the plant, will simplify the planning and implementing of decontamination procedures. This, in turn, will shorten the durations that workers are exposed to radiation.

- Worker access to contaminated areas is controlled to assure that workers wear proper protective equipment and limit their time in the areas.

10.3.1.2   Decommissioning Steps

The plan for decommissioning is to promptly decontaminate or remove all materials, from the site, that prevent release of the facility for unrestricted use. This approach, referred to in the industry as DECON, avoids long-term storage and monitoring of wastes on site.

The applicant has briefly described the decommissioning approach to be employed at the facility. The applicant states that implementation of the DECON alternative for decommissioning the facility may begin immediately after Separations Building Module equipment shutdown. The applicant estimates that the DECON alternative will take approximately 9 years to complete in three phases (3 years/module).

Decommissioning activities will generally include: (1) installation of decontamination facilities; (2) purging of process systems; (3) dismantling and removal of equipment; (4) decontamination and destruction of Confidential and Secret Restricted Data material; (5) sales of salvaged materials; (6) disposal of wastes; and (7) completion of a final radiation survey. Credit is not taken for any salvage value that might be realized from the sale of potential assets (e.g., recovered materials or decontaminated equipment) during or after decommissioning.

Overview

Decommissioning, using the DECON approach, requires residual radioactivity to be reduced below specified levels so the facilities may be released for unrestricted use. Current NRC guidelines for unrestricted release serve as the basis for decontamination costs estimated by the applicant. The applicant intends to remove all enrichment-related equipment from the buildings in such a manner that only the building shells and site infrastructure remain. The equipment to be removed by the applicant will include: all piping and components from systems providing uranium hexafluoride (UF$_6$) containment; systems in direct support of enrichment (such as refrigerant and chilled water); radioactive- and hazardous-waste-handling systems; and contaminated filtration systems, etc. The remaining site infrastructure will include services such as: electrical power supply; treated water; fire protection; ventilation systems; plant cooling water; communications; and sewage treatment.

The applicant will install two new facilities dedicated for decontamination of plant components and structures. Existing plant buildings are assumed to house the facilities. One facility will be specially designed to accommodate repetitive cleaning of thousands of centrifuges, and the other will serve as a general-purpose facility used primarily for larger components. The two new facilities will be the primary locations for decontamination activities. The small decontamination area in the Technical Services Building (TSB), used during normal operation, may also handle small items at decommissioning. The applicant estimates that the time for installation is approximately 1 year. The applicant provided details of the facilities in Section 10.1.7 of the SAR (LES, 2005).

The applicant states, in Section 10.1.6.1 of the SAR (LES, 2005), that decontaminated components may be reused or sold as scrap. The applicant will decontaminate all equipment that is to be reused or sold as scrap to a level at which further use is unrestricted. Materials that cannot be decontaminated will be disposed of by the applicant in a licensed radioactive-waste-disposal facility.

The applicant states in Section 10.1.6.1 of the application that contaminated portions of the buildings will be decontaminated as required. When decontamination is complete, the applicant will survey all areas and facilities on the site, to verify that further decontamination is not required. The applicant will continue decontamination activities until the entire site is demonstrated to be suitable for unrestricted use. NRC will independently confirm that the site is suitable to be released for unrestricted use. NRC will not authorize unrestricted release of materials and equipment unless all release criteria applicable at the time of decommissioning have been met. NRC will not authorize release of the site for unrestricted use until the applicant adequately demonstrates that all decommissioning criteria applicable at the time of decommissioning have been met.

Process System

The applicant states, in Section 10.1.6.3 of the SAR (LES, 2005), that at the end of the useful life of each Separations Building Module, the enrichment process will be shut down and UF$_6$ will be removed to the fullest extent possible by normal process operation. This will be followed by evacuation and purging with nitrogen. The applicant estimates that the shutdown and purging portions of the decommissioning process will take approximately 3 months.

Dismantling

Dismantling involves cutting out, disconnecting, etc., all components requiring removal. The operations themselves may be simple, but very labor-intensive. Depending on the level of contamination, the use of protective clothing may be required. The applicant states, in Section 10.1.6.4 of the SAR (LES, 2005), that the work process will be optimized, considering the following:

- Minimizing the spread of contamination and the need for protective clothing;

- Balancing the number of cutting and removal operations with the resultant decontamination and disposal requirements;

- Optimizing the rate of dismantling with the rate of decontamination facility throughput;

- Providing storage and laydown space required, as affected by retrievability, criticality safety, security, etc.; and

- Balancing the cost of decontamination and salvage with the cost of disposal.

The applicant states, in Section 10.1.6.4 of the SAR (LES, 2005), that the details of the complex optimization process will necessarily be decided near the end of plant life, taking into account specific contamination levels and available waste disposal sites. To avoid laydown space and contamination problems, dismantling will likely be allowed to proceed no faster than the downstream decontamination process. The applicant estimates that dismantling and decontamination will take approximately 3 years per Separations Building Module.

Sale/Salvage

Items to be removed from facilities can be categorized as potentially reusable equipment, recoverable scrap, and wastes. However, based on a 30-year operating life, the applicant does not assume that the facility operating equipment has any reuse value. According to the applicant, wastes will also have no salvage value. With respect to scrap, the applicant states, in Section 10.1.6.6 of the SAR (LES, 2005), that a significant amount of aluminum will be recovered, along with smaller amounts of steel, copper, and other metals. For security and convenience, the uncontaminated material will likely be smelted to standard ingots, then sold at market price. However, the applicant has not assigned salvage value to scrap, in estimating its decommissioning funding requirements. Contaminated material will be disposed of as low-level radioactive waste.

Disposal

The applicant states, in Section 10.1.6.7 of the SAR (LES, 2005), that all wastes produced during decommissioning will be collected, handled, and disposed of in a manner similar to that described for those wastes produced during normal operation. According to the applicant, wastes will consist of normal industrial trash, non-hazardous chemicals and fluids, small amounts of hazardous materials, and radioactive wastes. The radioactive waste will primarily be crushed centrifuge rotors, trash, and citric cake. Citric cake will consist of uranium and metallic compounds precipitated from citric acid decontamination solutions. The applicant estimates approximately 5000 cubic meters (m^3) [180,000 cubic feet (ft^3)] of radioactive waste to be generated over the 9-year period of facility decommissioning activities. This waste will be subject to further volume reduction processes before disposal.

The applicant states, in Section 10.1.6.7 of the SAR (LES, 2005), that radioactive wastes will ultimately be disposed of in licensed low-level radioactive-waste disposal facilities. Hazardous wastes will be disposed of in permitted hazardous waste-disposal facilities. Non-hazardous and non-radioactive wastes will be disposed of in a manner consistent with good industrial practice and in accordance with all applicable regulations. A complete estimate of the wastes and effluents to be generated during decommissioning will be provided in the applicant's plan for completion of decommissioning, to be submitted to NRC at the time of decommissioning.

The applicant states, in Section 10.1.6.7 of the SAR (LES, 2005), that Confidential and Secret-Restricted Data components and documents on site will be disposed of in accordance with the requirements of 10 CFR Part 95. Classified portions of the centrifuges will be destroyed; piping will likely be smelted; documents will be destroyed; and other items will be handled in an appropriate manner. Details will be provided in the facility "Standard Practice Procedures Plan for the Protection of Classified Matter and Information," submitted separately, in accordance with Part 95.

Final Radiation Survey

The applicant states, in Section 10.1.6.8 of the SAR (LES, 2005), that it will perform a final radiation survey, to verify proper decontamination, to allow the site to be released for unrestricted use. The initial radiation survey performed before initial operation will provide data on the natural background radiation of the area that can be used to determine any increase in levels of radiation. The applicant states, in Section 10.1.6.8 of the SAR (LES, 2005), that radioactivity over the entire site will systematically be measured in the final survey. The intensity of the survey will vary depending on the location (i.e., the buildings, the immediate area around the buildings, the controlled fenced area, and the remainder of the site). The survey procedures and results will be documented in a report. The report will include, among other things, a map of the survey site, measurement results, and the site's relationship to the surrounding area. If the results are above allowable residual radioactivity limits, further decontamination will be performed until the results are determined to be below limits.

10.3.1.3   Management/Organization

The applicant states, in Section 10.1.5.4 of the SAR (LES, 2005), that management of the decommissioning program will ensure that proper training and procedures are provided to protect worker health and safety. The programs will focus heavily on minimizing waste volumes and worker exposure to hazardous and radioactive materials. Contractors assisting with decommissioning will likewise be subject to facility training requirements and procedural controls. The NRC staff finds the applicant's general plans acceptable regarding:

- Responsibilities of management of the decommissioning program;

- Minimization of waste volumes and worker exposures; and

- Procedural control and training requirements for contractors assisting in decommissioning.

Details related to these three items are typically provided in the detailed decommissioning plan at the time of decommissioning.

10.3.1.4   Health and Safety

The applicant states, in Section 10.1.5.5 of the SAR (LES, 2005), that, as with normal operation, during decommissioning, the policy will be to keep individual and collective occupational radiation exposures as low as is reasonably achievable (ALARA). A health physics program will identify and control sources of radiation, establish worker-protection requirements, and direct the use of survey and monitoring instruments.

10.3.1.5   Waste Management

The applicant states, in Section 10.1.5.6 of the SAR (LES, 2005), that radioactive and hazardous wastes produced during decommissioning will be collected, handled, and disposed of in accordance with all regulations applicable to the facility at the time of decommissioning. Generally, procedures will be similar to those described for wastes produced during normal operation. These wastes will ultimately be disposed of in licensed radioactive or permitted hazardous-waste-disposal facilities located elsewhere. Non-hazardous and non-radioactive wastes will be disposed of consistent with good industrial practice and in accordance with applicable regulations.

10.3.1.6   Security/Nuclear Material Control

The applicant states, in Section 10.1.5.7 of the SAR (LES, 2005), that requirements for physical security and for material control and accounting (MC&A) will be maintained, as required, during decommissioning, in a manner similar to the programs in force during operation. The plan for completion of decommissioning, submitted near the end of plant life, will provide a description of any necessary revisions to these programs.

10.3.1.7   Recordkeeping

The applicant states, in Section 10.1.5.8 of the SAR (LES, 2005), that records important for safe and effective decommissioning of the facility will be kept in the applicant's files. Information maintained in these records will include:

1.   Records of spills or other unusual occurrences involving the spread of contamination in and around the facility, equipment, or site. These records may be limited to instances when contamination remains after any cleanup procedures or when there is reasonable likelihood that contaminants may have spread to inaccessible areas, as in the case of possible seepage into porous materials such as concrete. These records will include any known information on identification of involved nuclides, quantities, forms, and concentrations.

2.   As-built drawings and modifications of structures and equipment, in restricted areas, where radioactive materials are used or stored, and of locations of possible inaccessible contamination, such as buried pipes, which may be subject to contamination. Required drawings will be referenced as necessary, although each relevant document will not be indexed individually. If drawings are not available, appropriate records of available information concerning these areas and locations will be substituted.

3.	Except for areas containing only sealed sources, a list contained in a single document and updated every 2 years, of the following:

a.	All areas designed and formerly designated as Restricted Areas, as defined under 10 CFR 20.1003;

b.	All areas outside of Restricted Areas that require documentation specified in item 1 above;

c.	All areas outside of Restricted Areas where current and previous wastes have been disposed of, as documented under 10 CFR 20.2108; and

d.	All areas outside of Restricted Areas that contain material such that, if the license expired, the licensee would be required to either decontaminate the area to meet the criteria for decommissioning, in 10 CFR Part 20, Subpart E, or apply for approval for disposal under 10 CFR 20.2002.

4.	Records of the cost estimate performed for the DFP and records of the funding method used for assuring funds.

10.3.1.8	Decontamination

The following paragraphs discuss the facilities, procedures, and expected results of decontamination, as described by the applicant in Section 10.1.7 of the SAR (LES, 2005).

The applicant states, in Section 10.1.7.1 of the SAR (LES, 2005), that the primary contamination throughout the plant will be in the form of small amounts of UO_2F_2, with even smaller amounts of UF_4 and other compounds. Radiological contamination will be characterized by the applicant, before beginning decommissioning activities. NRC staff review of the final decommissioning plan will include a review of the nature and extent of contamination present at the time of decommissioning.

Facilities

The applicant states, in Section 10.1.7.2 of the SAR (LES, 2005), that a decontamination facility will be required to accommodate decommissioning. A specialized facility is needed for optimal handling of the thousands of centrifuges to be decontaminated, along with the UF_6 vacuum pumps and valves. Additionally, a general purpose facility is needed for handling the remainder of the various plant components. The applicant will most likely install these facilities in existing plant buildings (such as the Centrifuge Assembly Building).

The applicant describes the specialized facility as having four functional areas: a disassembly area; a buffer stock area; a decontamination area; and a scrap storage area for cleaned stock. The general purpose facility may share the specialized facility decontamination area. However, because of various sizes and shapes of other plant components needing handling, the disassembly area, buffer stock areas, and scrap storage areas may not be shared.

The applicant assumes that equipment in the decontamination facilities will include:

- Transport and manipulation equipment;
- Dismantling tables, for centrifuge externals;
- Sawing machines;
- Dismantling boxes and tanks, for centrifuge internals;
- Degreasers;
- Citric acid and demineralized water baths;
- Contamination monitors;
- Wet-blast cabinets;
- Crusher, for centrifuge rotors;
- Smelting and/or shredding equipment; and
- Scrubbing facility.

The applicant states, in Section 10.1.7.2 of the SAR (LES, 2005), that the decontamination facilities provided in the TSB for normal operational needs would also be available for cleaning small items during decommissioning.

Procedures

The applicant states, in Section 10.1.7.3 of the SAR (LES, 2005), that procedures for decontamination will be developed and approved by plant management to minimize worker exposure and waste volumes, and to assure work is carried out in a safe manner. If, as expected, European gas centrifuge enrichment facilities are decommissioned before the proposed facility, then the experience gained will be incorporated into the procedures to be developed by the applicant. NRC staff will assess the procedures used and the results of the final survey. A confirmatory survey is expected to be part of NRC's assessment of the final survey. The facility and site will not be released for unrestricted use unless it is demonstrated that the residual contamination is within the limits and criteria in place at the time of decommissioning.

The applicant states, in Section 10.1.7.3 of the SAR (LES, 2005), that contaminated plant components will be cut up or dismantled, then processed through the decontamination facilities.

The applicant states, in Section 10.1.7.3 of the SAR (LES, 2005), that the centrifuges will be processed through the specialized facility. The following operations will be performed:

- Removal of external fittings;

- Removal of bottom flange, motor and bearings, and collection of contaminated oil;

- Removal of top flange, and withdrawal and disassembly of internals;

- Degreasing of items as required;

- Decontamination of all recoverable items for smelting; and

- Destruction of other classified portions by shredding, crushing, smelting, etc.

10-9

Results

The applicant states, in Section 10.1.7.4 of the SAR (LES, 2005), that conventional decontamination techniques are effective for all plant items, based on their experience in decommissioning Urenco facilities in Europe. Recoverable items will be decontaminated and suitable for reuse except for a very small amount of intractably contaminated material. Material requiring disposal will primarily be centrifuge rotor fragments, trash, and residue from the effluent treatment systems. The applicant does not anticipate problems that will prevent the site from being released for unrestricted use.

10.3.1.9   Decommissioning Costs

The applicant submitted decommissioning cost information consistent with the recommendations in NUREG-1757, Volume 3, "Consolidated NMSS Decommissioning Guidance - Financial Assurance, Recordkeeping, and Timeliness" (NRC, 2003). The applicant presented its decommissioning cost estimate breakdown in SAR Tables 10.1-1 through 10.1-14 (LES, 2005). Decommissioning cost information included labor costs, proposed decontamination methods and unit costs, waste disposal costs, final survey costs, and costs for dispositioning depleted uranium tails. The decommissioning costs were based on the decommissioning experience of Urenco, the applicant's principal general partner, in decommissioning gas centrifuge enrichment plants in Europe.

The applicant estimates the cost of decommissioning the facility to be approximately $942 million, in 2004 dollars, which includes an estimated cost of $131 million to decommission the supporting structures, an estimated tails-disposition cost of $622 million, and a 25 percent contingency factor, equal to $188 million. More than 97 percent of the cost to decommission the structures are attributed to the dismantling of the centrifuges and other equipment in the Separation Building Modules.

The cost analysis for decommissioning the centrifuges and equipment supporting the Separation Building is classified. The staff reviewed the number of centrifuges, the estimated man-hours to decontaminate the centrifuges, and the estimated volume of material resulting from the disposal of the centrifuges and supporting equipment; confirmed that the estimate includes a 25 percent contingency; and also confirmed that no credit is taken for salvage of materials or equipment. The use of a 25 percent contingency factor and taking no credit for salvage value is consistent with staff guidance in NUREG-1757, Volume 3 (NRC, 2003).

The unclassified estimate for decommissioning the remainder of the facility was reviewed considering labor costs, decontamination methods and unit costs, waste disposal costs, depleted uranium disposition costs, and final survey costs. Both the classified and non-classified estimates were evaluated and found to be reasonable and consistent with estimates provided in NUREG/CR-6477, "Revised Analyses of Decommissioning Reference Non-Fuel-Cycle Facilities" (NRC, 2002b). While the cases in NUREG/CR-6477 (NRC, 2002b) do not include a case for decommissioning a gas centrifuge enrichment plant, some information can be compared with the cost estimates provided by the applicant. The following other considerations were also included in comparing the applicant's estimates with those in NUREG/CR-6477:

1. The proposed facility operates at subatmospheric pressures that minimize the spread of contamination throughout the plant.

2. The plant has design features and operating procedures to minimize releases of uranium hexafluoride (e.g., cylinder feed, withdrawal, blending, and sampling systems). These design and operating features are principally intended to minimize worker chemical exposures, but also result in low contamination levels in occupied areas of the plant.

3. Feed material will be restricted to material specifications meeting the requirements of American Society for Testing and Materials C787, "Standard Specification for Uranium Hexafluoride for Enrichment" (ASTM, 2003). Through this specification, the applicant will control the entry of other radioactive contaminants into the process systems.

4. Most of the process equipment in the plant is aluminum, which can be more easily cut and processed than the steel components assumed in the cases described in NUREG/CR-6477 (NRC, 2002b).

Based on the staff's review of the classified and unclassified information, the staff found the cost estimate for decommissioning the facility to be reasonable.

The applicant conservatively estimated that the facility will generate 132,942 MT of depleted uranium over a nominal 30 years of production, and did not reduce the estimate of depleted uranium based on the planned operations approach where production would actually end 5 years earlier. The applicant estimated the waste processing and disposal cost of UF_6 tails at $4.68 per kilogram of uranium (kg U) or $4,680 per metric ton of uranium (MTU). This cost is based on the total of the three cost components that make up the total disposition cost for DUF_6 (i.e., deconversion, disposal, and transportation). The staff reviewed the basis of each of these three costs components, and has concluded that they are reasonable.

The deconversion cost was based on proprietary information on a previously proposed private deconversion plant using the Cogema dry conversion process producing U_3O_8 and aqueous hydrogen fluoride (HF) (NRC, 2005). The proposed process was the same as the plant Cogema has been operating in Pierrelatte, France for 20 years. The cost estimate was adjusted to account for differences in planned operating capacities, Euros-to-dollars conversion, and other costs associated with "Americanization." "Americanization" refers to costs to obtain regulatory approval and costs to convert European equipment standards to standards used in the United States. These cost estimates used a proprietary Urenco business study of a proposed 3,500 Metric Tons (MT) U/year deconversion plant for the Capenhurst site. The study was based on a Cogema response to a Urenco request for proposal. The applicant modified the Cogema information to reflect a 7,000 MT U/year capacity by doubling the operating costs and by adding funds to reflect the increased capital and construction costs of a larger capacity plant considering the shared nature of some systems. Additional funds were also added for Americanizing the design and for licensing.

The Cogema proposal assumed that HF would be sold commercially and did not include the costs to neutralize aqueous HF to calcium fluoride. Staff consider that neutralization would have no effect on the overall deconversion costs because those costs would be balanced by the elimination of costs for equipment for storing HF prior to commercial sale. The cost of disposing the calcium fluoride ($0.02/kg U) was included in the estimate.

The transportation and disposal costs were based on estimates provided by vendors of transportation and disposal services (LES, 2005a). Transportation costs were based on an estimate from Transportation Logistics International. This transportation estimate ($0.85/kg U) was independent of distance. The disposal cost of $1.14/kg U for depleted uranium oxides was based on an estimate provided by Waste Control Specialists. Staff compared the Waste Control Specialists estimate to an estimate for disposal of decommissioning wastes the applicant had obtained from Envirocare of Utah and found it to be consistent. The Envirocare disposal estimate for decommissioning waste was $2.12/m^3 ($75/ft^3) (LES, 2004). For the disposal of U_3O_8, the equivalent disposal cost at Envirocare is $1.07/kg U.

Further, the applicant submitted an estimate for tails disposition from the U.S. Department of Energy (DOE) (DOE, 2005) as additional evidence of the reasonableness of their estimate. The DOE estimate included conversion, transportation, storage, disposal, and decommissioning costs of the conversion facility and totaled $3.34/kg DUF_6 ($4.91/kg U) in 2004 dollars. This is less than 5 percent of the difference in the applicant's estimate of $4.68/kg U. Staff considers that the DOE estimate provides additional assurance that the applicant's estimate of depleted uranium disposition costs is reasonable.

Based on the staff's review of the classified and unclassified information, the staff found that the cost estimate for decommissioning the facility is reasonable and the cost estimate fulfills the requirements of 10 CFR 30.35(e), 10 CFR 40.36(d), and 10 CFR 70.25(e) and the evaluation criteria in Section 4.1 of NUREG-1757, Volume 3 (NRC, 2003) for the following reasons:

- The cost estimate is based on documented and reasonable assumptions;

- The cost estimates for individual facility activities and components are reasonable and, to the extent possible, consistent with NRC cost estimation reference documents;

- The cost estimate reflects decommissioning under appropriate facility conditions;

- The cost estimate includes costs for labor, equipment and supplies, overhead and contractor profit, sampling, and miscellaneous expenses;

- The cost estimate includes costs for all major decommissioning activities, including planning and preparation; decontamination or dismantling facility components; packaging, shipping, and disposal of wastes; restoration of facility grounds; and the final radiation survey.

- The computations are correct;

- No credit is taken for salvage value;

- The decommissioning cost estimate includes an adequate contingency factor of 25 percent; and

- The decommissioning cost estimate provides a description of how it will be adjusted periodically over the life of the facility.

10.3.1.10 Financial Assurance for Decommissioning

The applicant stated it will utilize a surety bond method to provide reasonable assurance of decommissioning funding as required by 10 CFR 30.35(f)(2), 10 CFR 40.36(e)(2), and 10 CFR 70.25(f)(2). The applicant provided draft copies of the surety bond and standby trust language. Finalization of the specific financial instruments to be utilized will be completed, and signed originals of those instruments will be provided to the NRC for final confirmation of the instrument prior to the applicant receiving licensed material at the facility. In addition, the applicant committed to provide continuous financial assurance through the completion of decommissioning and termination of the licenses. Although the applicant plans to sequentially install and operate the separations buildings modules over time, financial assurance for decommissioning of the full-size facility will initially be set aside, as well as an estimated 3 years of tails disposition. Thereafter, funding for tails disposition will be provided at a rate in proportion to the amount of accumulated tails onsite up to the maximum amount of the tails as described in Sections 10.2.1 and 10.2.2 of the SAR. If the applicant's schedule for future enrichment module phase-in changes, the applicant indicated it may reduce the funding for facility decontamination and decommissioning cost estimate to reflect the actual number of operating modules.

The surety bond method to be adopted by the applicant will provide a guarantee that decommissioning costs will be paid in the event the applicant is unable to meet its decommissioning obligations at the time of decommissioning. The surety bond will be structured consistent with applicable NRC requirements and in accordance with NRC regulatory guidance contained in NUREG-1757, Volume 3 (NRC, 2003). Accordingly, the applicant stated that its surety bond will contain, but not be limited to, the following attributes:

- The surety bond will be open-ended or, if written for a specified term, such as 5 years, will be renewed automatically unless 90 days or more prior to the renewal date, the issuer notifies the NRC, the trust to which the surety is payable, and the applicant of its intention not to renew. The surety bond will also provide that the full face amounts are paid to the beneficiary automatically prior to the expiration without proof of forfeiture if the applicant fails to provide a replacement acceptable to the NRC within 30 days after receipt of notification of cancellation.

- The surety bond will be payable to a standby trust established for decommissioning costs. The trustee and trust will be ones acceptable to the NRC. For instance, the trustee may be an appropriate State or Federal government agency or an entity which has the authority to act as a trustee and whose trust operations are regulated and examined by a Federal or State agency.

- The surety bond and standby trust will remain in effect until the NRC has terminated the license.

In accordance with 10 CFR 30.35(e), 10 CFR 40.36(d), and 10 CFR 70.25(e), the applicant will update the decommissioning cost estimate for the facility and the associated funding levels, over the life of the facility. These updates will take into account changes resulting from inflation or site-specific factors, such as changes in facility conditions or expected decommissioning procedures. These funding level updates will also address anticipated accumulated tails. As required by 10 CFR 30.35(e), 10 CFR 40.36(d), and 10 CFR 70.25(e), such updates will occur at least every 3 years.

In Sections 1.2.5, 10.2.1, and 10.2.2 of the SAR (LES, 2005), the applicant described its approach to funding the surety bond financial assurance instrument to be used and for updating the DFP over time. Financial assurance for decommissioning will be provided during the operating life of the facility. Initially, the applicant will provide funding for decommissioning the facility and the tails expected to be generated during the first three years of operation. Funding for tails dispositioning will thereafter be provided at annual intervals. In Section 1.2.5 of the SAR (LES, 2005), the applicant requested an exemption to fund decommissioning on an incremental basis. Section 1.2.3.6 of this SER discusses the approval of this exemption request as required in 10 CFR 40.14 and 10 CFR 70.17.

Updates of the DFP and the financial assurance instrument will be provided as follows:

- In the initial executed financial assurance instrument submitted prior to receipt of licensed material, the applicant will provide full funding for decontamination and decommissioning of the full-size facility.

- In the initial executed financial assurance instrument submitted prior to receipt of licensed material, the applicant will provide funding for the disposition of depleted uranium tails in an amount needed to disposition the first three years of depleted uranium tails generation.

- Subsequent updated decommissioning funding estimates and revised funding instruments for facility decommissioning will be provided at least every three years. If the applicant reduces the amount of funding for the facility because of a change in module phase-in, the revisions will be submitted prior to the operation of each facility module. This will allow the applicant to modify its initial facility decommissioning funding approach to reflect changes in future enrichment module phase-in schedules.

- Subsequent updated decommissioning cost estimates and revised funding instruments for depleted uranium disposition will be provided annually on a forward-looking basis to reflect projections of depleted uranium byproduct generation.

The above DFP update schedule will provides updates at a frequency of at least every 3 years in accordance with 10 CFR 30.35(e), 10 CFR 40.36(d), and 10 CFR 70.25(e).

The initial financial obligation will be the entire facility decommissioning cost ($131 million), the cost for dispositioning the first three years of generation of depleted uranium ($22.7 million based on generating 4,861 MT of depleted uranium in the first 3-year period), and a 25 percent contingency of $38.5 million giving a total decommissioning obligation for this period of $192 million. These estimates are in 2004 dollars. This approach to funding the financial assurance instrument is acceptable to NRC staff because the amount of financial assurance will be sufficient to cover the decommissioning obligation of the licensee at any point in time in the event that the licensee is unable to complete decommissioning for any reason.

Because final executed copies of the financial assurance mechanism will not be provided to NRC until prior to receipt of licensed material, NRC staff is imposing the following license conditions:

"1.   The licensee shall provide final copies of the proposed financial assurance instruments to NRC for review at least 6 months prior to the planned date for obtaining licensed material, and provide to NRC final executed copies of the reviewed financial assurance instruments prior to the receipt of licensed material.

2.   In addition, the Decommissioning Funding Plan cost estimate shall be updated as follows:

    a.   In the first executed financial assurance instrument submitted prior to receipt of licensed material, the licensee will provide full funding for decontamination and decommissioning of the full-size facility.

    b.   In the first executed financial assurance instrument submitted prior to receipt of licensed material, the licensee will provide funding for the disposition of depleted uranium tails in an amount needed to disposition the first three years of depleted uranium tails generation.

    c.   Subsequent updated decommissioning funding estimates and revised funding instruments for facility decommissioning will be provided, at a minimum, every three years. Any proposed reduction based on changes to module phase-in will be submitted 6 months prior to the scheduled operation of the facility module.

    d.   Subsequent updated decommissioning cost estimates and revised funding instruments for depleted uranium disposition will be provided annually on a forward-looking basis to reflect projections of depleted uranium byproduct generation.

3.   The Decommissioning Funding Plan cost estimates shall be provided to NRC for review, and subsequently, after resolution of any NRC comments, final executed copies of the financial assurance instruments shall be provided to NRC."

With the above proposed license conditions and the exemption discussed in Section 1.2.3.6, NRC staff finds the DFP and proposed surety bond method acceptable.

10.4   EVALUATION FINDINGS

The NRC staff has evaluated the applicant's decommissioning financial assurance plan in accordance with NUREG-1757 (NRC, 2003). On the basis of this evaluation, the NRC staff has determined that the applicant's financial assurance for decommissioning provides sufficient funding to ensure decommissioning and decontamination of the facility even if the licensee is unable to meet its financial obligations, and, therefore, provides reasonable assurance of protection for workers, the public, and the environment.

Because final executed copies of the financial assurance mechanism will not be provided to NRC until prior to receipt of licensed material, NRC staff is imposing the following license conditions:

"1.   The licensee shall provide final copies of the proposed financial assurance instruments to NRC for review at least 6 months prior to the planned date for obtaining licensed material, and provide to NRC final executed copies of the reviewed financial assurance instruments prior to the receipt of licensed material.

2.   In addition, the Decommissioning Funding Plan cost estimate shall be updated as follows:

a.   In the first executed financial assurance instrument submitted prior to receipt of licensed material, the licensee will provide full funding for decontamination and decommissioning of the full-size facility.

b.   In the first executed financial assurance instrument submitted prior to receipt of licensed material, the licensee will provide funding for the disposition of depleted uranium tails in an amount needed to disposition the first three years of depleted uranium tails generation.

c.   Subsequent updated decommissioning funding estimates and revised funding instruments for facility decommissioning will be provided, at a minimum, every three years. Any proposed reduction based on changes to module phase-in will be submitted 6 months prior to the scheduled operation of the facility module.

d.   Subsequent updated decommissioning cost estimates and revised funding instruments for depleted uranium disposition will be provided annually on a forward-looking basis to reflect projections of depleted uranium byproduct generation.

3.   The Decommissioning Funding Plan cost estimates shall be provided to NRC for review, and subsequently, after resolution of any NRC comments, final executed copies of the financial assurance instruments shall be provided to NRC."

10.5   REFERENCES

(ASTM, 2003) American Society for Testing and Materials (ASTM). ASTM C787, "Standard Specification for Uranium Hexafluoride for Enrichment," 2003.

(DOE, 2005) U.S. Department of Energy (DOE), letter to Louisiana Energy Services, "Conversion and Disposal of Depleted Uranium Hexafluoride (DUF6) Generated by Louisiana Energy Services, LP (LES)," March 1, 2005.

(LES, 2004) Louisiana Energy Services (LES), letter to U.S. Nuclear Regulatory Commission, "Response to NRC Request for Additional Information Regarding Decommissioning Funding Plan," December 10, 2004.

(LES, 2005a) Louisiana Energy Services (LES), letter to U.S. Nuclear Regulatory Commission, "Clarifying Information Related to Depleted Uranium UF_6 Disposition Costs and Request for License Condition," March 29, 2005.

(LES, 2005b) Louisiana Energy Services (LES). "National Enrichment Facility Safety Analysis Report," Revision 6, 2005.

(NRC, 2002a) U.S. Nuclear Regulatory Commission (NRC). NUREG-1520, "Standard Review Plan for the Review of a License Application for a Fuel Cycle Facility," 2002.

(NRC, 2002b) U.S. Nuclear Regulatory Commission (NRC). NUREG/CR-6477, "Revised Analyses of Decommissioning Reference Non-Fuel-Cycle Facilities," 2002.

(NRC, 2003) U.S. Nuclear Regulatory Commission (NRC). NUREG-1757, Volume 3, "Consolidated NMSS Decommissioning Guidance - Financial Assurance, Recordkeeping, and Timeliness," 2003.

(NRC, 2005) Johnson, T.C., U.S. Nuclear Regulatory Commission (NRC). April 8, 2005, "In-Office Review Summary: Louisiana Energy Services Decommissioning Funding," April 29, 2005.

11.0 MANAGEMENT MEASURES

Management measures are functions that the applicant performs, generally on a continuing basis, which are applied to items relied on for safety (IROFS), to ensure compliance with established performance requirements that the IROFS are available and reliable. Management measures shall be implemented to assure compliance with performance requirements, and the degree to which they will be applied will be a function of the item's importance in terms of meeting performance requirements as evaluated in the Integrated Safety Analysis (ISA). This chapter addresses each of the management measures included in the 10 CFR Part 70 definition of management measures, including: (a) configuration management (CM); (b) maintenance; (c) training and qualifications; (d) procedures; (e) audits and assessments; (f) incident investigations; (g) records management; and (h) other quality assurance (QA) elements.

The purpose of this review is to verify whether the applicant provided conclusive information to ensure that the management measures applied to IROFS, as documented in the ISA Summary, provide assurance that the IROFS will be available and reliable, consistent with the performance requirements of 10 CFR 70.61. If a graded approach is used, the review will also determine whether the measures are applied to the IROFS in a manner commensurate with the IROFS' importance to safety.

11.1 REGULATORY REQUIREMENTS

The requirements for fuel cycle facility management measures are specified in 10 CFR Part 70, "Domestic Licensing of Special Nuclear Material."

1. 10 CFR 70.4 states that management measures include: CM; maintenance; training and qualifications; procedures; audits and assessments; incident investigations; records management; and other QA elements.

2. 10 CFR 70.62(a)(3) states that records must be kept for all IROFS failures; describes required data to be reported; and sets time requirements for updating the records.

3. 10 CFR 70.62(d) requires an applicant to establish management measures, for application to engineered and administrative controls and control systems that are identified as IROFS, pursuant to 10 CFR 70.61(e), to ensure they are available and reliable.

4. 10 CFR 19.12 states requirements for workers instructions that are applicable to personnel training and qualifications.

5. 10 CFR 70.22(a)(8) states requirements for license applications to address proposed procedures to protect health and minimize danger to life and property.

6.	10 CFR 70.72 requires a licensee to establish a CM program to evaluate, implement, and track changes to the facility; structures, systems and components (SSCs); processes; and activities of personnel.

7.	10 CFR 70.74(a) and (b) state requirements for incident investigation and reporting.

## 11.2	REGULATORY ACCEPTANCE CRITERIA

The acceptance criteria for the U.S. Nuclear Regulatory Commission's (NRC's) review of the applicant's management measures program are outlined in Section 11.4.3 of NUREG-1520 (NRC, 2002).

## 11.3	STAFF REVIEW AND ANALYSIS

11.3.1 Configuration Management Program (CM)

The applicant's CM function is described in Section 11.2 of the Safety Analysis Report (SAR) (LES, 2005).

11.3.1.1	CM Policy

The goal of the CM program is to ensure that accurate and current documentation matches the facility's physical/functional configuration, and to ensure that IROFS are available and reliable, and comply with regulatory requirements. The CM program will be implemented throughout facility design, construction, testing, and operation. The applicant has defined the CM policy for the facility in Section 11.1.1 of the SAR (LES, 2005). During design and construction, the Engineering and Contracts organization has primary responsibility for the implementation of the CM program, through the design control process. As the project transitions from the design and construction phase to the operations phase, the responsibility for CM transitions to the Technical Services organization. In all cases, the applicant's CM program will provide for the control of key documentation, including the ISA, in accordance with design control, document control, and records management procedures. All design changes will undergo formal review, including interdisciplinary reviews, as appropriate. The applicant's CM program policy statement addresses the following topics: (1) the scope of SSCs and administrative items that will be covered under the CM program; (2) CM interfaces with other management measures and QA; (3) the objectives of the CM program; (4) a description of CM activities; and (5) organizational descriptions of duties and responsibilities.

The CM program will be applied to all SSCs that the ISA identifies as IROFS, and any items that impact the function of IROFS. The CM program will also include all calculations, safety analyses, design criteria, engineering drawings, system descriptions, technical documents, and specifications that establish specific design requirements for IROFS. The scope of documents under CM expands to include relevant data and procedures applicable to IROFS during construction, initial startup, and operations. CM procedures will provide for evaluation, implementation, and tracking of changes to IROFS, as well as processes, equipment, computer programs, and personnel activities that impact the reliability of IROFS.

The applicant describes how the CM function relates to other management measures, including QA; records management; maintenance; training and qualifications; incident investigation, audits and assessments; and procedures. Specifically, the QA Program establishes the framework for CM, and the records associated with IROFS or items affecting IROFS. Records are generated and processed in accordance with QA Program requirements. The design basis, which will be controlled under CM, establishes maintenance requirements.

The applicant clearly states the objectives of the CM function, which will be to ensure design and operation within the design bases of IROFS. This will be accomplished by identifying and controlling documentation associated with IROFS, controlling changes to IROFS, and maintaining the physical configuration of the facility consistent with the approved design. Changes to the approved design are subject to review and approval, to ensure consistency within the design bases of IROFS. The CM function and the design will be subject to periodic audits and assessments to confirm that the design and physical configuration will be consistent with the design basis. Identified problems will be subject to the applicant's corrective action process which is identified in the applicant's QA Program Description (QAPD) and associated implementing procedures. Prompt corrective actions will be developed as a result of incident investigations or in response to unfavorable audit or assessment results.

The applicant provides a description of CM activities. CM activities will be conducted and documented under design control provisions and involve the systematic preparation, review, and approval of design and construction documentation for ensuring that consistency is maintained between design and design bases for IROFS. CM activities also provide for operation of IROFS within the limits specified in the ISA and control changes to the facility in accordance with 10 CFR 70.72. Finally, CM activities provide records for certifying that personnel conducting activities relied on for safety are appropriately trained and qualified and revisions for ensuring only formally reviewed and approved procedures, specifications, and drawings are used for activities associated with IROFS.

During design and construction, the CM program will be implemented by the Engineering and Contracts organization, which is under the direction of the Engineering and Contracts Manager. The engineering organization is organized by discipline. Responsible lead engineers will be in charge of each discipline. In addition to having primary technical responsibility for work performed by their disciplines, lead engineers will be responsible for conducting interdisciplinary reviews of design changes. Lead engineers in each discipline report to design managers or project managers, who report to the Engineering and Contracts Manager. Work performed within the Engineering and Contracts organization is also subject to reviews, as appropriate, by construction management, operations, QA, and procurement personnel. The applicant's various departments and contractors will perform quality-related activities. The QA Programs developed by the applicant's primary contractors will be consistent with the requirements of the applicant's QA Program for those contracted activities determined to be within the scope of the QA Program. All applicant and contractor personnel are responsible for identifying quality problems. If a disagreement exists among personnel concerning a quality-related issue, the issue will be elevated to the appropriate level of management, up to and including the applicant's President.

11.3.1.2   Design Requirements

During design and construction, the Engineering and Contracts organization establishes and maintains design requirements and associated design bases. Whereas, during the operations phase, this function transitions to the Technical Services Organization. Design requirements will be documented and maintained in a design requirements document, which also contains a listing of IROFS. The design requirements documentation and the design-basis documents will be controlled in accordance with the design control provisions of the CM program. The design documents associated with IROFS, items that affect the function of IROFS, and items required to satisfy regulatory requirements, as well as analyses that constitute the ISA of the design bases, will be subject to interdisciplinary reviews and design verification. Design changes are evaluated to ensure consistency within the design basis. Qualified individuals will prepare and review design documents, and the responsible manager approves the corresponding design documentation. The QA director conducts audits on the design-control process, using independent, technically qualified staff.

The applicant describes a design-verification process for design documents where emphasis will be placed on assuring conformance with applicable codes, standards, and license application design criteria. Design verification ensures that design documents are consistent with the design criterion used for IROFS. Design verification will be accomplished through design review, alternative calculations, or qualification testing. Engineering personnel assigned to perform inter-disciplinary checks and reviews of design documents have the authority to withhold approval pending resolution of all questions concerning the design. Design bases will be appropriately referenced in the design document, and the reviewer will verify proper application. Design-verification testing will demonstrate adequate performance under the most adverse design conditions. Independent design verification will be accomplished before other organizations use the design document, to support other activities. In all cases, design verification for an item must be complete, before the item may be relied on to perform its intended function. After verification and approval, the design document will be distributed to the appropriate parties through the document control center.

Configuration control will be accomplished, during design, through the use of design-control procedures, which address preparation, review, verification, approval, release, and distribution of design documents for use. Engineering documents will be assigned a QA-Level verification that is based on safety significance. Configuration verification will be accomplished through the design verification process. All acceptance criteria the designer establishes will be incorporated into documents used to perform work associated with the design. Documentation will also be maintained demonstrating that work has been performed and configuration managed.

11.3.1.3   Document Control

The applicant will be required to establish a document control program that will be implemented and controlled through procedures. Document control procedures govern the preparation and issuance of, as well as changes to, key documents. In addition to document control procedures, the applicant will utilize an electronic document management system that will be used to file project records and to make available the latest controlled copy of design documents. Controlled documents will be maintained in the system until cancelled or superseded. The document control and records management procedures will be a part of the CM program and as such, they will capture the following documents, which are listed in approved procedures: (a) design-requirements documents; (b) design-bases documents; (c) the ISA of the design bases of IROFS; (d) nuclear criticality safety evaluations and analyses; (e) as-built drawings; (f) specifications; (g) procedures that are IROFS; (h) training procedures; (i) QA documents; (j) maintenance documents; (k) audit and assessment reports; (l) emergency operating procedures;

11-4

(m) emergency response plans; (n) system-modification documents; and (o) engineering documents.

11.3.1.4   Change Control

The applicant is mandated by 10 CFR 70.72 to implement a change-control function, and associated procedures that control changes to the technical baseline, as part of it's CM program. Review of the applicant's change control process, the applicant stated that for each change to the facility or to the activities carried out by personnel, an evaluation of the proposed change will be performed in accordance with the requirements of 10 CFR 70.72, as applicable. For every modification to the facility, the applicant has described their process of evaluation of the modification and how the modification will be performed to determine the need for any required changes or additions to the facility's processes, procedures, personnel training, testing program, or regulatory documents. The applicant provided a descriptive review process which will be implemented to ensure changes to the facility's physical/functional configuration, procedures, and controlled documents are implemented in accordance with the provisions of 10 CFR 70.72. Changes to the above items will be controlled through procedures. Before a change is implemented, the change-control process will require an appropriate level of technical, management, and safety review and approval.

During the design phase, changes to the design will include a systematic review of the design bases. Consistency between documents will be ensured through the interdisciplinary review process, which will ensure that design changes do not impact the ISA; or are accounted for in subsequent changes to the ISA; or such design changes will not be approved and/or implemented. Before the issuance of the license, the applicant has committed to notify the NRC of potential changes that reduce the level of commitments or the safety margin in the design bases of IROFS.

During the construction phase, the applicant has described the documentation, review, and approval processes that will be used to control all changes to documents issued for construction, fabrication, and procurement. The review of the applicant's processes, include the evaluation of changes to design documents and an evaluation of the design basis of IROFS and the ISA which will be required to be posted against each affected design document. Vendor drawings and data will undergo interdisciplinary reviews to ensure compliance with engineering and procurement specifications. During construction, design changes will continue to be evaluated against the approved design bases. The applicant is mandated by 10 CFR 70.72 to have a fully implemented CM program which will be inclusive of reporting and tracking changes made without prior NRC approval.

The applicant's description of design control used during the operations phase, describes the methodologies used to document, review, and approve design changes before implementation. During facility operation, the applicant describes measures that will be implemented to ensure that the quality of facility SSCs are not compromised by planned changes and modifications. Modification procedures will be implemented that state the requirements which will be met to implement a modification, as well as the requirements for initiating, approving, monitoring, designing, verifying, and documenting modifications. Modification procedures will be written to ensure that policies are maintained to satisfy the applicant's QA requirements. Each modification will be evaluated for any required changes to the facility's procedures, training programs, testing programs, or regulatory requirements described in 10 CFR 70.72. A nuclear criticality safety evaluation will be prepared and approved for any change that involves or could affect uranium on-site. The evaluation of modifications will take into consideration such items as: (a) radiation exposure; (b) cost; (c) lessons-learned; (d) QA aspects; (e) operability issues; (f) maintenance issues; (g) constructability issues; (h) testing requirements; (i) environmental concerns; and (j) human factors.

11.3.1.5 Assessments

Review of the applicant's CM assessment function included description of periodic assessments of the CM program and design to determine the effectiveness of the CM program and to correct deficiencies. This review of Assessments will be carried out in accordance with the applicant's QAPD, Appendix A, Section 18, "Audit Schedules." The applicant describes the implementation strategy used for scheduling internal and external audits/assessments in a manner consistent with the activities status and importance to on-going work activities. These assessments will be based upon thorough reviews of the adequacy of documentation and effectiveness of the quality assurance program and system walk-downs of the as-built facility. As stated in the applicant's QAPD, Appendix A, Section 18, "Audits" these periodic assessments will confirm the implementation of the CM programmatic elements and verification of compliance and consistency of design with the design bases. If problems or deficiencies are encountered, and have been evaluated as significant conditions adverse to quality for reportability to the NRC, prompt corrective actions will be developed in accordance with the applicant's QAPD, Appendix A, Section 16, Corrective Action Procedures and 10 CFR Part 21, "Reporting of Defects and Noncompliance."

11.3.2 Maintenance

The maintenance and functional testing programs is described in Section 11.2 of the applicant's SAR (LES, 2005). The applicant has outlined those planned and scheduled maintenance and functional testing programs, for IROFS, that will ensure that equipment and controls will be maintained in a condition of readiness, and will perform their safety functions when required. The applicant has described the performance measures and use of surveillance/monitoring activities to detect degradation or adverse trends in the performance of IROFS, so that proactive action may be taken before failure. The applicant has described methods and practices in correlation with corrective maintenance and the applicant's Quality Assurance Program function Section 10, of the applicant's QAPD, " Inspection" will be implemented to ensure that equipment which has unexpectedly degraded or failed will be repaired or replaced. Finally, the applicant has defined a functional testing program that defines how functional testing will be performed under pre-operational and operational conditions, and measures for ensuring that IROFS will be capable of performing their safety functions, when required. The maintenance organization plans, schedules, and tracks maintenance activities, and maintains records for these activities.

The applicant has described implementing measures that ensure that the quality of facility SSCs will be not compromised by planned modifications and/or maintenance activities. Modifications to facility SSCs will be controlled through the use of facility administrative procedures that the Technical Services Manager approves, and the QA Manager concurs. Revisions to these procedures for controlling modifications must also receive approval and concurrence from these two respective managers. The administrative procedures will contain requirements that will implement modification requirements, as well as requirements for initiating, approving, monitoring, designing, verifying, and documenting modifications. Administrative procedures will satisfy the QA standards specified in the applicant's QAPD.

The applicant has described a set of methods and practices that will be applied to it's corrective maintenance, preventive maintenance, and functional-test maintenance elements. The applicant's descriptive programs will include the preparation of authorized work instructions that will contain detailed steps of work sequencing and written procedures associated with the performance of these methods and practices. The applicant has stated that the methodology and practices include: (a) parts lists; (b) as-built or redlined drawings; (c) notification to the Operations organization before conducting repairs and removing IROFS from service; (d) radiation work permits; (e) replacement with like-kind parts and the control of new or replacement parts, to ensure compliance with 10 CFR Part 21; (f) compensatory measures while performing work on IROFS; (g) procedural control of the removal of components from service and for the return of components to service; (h) ensuring safe operations during the removal of IROFS from service; and (i) notification to Operations personnel that repairs have been completed. Written procedures for the performance of maintenance activities will include the methods and practices listed above. The applicant will require contractors performing maintenance work activities involving IROFS to follow the same maintenance procedures that the applicant's personnel are required to use.

When maintenance activities involve IROFS, the applicant has described the methods listed below for corrective maintenance, preventive maintenance (PM), functional testing after maintenance, and surveillance/monitoring activities:

- Pre-maintenance activities require reviews of the work to be performed, including procedure reviews for accuracy and completeness;

- New procedures or work activities that involve or could affect uranium on-site require preparation and approval of a Nuclear Criticality Safety evaluation (NCS) and if required an NCS analysis;

- Procedure steps will require notification of all affected parties (operators and appropriate managers) before performing work, as well as after work is completed (the notification will include potential degradation of IROFS during the planned maintenance); and

- Work control will be implemented through comprehensive procedures that maintenance workers will follow.

The applicant has established protocols for the review of these procedures which include review concurrence obtained from the various safety disciplines, including nuclear criticality safety, fire protection, radiation safety, industrial safety, and chemical process safety. The applicant has structured these procedures to describe as a minimum the following: (a) the qualifications of personnel authorized to perform maintenance, functional testing, or surveillance activities; (b) controls and specifications on the use of replacement components or materials; (c) post-maintenance testing requirements to verify component operability; (d) tracking and records management of maintenance activities; and (e) safe work practices (e.g., component tag-out; confined-space entry requirements; moderation control or exclusion area; radiation or hot work permits; and criticality, fire, chemical, and environmental safety for workers).

11-7

11.3.2.1   Corrective Maintenance

The applicant's corrective maintenance function will involve the repair or replacement of equipment that has unexpectedly degraded or failed. For IROFS, corrective maintenance will restore the equipment to acceptable performance through the use of planning, controlling, and documentation of repair and replacement activities. After performing corrective maintenance on IROFS, and before the applicable IROFS are declared operable, the necessary functional testing of the IROFS will be performed to ensure that the IROFS perform their intended safety functions. The applicant has indicated that, with respect to corrective maintenance, the applicant's Corrective Action Program (CAP) will require facility personnel to determine the cause of conditions adverse to quality and will also require facility personnel to act to correct these conditions.

11.3.2.2   Preventative Maintenance (PM)

The applicant describes their Preventative Maintenance program activities as containing activities that will be inclusive of preplanned, scheduled periodic refurbishment, partial or complete overhaul or replacement of IROFS to ensure that they are reliable and available to perform their intended functions. Planning for PM will take into consideration the results of surveillance and monitoring, including failure history as well as, instrument calibration and testing. PM tasks, frequencies, and procedures will be initially determined using industry experience, industry standards, and vendor recommendations, taking into consideration the appropriate balance between the objectives of equipment failure prevention through PM and the objective of minimizing unavailability of IROFS, because of PM activities. Feedback from acquired experience in maintenance activities and incident investigations will be used, as appropriate, to modify the frequency or scope of PM activities. Deviations from manufacturer recommendations or industry standards will be documented. All records pertaining to PM will be maintained in accordance with the applicant's Records Management System and in accordance with 10 CFR § 70.62 the results of the applicant's PM activities related to IROFS reliability and failure will be evaluated by all safety disciplines to determine any ISA impact and the need for any updates.

The applicant will develop and implement the necessary post-maintenance functional testing of SSCs after conducting PM on IROFS and before returning IROFS to service. Post-maintenance testing will ensure that IROFS that have undergone PM are capable of performing their safety functions.

11.3.2.3   Surveillance and Monitoring

The applicant has described their surveillance and monitoring system as being proactive, which will be utilized to detect degradation and adverse trends of IROFS and SSCs identified as IROFS, prior to component failure. This proactive system primarily consists of periodic surveillance tests, plant computer monitoring, operator rounds, and walk-downs, and/or inspections performed at specified intervals for all IROFS, as well for all items that could impact the function of IROFS and SSCs identified as IROFS. The applicant's surveillance and monitoring system utilizes the selection of measurements of monitored performance specifications to detect degradation or adverse trends in IROFS, whereas, proactive corrective action may be taken prior to failure. Results of surveillance monitoring will be trended to detect signs of performance degradation. Any signs of potential performance degradation will be addressed by appropriate corrective actions, such as corrective maintenance or adjustment of preventive maintenance frequencies. Parameters to be monitored or evaluated during surveillance/monitoring activities will be selected, based on their ability to detect the predominate failure modes of critical components. Plant performance criteria to be monitored will be established using industry operating experience, plant equipment operating experience, operating data, and surveillance data.

The applicant has stated that the surveillance program will adhere to "Inspection, Testing, and Maintenance Baseline Design Criteria," as stated in 10 CFR 70.64. The surveillance(s) will be included in the work control process, and the results will be trended. Frequencies will be adjusted or other corrective actions taken as necessary.

11.3.2.4 Functional Testing

The applicant's description of provisions describing their functional testing program will be applied to IROFS after initial installation, as part of periodic surveillance testing, and after the performance of corrective maintenance or PM, or calibration to verify that IROFS are capable of performing safety functions after these maintenance activities. The applicant's overall functional testing program will consist of a two primary sets of test programs; pre-operational testing program, and an operational testing program. The Pre-operational testing program is comprised of two subsets of tests which include functional tests and initial startup testing. Whereas, the Operational testing program is comprised of two different subsets of tests, which include periodic and special testing. The applicant describes the objectives of the preoperational and operational testing programs are to provide assurance that IROFS meet applicable contractual, regulatory, licensing, and design requirements, and can be relied on to perform design functions required to protect the health and safety of workers and the public. Test procedures will be sufficiently detailed to ensure that qualified personnel can perform the required functions without direct supervision. Test procedure content will be consistent and will consist of the following sections: (a) Title; (b) Purpose; (c) References; (d) Time Required; (e) Prerequisites; (f) Test Equipment; (g) Limits and Precautions; (h) Required Plant Unit Status; (i) Prerequisite System Conditions; (j) Test Method; (k) Data Required; (l) Acceptance Criteria; (m) Procedure; and (n) Enclosures. Review of the above noted items (a through n) the applicant has provided broader detail and elaboration on specific attributes for each test procedure section in the SAR (LES, 2005).

Preoperational Testing Program

The applicant describes their preoperational testing program primarily consisting of functional testing and initial start-up testing. All preoperational functional tests associated with IROFS will be completed before the introduction of licensed material at the facility. Before the start of pre-operational functional testing, the facility will undergo a constructor turnover process, where systems or portions of systems will be turned over to the applicant following as-built drawing verification, purging, cleaning, and initial instrument calibration by the constructor. Functional testing will begin as systems or portions of systems are turned over to the applicant. Pre-operational functional testing at the facility will be conducted to initially determine various facility parameters and to verify that IROFS are capable of performing their intended functions and that facility SSCs are capable of meeting performance requirements. These tests are primarily associated with IROFS (QA Level 1) and certain QA Level 2 SSCs. For SSCs that are not QA Level 1, acceptance criteria will be established to ensure worker safety and reliable system operation. Initial-start-up testing will involve the initial introduction of licensed material and all subsequent testing through completion of Enrichment Setting Verification. Initial-start-up testing begins with the initial introduction of licensed material and ends with the start of commercial operation. Its purpose is to ensure safe introduction of licensed material and to verify parameters assumed in the ISA.

The applicant states that only properly reviewed and approved test procedures and plans will be used for all pre-operational tests. The applicant ensures that copies of approved test procedures will be made available to NRC staff approximately 60 days before implementation, and not less than 60 days before the scheduled introduction of licensed material for initial-start-up testing. The pre-operational test plan will be made available to NRC at least 90 days before the start of testing. Records of testing schedules and test results for all IROFS will be maintained for all required pre-operational testing. The responsible department manager will review and approve the results of each pre-operational test before the test process continues, where test results form the basis for continuing the scheduled test program. Retesting will be conducted as necessary. All modifications to IROFS that the applicant considers necessary during the testing phase will be evaluated in accordance with 10 CFR 70.72, and the impact of these modifications will be evaluated during the 10 CFR 70.72 evaluation process. To ensure the adequacy of operating, emergency, and surveillance procedures under actual or simulated operating conditions, and before beginning actual plant operations, the applicant has agreed to the trial use of these procedures throughout the testing program phases, as practicable.

Operational Testing Program

The applicant's operational testing program will consist of periodic testing and special testing. Periodic testing consists of testing conducted on a periodic basis, to provide continuing verification that IROFS are capable of meeting performance requirements. Periodic testing also includes surveillance tests and post-maintenance tests. Special testing is defined as testing that is of a nonrecurring nature and does not fall under any of the other testing programs. The Maintenance Manager will have the overall responsibility for the development and conduct of the operational testing program. The operational testing program will use test coordinators for various tests to be performed. The test coordinators are responsible for: (a) verification of all testing prerequisites; (b) observance of all testing limits and precautions during the test evolution; (c) compliance with facility license requirements and other facility directives; (d) identification and implementation of corrective actions necessary to resolve system conditions observed during the test evolution; (e) verification of proper data acquisition and evaluation; (f) ensuring the observance of personnel safety precautions during test evolutions; and (g) coordinating additional test support from other organizations. Periodic and special testing procedures will be sufficiently detailed such that qualified personnel can perform the required functions without direct supervision. Procedures will be reviewed and approved before implementation.

The applicant's periodic testing program, will be a component of the operational testing program and will be used to verify that the facility is capable of operating in a reliable manner and will ensure that the health and safety of workers, the public, the environment, and compliance with all regulatory requirements is achieved. The periodic testing program begins during the pre-operational testing phase and continues throughout the life of the facility. A periodic testing schedule will be established to ensure timely performance of testing requirements and evaluation of testing results. The applicant describes the use of a computer database within the

SAR (LES 2005) which will be used for scheduling and organizing surveillance test requirements, procedures, and activities, and for ensuring that surveillance requirements are met and are within the required test interval. The applicant's database will include all periodic testing, calibrations, or inspections required by regulations or license conditions. For each surveillance requirement, the applicant's database will provide information concerning the specific test to be performed, test equipment, plant cascade, test frequency, and testing dates for the preceding and next required surveillance test. Testing will be scheduled such that the safety of the plant is not dependent upon the performance of IROFS that have not been tested within specified testing intervals. If system or plant-unit conditions preclude the performance of a test within its required interval, the applicant's Health, Safety, and Environment (HS&E) Manager, Operations Manager, and Maintenance Manager will be notified, in writing, by the responsible department, and the responsible department will be required to ensure that the test will be performed as soon as practical, once the required conditions are met. All periodic testing associated with QA Level 1 and QA Level 2 SSCs will be performed in accordance with approved written procedures.

The applicant's special testing program encompasses all testing that is not pre-operational testing, periodic testing, post-modification testing, or post-maintenance testing. Special testing will be non-recurring testing conducted to determine facility parameters and/or to verify the capability of IROFS to meet performance requirements. Special testing will be conducted for such purposes as data acquisition and/or information-gathering, for special analyses or incident investigations, verification of the effectiveness of corrective actions, and confirmation of results expected from facility modifications, where such confirmatory testing falls outside the realm of routine post-modification or post-maintenance testing. The applicant's Plant Manager will have the authority to determine when certain tests will be conducted as special tests. This determination will be made based on evaluations that consider such factors as: (a) whether the proposed testing evolution involves unusual operational configurations for which no prior experience exists; (b) whether a potential exists for conducting an improper test that could have an impact on primary plant parameters; and (c) whether the test activity involves seldom-performed evolutions, where any previous experience with such evolutions is no longer relevant to the overall test activity.

11.3.3 Training and Qualifications

The applicant discusses training and qualification programs for the operations phase of the facility in Section 11.3 of the SAR (LES, 2005). Operations phase training programs include training for pre-operational functional testing and initial-start-up testing. The applicant's purpose and objective of the training program is to ensure that training is administered and job proficiency is maintained for all facility personnel who perform activities relied on for safety. The applicant states that qualification will be indicated by both successful completion of prescribed training, demonstration of the ability to perform assigned tasks, and, where required by regulation, maintaining an active and valid license issued by the regulatory agency. Training will involve formal training to establish knowledge of the principles in the relevant job field and on-the-job training to develop work performance skills. Continuing training or periodic retraining will be provided to maintain the knowledge, skills, and proficiency of personnel in the relevant fields. The applicant's QAPD, Appendix A, "Quality Assurance Training" provides training and qualification requirements for QA personnel during the design, construction, and operations phases. The Quality Assurance Training detailed by the applicant in Appendix A of the QAPD identifies QA Indoctrination Training provisions for personnel who perform QA Level 1 activities which are applied exclusively to IROFS or any items which are determined to affect the function of IROFS. The applicant's QA Indoctrination and Training provisions include general criteria inclusive of applicable codes, standards, QA implementing procedures, QA Program elements and job responsibilities and authorities. In addition to this training, personnel who perform QA Level 1 activities are required to complete training in job-specific QA activities prior to the performance of QA Level 1 work.

11.3.3.1 Organization and Management of the Training Function

In Section 11.3.1 of the SAR (LES, 2005), the applicant states that line management will be responsible for the content and for conducting training of personnel. The applicant describes their organization and management structure in detailed position descriptions in Chapter 2 of the SAR (LES, 2005). The training organization has been delineated to provide support to line management through planning, directing, analyzing, developing, conducting, evaluating, and controlling of systematic performance-based training. For personnel performing activities relied on for safety, the relevant indoctrination and training requirements are established through facility administrative procedures. Exceptions from training requirements as stated by the applicant, will require formal documented justification and approval by appropriate management. The Human Resources Manager will be responsible for training programs at the facility. The applicant will develop, implement and use lesson plans for all classroom, on-the-job training, and maintenance of training records on each employee's qualifications, and experience.

11.3.3.2 Analysis and Identification of Functional Areas Requiring Training

The applicant has described the methods that will be employed to perform an analysis of job training needs and to ensure that personnel who work on tasks related to IROFS are provided the appropriate training. Furthermore, the applicant states that employees will participate in Job Hazard Analysis (also referred to as Job Safety Analysis), which will be used as part of on-the-job training, and will provide employees with the skills required to safely conduct job activities. The Job Hazard Analysis process will be used to enhance personnel recognition for defining the work scope, identifying the hazards, mitigation of the hazard, completion of the task and providing feedback. In developing the list of training tasks for specific job training activities, the training organization will consult the appropriate technical and management personnel. During the consultation process, the applicant will employ and list specific tasks pre-selected for training which are reviewed and compared to training materials used to systematically evaluate training. The task list will be refined and updated when changes occur to procedures, processes, plant systems, equipment or job scope.

11.3.3.3 Position Training Requirements

The applicant's Position Training Requirements provide a description for developing minimum training requirements for positions that involve activities which are relied on for safety or who perform functions that prevent or mitigate an accident sequence described in the ISA Summary . The initial identification of job-specific training requirements will be based on experience, and the level at which an employee initially enters the training program will be based on the employee's past experience, level of ability, and qualifications.

The applicant has stated that facility personnel may receive training in the areas of General Employee Training, Technical Training, and/or Employee Development/Management-

Supervisory training. The applicant's General Employee Training will encompass: (a) QA policies and procedures, (b) nuclear safety (inclusive of dosimetry, protective clothing and equipment), (c) industrial safety, (d) health and first aid, (e) facility security, (f) emergency planning and implementing procedures, (g) administrative controls and procedure use, (h) facility systems and equipment, (i) chemical safety, (j) fire protection and fire brigade, (k) new employee orientation, and (l) applicable regulations. General Employee Training will be required for all personnel and contractors who report to facility management. Technical training will help facility employees gain a technical understanding of the fundamental concepts, practices, and procedures common to a gas centrifuge uranium enrichment facility and will develop the skills necessary to competently perform assigned work. Technical training will consist of the following: (a) Initial Training for providing an understanding of fundamental principles, concepts, and procedures; (b) On-the-Job Training and Qualifications for providing the necessary job-related skills for the position; (c) Continuing Training for maintaining and improving job-related knowledge and skills; and (d) Special Training, which involves subjects, required for a specific job, that are unique in nature.

Nuclear Safety Training

The applicant 's Nuclear Safety Training Programs will be broad based and encompass various types of job functions such as, (production operator, radiation protection technician and contractor personnel) which will be used in conjunction with criticality safety and radiation safety responsibilities associated with each position. The applicant describes their nuclear safety training program as being structured in a manner that emphasizes the high level of importance placed on radiological, criticality and chemical safety of plant personnel and the public. Nuclear Safety Training will be structured as follows:

Personnel access procedures are developed to ensure the completion of formal nuclear safety training prior to permitting unescorted access into Controlled Access Areas. Comprehensive training sessions will be administered covering such topics as criticality safety, radiation protection, and emergency procedures. The applicant's training sessions include the following course topics and subject matter:

a.   Notices, reports and instructions to workers,
b.   Practices designed to keep radiation exposures ALARA
c.   Methods of controlling radiation exposures
d.   Contamination control methods (including decontamination)
e.   Use of monitoring equipment
f.   Emergency procedures and actions
g.   Nature and sources of radiation
h.   Safe use of chemicals
i.   Biological effects of radiation
j.   Use of personnel monitoring devices
k.   Principles of nuclear criticality safety
l.   Risk to pregnant females
m.   Radiation protection practices
n.   Protective clothing (donning and doffing practices)
o.   Respiratory protection
p.   Personnel surveys

The applicant's Criticality Safety Training is described as being developed and implemented in accordance with ANSI/ANS 8.19, "Administrative Practices for Nuclear Criticality" (ANSI/ANS, 1996) and ANSI/ANS 8.20, "Nuclear Criticality Safety Training" (ANSI/ANS, 1991). The applicant's nuclear safety training program will be evaluated and updated annually through roundtable safety meetings held by supervisors and managers.

<u>Fire Brigade Training</u>

The applicant's descriptive scope and intent of Fire Brigade Training is to develop a core group of facility personnel skilled at fire prevention, fire fighting techniques, first aid procedures, and emergency response. Fire Brigade Training Program requirements will be used for initial training of all new fire brigade members, semi-annual class room training and emergency response drills, annual classroom training and drills, annual practical training, and leadership training for fire brigade leaders.

<u>Technical Training</u>

Technical Training as described by the applicant will be designed, developed and implemented to provide assistance to facility employees in retaining a comprehensive knowledge base of the fundamentals, procedures and practices employed in a gas centrifuge uranium enrichment facility. Technical training will be also used to develop manipulative skills that are required to demonstrate competency. Technical training will be divided into four segments: Initial Training, On-the-Job Training and Qualifications, Continuing Training, and Special Training.

<u>Initial Training</u>

The applicant describes initial training and qualification programs as being used to provide an understanding of the fundamentals, basic principles, and procedures used to control work evolution or activity. Initial training will be administered in various formats, which include live lectures, taped and filmed lectures, self-guided study, demonstrations, laboratories and workshops, and on-the-job-training. Initial job training and qualification programs will be developed for operations, maintenance and technical services classifications. Logical sequencing or block modules will be presented in a objective manner that human behavioral objectives will be identified. The applicant has identified initial training programs which will be inclusive of; Operations Initial Training, Mechanical Maintenance Initial Training, Instrumentation and Electrical and Maintenance Initial Training, Health Physics and Chemistry Initial Training, and Engineer/Professional Initial Training

<u>Operations Initial Training (General/Specific Systems, Nuclear Preparatory, Plant Familiarization)</u>

General Systems training, as described by the applicant, is comprised of training and instruction that will be developed to provide a trainee with basic knowledge on concepts and fundamentals of mathematics, physics, chemistry, heat transfer, and electrical theory. The applicant has augmented this training with on-the-job-training associated with detailed systems and component operation supplemented by elementary process instrumentation and control.

Specific Systems training is described by the applicant as providing basic instruction in system and component identification and basic system operating characteristics. This training module

11-14

will provide a general overview of enrichment plant equipment, nomenclature, and instruction describing basic system operations.

Nuclear Preparatory training will be provided by the applicant to operations personnel preceding completion of Specific Systems training. Training objectives described in this training module include basic nuclear physics, plant chemistry, thermodynamics, radiation protection, and enrichment theory.

Plant Familiarization will be provided by the applicant as orientation of employees to plant layout, systems, and practical laboratory and equipment work at the facility.

Mechanical Maintenance Initial Training (General Systems, Fundamental Shop Skills, Plant Familiarization)

General Systems Training is the same course module as described in Operations Initial Training and will also be provided to Mechanical Maintenance personnel during initial training.

Fundamental Shop Skills as described by the applicant will provide instruction in fundamental mechanical maintenance performance characteristics. This training module will emphasize academic instruction combined with hands-on-training familiarizing trainees with design operational and physical characteristics of enrichment facility components and the basic skill sets used to perform mechanical repairs or equipment replacement. Factors applied by the applicant include task training lists, which will be integrated into the module to provide assurance that each prospective trainee attains a minimum level of performance. Practical tasks will be assigned to trainees and correlated to the use of work procedures as a guide to task completion with emphasis placed on radiological and industrial safety.

Plant Familiarization Training is the same course as previously noted, the exception as noted by the applicant is that this training module will also be provided to Mechanical Maintenance personnel during initial training.

Instrumentation and Electrical and Maintenance Initial Training (General Systems, Basic Instrument and Electrical, Basic Performance, Plant Familiarization)

General Systems Training is the same course module as described in Operations Initial Training and will also be provided to Instrumentation and Electrical and Maintenance personnel during initial training.

Basic Instrument and Electrical training is described by the applicant as providing prospective trainees with refresher training in electrical and electronic fundamentals, digital techniques and application, instrumentation and control theory and application and instruction in measuring and test equipment use. The applicant describes this training module as providing a general overview and working knowledge of nuclear and non-nuclear instrumentation systems, overall integrated plant operation and control, and the hazards of calibration errors and calibration during plant operation.

Basic Performance Training as explained by the applicant, describes fundamental performance familiarization and instruction provided to trainees regarding plant test procedures, test equipment, testing, plant records and reports, as well as data collection during operations.

11-15

Training will be provided by the applicant in the basic application and use of thermodynamics utilized in plant heat transfer.

Plant Familiarization Training is the same course as previously noted, the exception as noted by the applicant is that this training module will also be provided to Instrumentation and Electrical Maintenance personnel during initial training.

Health Physics and Chemistry Initial Training (General Systems, Fundamental Health Physics, Fundamental Chemistry, Plant Familiarization)

General Systems Training is the same course module as described in Operations Initial Training and will also be provided to Health Physics and Chemistry personnel during initial training.

Fundamental Health Physics training is described by the applicant as providing prospective trainees with comprehensive and theoretical understanding of nuclear processes. Training objectives include practical presentations demonstrating the proper techniques and application of theory in the use of non-automated counting and spectrographic equipment and use of portable survey instruments . The applicant describes this training module as providing a detailed overview of administrative materials.

Fundamental Chemistry training is provided by the applicant to familiarize trainees with training pertaining to chemistry, chemistry theory, techniques, and procedures. The applicant's emphasis will be placed on familiarization of trainees with chemical safety hazards in the laboratory and during plant operation.

Plant Familiarization Training is the same course as previously noted, the exception as noted by the applicant is that this training module will also be provided to Health Physics and Chemistry personnel during initial training.

Engineer/Professional Initial Training (Facility Orientation, Basic Engineer/Professional Training, Enrichment/Chemical Engineer/ Professional Training, Engineer/ Professional Systems Training)

The applicant's Engineer Professional training is for the applicant's technical staff and managers. Professional Engineer training is described as follows:

Facility Orientation training is provided by the applicant as an orientation to each section of the enrichment facility. Training will be provided based upon on-the-job-task lists, which detail training objectives that will be accomplished while working in the section.

Basic Engineer/Professional Training is described by the applicant as providing prospective trainees with a basic understanding of how uranium is enriched. Training objectives will include practical presentations demonstrating the systems and components required for producing the final product and the interrelationship and interface requirements with the various facility organizations used to achieve facility objectives.

Enrichment/Chemical Engineer/ Professional Training is described by the applicant as providing chemical engineer's with comprehensive and theoretical understanding of enrichment operations. Training objectives which will be addressed specifically by the applicant include; thermal science, nuclear physics, and application of theory in an enrichment facility.

11-16

Engineer/ Professional Systems Training will be provided by the applicant as an overview of plant systems, components and procedures necessary to operate an enrichment facility safely and efficiently.

<u>On-the Job Training and Qualifications</u>

On the Job Training and Qualifications has been described by the applicant in a systematic manner which takes into consideration job related skills and knowledge of the position. The foundation of the OJT/qualifications program will comprise applicable tasks and related procedures for each respective technical area which the applicant states will be designed to supplement and compliment training received through formal classroom, laboratory and simulator training. The applicant concludes by stating that the objective of the OJT Training program is to assure the trainee's ability to perform job specific tasks as described in detailed task descriptions and Training and Qualification Guides.

<u>Continuing Training</u>

Continuing Education is defined and described by the applicant as any training that is not provided as initial qualification and basic training which maintains and improves job-related knowledge and skills. Continuing training will be used by the applicant to supplement and provide additional training when warranted. Categorization of continuing training will be as follows:

a.   Facility systems and component changes;

b.   OJT/Qualification program testing;

c.   Policy and procedure changes;

d.   Operating experience program documents review to include Industry and in-house operating experiences;

e.   Continuing training required by regulation (e.g., emergency plan training);

f.   General employee, special, administrative, vendor, and/or advanced training topics supporting tasks that are elective in nature;

g.   Training identified to resolve deficiencies (task-based) or reinforce seldom used knowledge skills;

h.   Refresher training on initial training topics;

i.   Structured pre-job instruction, mock-up training, walk-throughs; and

j.   Quality awareness.

Components of the applicant's continuing training process is comprised of both formal and informal delivery systems which will be administered at frequencies based on the need to maintain on the job proficiency. A systematic approach will be developed by department or

11-17

section and results from data gathered from job performance and safe facility operation will be used for determining the content and context of continuing training.

<u>Special Training</u>

Special Training has been described and defined by the applicant as involving training subjects of unique nature required for a particular area of work. The applicant only administers special training to select personnel based on specific needs not directly related to discipline specific qualifications.

11.3.3.4   Basis for Training and Training Objectives

The applicant's Basis for Training and Training Objectives program content will be established based on an analysis of job training needs and position-specific requirements. Overall training objectives will include the demonstration of knowledge, skills, and abilities, as well as the achievement of performance standards. The applicant states that the basis for safety training will be derived from compliance with applicable sections of the Occupational Safety and Health Administration (OSHA) regulation 29 CFR 1910 (Occupational Safety and Health Standards), 1910.1200 (Hazard Communication), 10 CFR Part 20 (Standards for Protection Against Radiation) and 10 CFR 19 (Notices, Instructions, and Reports to Worker: Inspection and Investigations). The applicant will provide continuing training in these areas to maintain employee proficiency. Training basis and the topics covered will be provided to all persons under the direct supervision of facility management (including contractors) and will be required to participate in General Employee Training. Provisions have been stipulated for certain facility support personnel which is dependent upon normal work assignment, which will exclude participation in all topics of some training. Temporary maintenance and service personnel receive General Employee Training to the extent necessary to assure safe execution of their assigned work duties. In some cases, the applicant may include portions of General Employee Training with New Employee Orientation.

11.3.3.5   Lesson Plans and Other Training Guides

Lesson plans will be used for classroom training and on-the-job training, as required, and include trainee performance standards. Lesson plans will be developed from the learning objectives, which will be based on job performance standards. Lesson plans and training guides will be developed and reviewed by the training function organization, as well as the organization that is cognizant in the subject matter.

11.3.3.6   Evaluation of Trainee Accomplishment

The applicant states that observation/demonstration or oral or written tests will be used, as appropriate, to evaluate and measure a trainee's understanding and command of learning objectives and knowledge and skill sets acquired for job performance. Evaluations will be performed by individuals qualified in the training subject matter.

11.3.3.7   Conduct of On-the-Job Training

On-the-job training will be used in conjunction with classroom training for activities identified as IROFS. Successful completion of on-the-job training requirements involves the performance of either the actual task or a simulation of the task. On-the-job training using task simulation will be carried out with the trainee explaining task actions using conditions encountered during the actual performance of the task.

11.3.3.8   Evaluation of Training Effectiveness

The applicant will periodically conduct systematic evaluations of the training program as a means to measure the program's effectiveness in ensuring the job proficiency and competency of all facility personnel is maintained. These evaluations will identify strengths and weaknesses in the training program and will determine whether the program meets current job needs. The need for any corrective actions will be determined. The applicant has outlined specific methods that may be used to collect training program evaluation data, including, among other things, testing to determine trainee accomplishment of objectives, trainee evaluation of the instruction, supervisory evaluation of job performance after training, and supervisory evaluation of the instruction. The QA Department will audit the facility training and qualification program. QA assessments of training program or topical areas may include the following elements:

a.   Management and administration of training and qualification programs;

b.   Development and qualification of the training staff;

c.   Position training requirements;

d.   Determination of training program content, including facility change control interface with configuration management systems;

e.   Design and development of training programs;

f.   Conduct of training;

g.   Trainee examinations and evaluations; and

h.   Training program assessments and evaluations.

The applicant describes the evaluation process and documentation of the results with emphasis placed on highlighted program strengths and weakness. Surveys, questionnaires, performance appraisals, staff evaluation, and overall effectiveness evaluations will be used as instruments for verification of implementation of the program.

11.3.3.9   Personnel Qualification

The applicant states that qualification and training requirements for process operator candidates will be established and implemented in plant procedures. The applicant describes the qualification requirements for key management positions in Chapter 2 of the SAR (LES, 2005). QA personnel also have specific training and qualification requirements, which are specified in the applicant's QAPD provided in Appendix A to Chapter 11 of the SAR (LES, 2005).

11.3.3.10   Periodic Personnel Evaluations

The applicant states that personnel who perform activities related to items relied on for safety (IROFS) will be evaluated at least biennially, to determine whether they are competent to continue performing these activities. The applicant's evaluation process will use written tests, oral tests, or on the job performance appraisals to conduct and document periodic personnel evaluations. Retraining will be provided when the results of periodic evaluation dictate the need. Retraining will also be administered when facility modification, procedure changes, and QA Program changes, result in new or revised information.

11.3.4 Procedure Development and Implementation

The applicant discusses it's programs for the development and implementation of procedures in Section 11.4 of the SAR (LES, 2005). The applicant states that all activities involving licensed materials or IROFS will be conducted in accordance with approved procedures, which will be used to control activities to ensure that they are conducted in a safe manner and in accordance with regulatory requirements. Procedures will be made available to NRC for inspection before the start of enrichment operations at the facility.

The applicant identifies four types of plant procedures that will be used to control activities at the facility: (1) operating procedures, (2) administrative procedures, (3) maintenance procedures, and (4) emergency procedures. Operating procedures will be used to directly control process operations at the facility. The applicant will implement a methodology for identifying, developing, approving, implementing, and controlling operating procedures. The identification of needed procedures will take into consideration the results of the ISA. Operating procedures will include, among other things: (a) the purpose of the activity; (b) governing regulations and policies; (c) steps for each operating process phase; (d) hazards and safety considerations; (e) operating limits; (f) precautions to prevent exposures; (g) measures to be taken in case of exposure; (h) IROFS associated with the process; and (i) the time frame for which the procedure is valid. The applicant states that applicable safety limits and IROFS will be clearly identified in the operating procedures.

Administrative procedures will be used to perform activities that support process operations. These procedures will control activities that include management measures such as: (a) CM; (b) nuclear criticality safety, radiation safety, chemical safety, and fire safety; (c) QA; (d) design control; (e) training and qualification; (f) audits and assessments; (g) incident investigations; (h) record-keeping and document control; (i) reporting; and (j) procurement. Administrative procedures will also be used for implementing facility security and emergency plans, including the Fundamental Nuclear Material Control Plan, the Emergency Plan, the Physical Security Plan, and the Standard Practice Procedures Plan for Protection of Classified Information.

Maintenance procedures will address: preventive and corrective maintenance of IROFS; surveillance; functional testing of IROFS; and pre-maintenance activity requirements. Emergency procedures will address the preplanned actions of operators and other plant personnel, in case of an emergency.

The applicant has described a process for creating, reviewing, and approving procedures. Procedure development will be carried out by the appropriate members of the facility staff or facility contractors. Initial "Draft Procedures" will be reviewed by other appropriate members of the facility staff, by personnel from the centrifuge supplier, and other vendors, as warranted. These reviews will verify the technical accuracy of the procedure and will include either a walk-down of the procedure in the field or a table-top walk-through. An independent review will be carried out for procedures that will be written for the operation of IROFS. The Plant Manager or designee will approve all procedures. If the procedure directly involves elements of QA, the QA Manager shall approve the procedure. The need for cross-disciplinary reviews will be considered during the procedure review process. The designated approving official will be responsible for determining the need for cross-disciplinary reviews.

The applicant states that procedures will be subject to re-verification and re-validation through periodic reviews, as a means to ensure continued accuracy and usefulness. Periodic reviews of operating procedures will be conducted at a minimum frequency of once every 5 years. Periodic reviews of radiation protection procedures and emergency procedures will be conducted at a minimum of once every year. Furthermore, in response to unusual incidents, including accidents, unexpected transients, operator errors, or equipment malfunction, the affected procedures will be reviewed. The applicable procedures will be reviewed after any modification to a system.

The applicant has defined a set of guidelines that describes how changes to procedures will be processed. The process for changes to procedures will first involve documentation of the proposed change, as well as the reason for the change. An evaluation of the proposed change will be performed in accordance with 10 CFR 70.72, as appropriate. If after the applicant's evaluation and determination is made that an amendment to the license is required to implement the proposed changes, the change will not be implemented until prior approval is received from NRC. The revised procedure, with the proposed changes, will then be reviewed by a qualified reviewer. The Plant Manager, a Department Manager, or a designee approved by the Plant Manager will be responsible for approving procedure changes, and for determining whether a cross-disciplinary review is necessary. The need for cross-disciplinary reviews will, as a minimum, be considered for proposed changes having a potential impact on chemical safety, radiation safety, and nuclear criticality safety, as well as material control and accounting.

The applicant states that the Document Control organization will distribute originally issued approved procedures and approved procedure revisions in a controlled manner. An index of the distribution of copies of all facility procedures will be established and maintained by document control. Revisions to procedures will be controlled and distributed in accordance with this index. Indexes will be reviewed and updated on a periodic basis or as required. Department managers or their designees will be responsible for ensuring that all personnel carrying out work activities, which require the use of procedures and have ready access to controlled copies of the procedures. On-the-job training and continuing training will be used to ensure that personnel are qualified to use the latest procedures. Both these training categories are components of the applicant's technical training program.

11.3.5 Audits and Assessments

In Section 11.5 of the SAR (LES, 2005), the applicant is required to implement a system of audits and assessments to help ensure that facility activities are conducted in a safe manner and for verifying compliance with regulatory requirements, licensing commitments, and procedures. The applicant describes a system of audits and assessments that consists of two distinct levels of activities: an audit activity structured to monitor compliance with regulatory requirements, licensing commitments, and facility procedures; and an assessment activity oriented toward determining the effectiveness of the activities in ensuring that IROFS are reliable and available to perform their intended safety functions. The applicant's approach to implementing the system of audits and assessments will involve performing audits and assessments on critical work activities associated with facility safety, environmental protection, and other areas.

Audits and assessments will be performed to assure that facility activities are conducted in compliance with regulatory requirements, licensing commitments, and written procedures, and that the processes and activities reviewed, effectively meet the applicant's specified objectives for ensuring the continued availability and reliability of IROFS. The applicant will implement a set of functions for which audits and assessments will be conducted. Audits and assessments will be conducted for the following areas: (a) nuclear criticality safety, radiation safety, and chemical safety; (b) industrial safety and fire protection; (c) environmental protection; (d) emergency management; (e) QA; (f) CM; (g) maintenance; (h) training and qualification; (i) procedures; (j) CAP and incident investigation; and (l) records management. The applicant has also specified actions that will be carried out in performing typical audits and assessments.

These specified actions include: (a) using approved audit and assessment checklists; (b) interviewing responsible personnel; (c) performing plant area walk-downs; (d) reviewing controlling plans and procedures; (e) observing work in progress; and (f) reviewing completed QA documentation.

The frequency of audits and assessments will be based on the status, safety significance, and history of the activities being performed. All major activities will be audited or assessed annually. Nuclear criticality safety audits will be conducted and documented quarterly, such that all aspects of the Nuclear Criticality Safety Program will be audited at least every 2 years. The operations organization will be assessed periodically to ensure that nuclear criticality safety procedures are being followed and the process conditions have not been altered to adversely affect nuclear criticality safety. Assessments will be conducted at least semiannually. A schedule will be established that identifies the audits and assessments to be performed and the responsible organization assigned to conduct the activity. This schedule will be reviewed periodically and revised as necessary to ensure the appropriate coverage, based on current and planned activities.

The applicant states that audits and assessments will be performed routinely by qualified staff who are not directly responsible for production activities. The QA department, under the direction of the QA Manager, will be responsible for conducting audits. The conduct of the audit will be carried out by an audit team, working under the direction of a lead auditor. The audit team will consist of one or more certified auditors. Audit team members will not have direct responsibility for the function and area being audited; however, audit team members will have technical expertise or experience in the area being audited. Auditors and lead auditors will be responsible for ensuring the performance of audits in accordance with applicable QA procedures and will hold certifications as required by the QA Program. The auditor certification process involves training in the following areas: (a) QA Program; (b) audit fundamentals, including audit scheduling, planning, performance, reporting, and follow-up action involved in conducting audits; (c) objectives and techniques of performing audits; and (d) on-the-job training.

Certification of auditors is based on the evaluation of the QA Director or Manager. This evaluation will take into consideration, among other things, such factors as: (a) education; (b) experience; (c) completion of the required training; (d) professional qualifications; (e) leadership; (f) sound judgment; (g) maturity; (h) analytical ability; and (l) past performance. A lead auditor must have participated in a minimum of five QA audits or audit equivalents within a period of time not to exceed 3 years before the date of certification.

Assessments will fall into two categories, which will both be owned and managed by line organizations. Management assessments will be conducted by line organizations that are responsible for the work activity. Independent assessments will be conducted by individuals not involved in the area being assessed. Personnel performing assessments will not require certification; however, they will be required to complete QA orientation training, as well as training on the assessment process. Nuclear criticality safety assessments will be performed under the direction of the criticality safety staff. The personnel who perform these assessments will not report to the production organization, and will have no direct responsibility for the function and/or area being assessed.

The applicant will use approved procedures for conducting internal and external audits and assessments. These procedures will meet the QA Program requirements and will provide for the following audit and assessment activities: (a) scheduling and planning of audits and assessments; (b) certification requirements for audit personnel; (c) development of audit plans and audit and assessment checklists, as applicable; (d) audit and assessment performance; (e) reporting and tracking of findings to closure; and (f) closure of audits and assessments.

The applicant will implement a system that captures the planned use of the results of audit and assessment activities, including the planning and implementation of corrective actions, based on audit and assessment findings and recommendations.

11.3.6 Incident Investigations

Section 11.6 of the SAR (LES, 2005) describes the applicant's incident investigation and corrective action process for investigating abnormal events and completing the appropriate corrective actions. The applicant will implement an incident investigation process for reporting deficiencies, abnormal events, and potentially unsafe conditions or activities. The applicant's overall incident investigation process will provide for incident identification, investigation, root-cause analysis, environmental protection analysis, recording, reporting, and follow-up. These activities will be performed according to written corrective action process procedures. Each event or condition will be evaluated to determine the level of investigation required. Guidance for evaluating the significance of occurrences will be contained in corrective action process procedures, and the extent of the investigation will depend on the significance of the incident, with respect to the levels of uranium released and/or the potential for exposure to workers, the public, or the environment.

The investigation process for abnormal events will implement the requirements for event notification to government agencies in accordance with regulations. The HS&E Manager will be responsible for maintaining a list of agencies to be notified, determining if a report to an agency is required, and notifying the agency when required. The applicant has described their process for reporting events to the NRC, as required by 10 CFR 70.50 and 70.74. Abnormal events that have the potential for actual HS&E consequences will be appropriately identified and reported to the HS&E Manager, who will be responsible for incident investigations into such events. The HS&E Manager or designee will be responsible for maintaining a record of corrective actions to be implemented as a result of incident investigations, in accordance with corrective action process procedures and for tracking corrective actions to completion. The applicant stated that it will monitor and document corrective actions through completion.

The applicant has described several specific features of its incident investigation process. These features include a commitment to establish processes to determine the specific or generic root cause(s) of abnormal events, the generic implications of such events, the recommendation of corrective actions, and reporting events to NRC, as required by 10 CFR 70.50 and 70.74. The investigation process will include a prompt risk-based evaluation of the event, to determine it's risk significance. Investigations will begin within 48 hours of the abnormal event, or sooner, depending on the event's safety significance.

The number of investigators for a given event will vary according to the complexity and severity of the event. Investigators will be independent from the line function(s) involved with the incident under investigation and investigators have been assured of no retaliation for participating in the investigations. Qualified internal or external investigators will be appointed to serve on investigating teams when required. The investigating teams will include at least one process expert and at least one team member trained in root-cause analysis.

The applicant will maintain auditable records and documentation related to abnormal events, investigations, and root-cause analyses, so that "lessons-learned" may be applied to other operations at the facility. For each abnormal event, an incident report will be written which includes: a description of the event; contributing factors; a root-cause analysis; findings; and recommendations. Relevant findings will be reviewed with all affected personnel. Details of the event sequence will be compared with accident sequences already considered in the ISA, and the ISA Summary will be modified to include evaluation of the risk associated with accidents of the type actually experienced.

The applicant will develop a corrective action process and implementing procedures for conducting incident investigations. These procedures will include the following elements: (a) a documented plan for investigating an abnormal event; (b) a description of functions, qualifications, and responsibilities of the investigative team manager and team members; (c) the scope of the investigative team's authorities and responsibilities; (d) assurance of management cooperation with the investigation effort; (e) assurance of the team's authority to obtain all the necessary information; (f) assurance of the investigative team's independence from the functional area involved in the incident being investigated; (g) retention of documentation relating to abnormal events for 2 years or for the life of the operation, whichever is longer; (h) guidance for investigative team members on how to apply a reasonable, systematic, structured approach to determining the root cause(s) and generic implications of the event; (i) requirements to make original investigation reports available to NRC, on request; and (j) a system for tracking the completion of appropriate corrective actions.

The applicant discusses a corrective action process where individuals involved in quality activities are responsible for identification and documentation of conditions adverse to quality, as well as, analysis of these conditions to determine the nature of corrective actions and the implementation of corrective actions in accordance with corrective action process procedures. The QA Program contains the corrective action process procedures for identifying, reporting, resolving, documenting, and analyzing conditions adverse to quality. Conditions adverse to quality may involve failures, malfunctions, deficiencies, deviations, defective materials and equipment, and nonconformances. Reports of conditions adverse to quality will be analyzed to identify trends in quality performance. Significant conditions and significant trends adverse to quality, along with root-cause analyses and corrective actions, will be documented and reported to senior management for review and assessment, in accordance with corrective action process procedures. The QA Manager will take follow-up action to verify proper and timely implementation of corrective action. The QA Program will require regularly scheduled audits and assessments of the corrective action process, to ensure that the needed corrective actions are identified.

All employees will have the authority and responsibility to initiate the corrective action process if they discover deficiencies.

11.3.7 Records Management

The applicant's records management function is described in Section 11.7 of the SAR (LES, 2005). Procedures for reviewing, approving, handling, identifying, retaining, distributing, retrieving, controlling, and maintaining QA records are required by the applicant's QA Program. QA records will include all completed records that furnish documentary evidence of the quality of items and/or activities affecting quality, including records related to health and safety. Records related to health and safety will be maintained in accordance with the requirements of Title 10, *Code of Federal Regulations*. The applicant has specified examples of the types of records that will be retained under records management procedures, including: (a) operating logs; (b) procedures; (c) supplier QA documentation; (d) nonconforming item reports; (e) test documentation/test results; (f) facility-modification records; (g) drawings; (h) specifications; (i) procurement documents; (j) nuclear material control and accounting reports;(k) maintenance activities including calibration records; (l) inspection documentation; (m) audit reports; (n) reportable occurrences and compliance records; (o) completed work orders; (p) license conditions records; (q) software verification records; (r) system descriptions; (s) as-built design-documentation packages; (t) regulatory reports; (u) corrective action documents; and (v) any other QA documentation required by specifications or procedures. These and other such records will be retained for at least the periods specified in the records management procedures. Other retention times will be specified for other facility records, as necessary, to meet applicable regulatory requirements.

That applicant states that for computer codes and computerized data used for activities relied on for safety, as specified in the ISA Summary, procedures will be established for maintaining the readability and usability of older codes and data as computing technology changes.

The applicant describes a Master File that will be maintained by the facility for storage and categorization of records and documents that fall within the scope of the applicant's records management function. The Master File storage system will provide for efficient and accurate retrieval of information. Access to and use of the Master File will be controlled by the facility. Documents in the Master File will be legible and identifiable by subject, and they will be considered valid only if stamped, initialed, signed or otherwise authenticated and dated by authorized personnel. Documents in the Master File may be originals or reproduced copies. Records stored in the Master File will be firmly attached in binders or placed in folders or envelopes; records will not be stored loosely. All records should be stored in steel file cabinets.

Computer storage of records will be permitted and will be done in a manner to preclude inadvertent loss and to ensure accurate and timely retrieval of data. Dual facility records storage will use an electronic data management system and storage of backup tapes in a fireproof safe. Records that are light-sensitive, pressure-sensitive, or temperature sensitive will be packaged and stored as recommended by the manufacturer of these media. Examples of such records include radiographs, photographs, negatives, and microfilm.

Written instructions will be prepared concerning the storage of records in the Master File, and the responsibility for implementing the requirements of the instructions will reside with a supervisor. These instructions will include, but will not necessarily be limited to, the following elements:

a.   description and identification of the locations of the Master File and the various record types within the Master File;

b.   the filing system to be used;

c.   a method for verifying that records received are in agreement with any applicable transmittal documents and are in good condition;

d.   a method for maintaining a record of the documents received;

e.   the criteria governing access to and control of the Master File;

f.   a method for maintaining control of and accountability for records removed from the Master File; and

g.   a method for filing supplemental information and for disposing of superseded records.

A qualified Fire Protection Engineer will evaluate record storage areas to assure records are adequately protected from damage.

11.3.8 Other QA Elements

The applicant addresses other QA elements that will be applied to IROFS and other management measures in Section 11.8 of the SAR (LES, 2005). The applicant has included a complete description of its application of QA elements to IROFS. The applicant's QA Program specifies mandatory requirements for performing activities affecting quality. These requirements are implemented through procedures. QA procedures and revisions to QA procedures are distributed on a controlled basis to organizations and individuals responsible for quality.

Before undertaking an activity, the applicant will establish implementation programs and procedures to document, approve, and implement the applicable portions of the QA Program. The applicant states that a management assessment of the QA Program will be performed at least 6 months before the scheduled receipt of licensed material at the facility. QA Program items identified as needing completion or modification as a result of this assessment will be entered into the applicant's corrective action process, and corrective action will be completed on such items before the scheduled receipt of licensed material at the facility. The QA Program will be monitored by management before the initial management assessment, through project review meetings and annual assessments. These management oversight and assessment activities will ensure that the QA Program is in place and effective before the receipt of licensed material.

The applicant states that the QA Program and supporting manuals and procedures will be reviewed periodically to ensure they are in compliance with applicable regulations, codes, and standards. New or revised regulations, codes, and standards will be incorporated into the QA Program and supporting manuals and procedures, as necessary. The applicant's President will assess the scope, status, adequacy, and regulatory compliance of the QA Program through regular meetings and correspondence with the Plant Manager and the QA organization. The QA Director will periodically inform the applicant's President and Plant Manager of quality concerns that require management resolution.

Personnel performing activities covered by the QA Program will perform work in accordance with approved procedures and will demonstrate suitable proficiency in their assigned tasks. The applicant commits to establishing formal training programs for QA policies, requirements, procedures, and methods. Ongoing training will be provided to ensure continuing proficiency as procedural requirements change. New employees will be required to attend an introductory QA class that covers QA authority, organization, policies, manuals, and procedures. Additional formal training will be conducted in specific topics such as NRC regulations and guidance, procedures, auditing, and applicable codes and standards. Supplemental training will be performed as required. On-the-job training will be performed by the employee's supervisor in QA area-specific procedures and requirements. Training records will be maintained for each person performing quality-related job functions.

The applicant states that the QA Program for design, construction, and pre-operational testing will continue simultaneously with the QA Program for the operations phase while construction activities are in progress. The applicant will develop and implement a system to plan and schedule system turnover as construction is completed. Before system turnover, written procedures will be developed for controlling the transfer of SSCs and associated documentation. These procedures will include checklists, marked drawings, documentation lists, system status, and receipt control. Major work activities performed by the applicant's contractors will be identified and controlled. Principal contractors will be required to comply with the applicable portions of 10 CFR Part 50, Appendix B, "Quality Assurance Criteria for Nuclear Power Plants and Nuclear Fuel Reprocessing Plants." The applicant will formally evaluate contractor performance, in accordance with the safety importance of the contracted activities.

The applicant will assign QA levels to facility SSCs and associated processes, based on their safety significance. Each component and document will receive a categorization of QA Levels 1, 2, or 3. The applicant's QA Program and it's supporting manuals, procedures, and instructions are applicable to items and activities designated as QA Levels 1 and 2. The QA-Level assignment will apply throughout the life of the facility, and will be based on the following definitions:

QA Level 1 Requirements

The applicant states that it's QA Level 1 Program will conform to the criteria established in 10 CFR Part 50, Appendix B. These criteria will be met by mandates to follow the guidelines of American Society of Mechanical Engineers (ASME) NQA-1-1994 (ASME, 1994), including supplements as revised by the ASME NQA-1a-1995 Addenda (ASME, 1995), as specified in the

11-27

applicant's QAPD, which is provided in Appendix A to the SAR (LES, 2005). The QA Level 1 Program will be applied to those SSCs and administrative controls that have been determined to be IROFS; items that affect the functions of IROFS; and items required to satisfy those regulatory requirements applicable to the QA Level 1 Program.

QA Level 2 Requirements

The applicant's QA Level 2 Program is an owner-defined QA Program that uses the ASME NQA-1-1994 standard (ASME, 1994), including supplements as revised by the ASME NQA-1a-1995 Addenda (ASME, 1995), as guidance. General QA Level 2 requirements are described in Section 2.0, "Quality Assurance Program for QA Level 2 Activities," in the applicant's QAPD, provided in Appendix A, to the SAR (LES, 2005) "Identification and Application of QA Controls." For contractors, the QA Level 2 Program will be described in documents that the applicant will approve. The QA Level 2 Program will be applied to owner-designated SSCs and activities. An International Organization for Standardization 9000 series QA Program may be acceptable for QA Level 2 applications provided it complies with QAPD requirements. The QA Director shall review and accept the QA Program manual.

QA Level 3 Requirements

The applicant's QA Level 3 Program is defined as standard commercial practice. A documented QA Level 3 Program is not required. As such, QA Level 3 will govern all activities not designated as QA Levels 1 or 2.

The QAPD, Appendix A to Chapter 11 of the applicant's SAR (LES, 2005), provides additional details and criterion to other QA elements that will be implemented to support the Management Measures described in Chapter 11. NRC approved the QAPD on April 9, 2004 (NRC, 2004).

11.4   EVALUATION FINDINGS

11.4.1 CM

The NRC staff has reviewed the CM function for the applicant's proposed uranium enrichment facility in accordance with the regulatory acceptance criteria of NUREG-1520 (NRC, 2002). The staff's evaluation found that the applicant's description of the overall CM program appropriately covered CM policy, design requirements, document control, change control, and assessments. Based on this evaluation, the NRC staff finds the applicant's CM program acceptable.

The applicant has suitably and acceptably described its implementation strategy for a CM program that meets the requirements of 10 CFR 70.72 and provides adequate assurance that facility changes are identified and controlled. IROFS are required to be identified and documented, organizational responsibilities are identified and defined, and administrative controls, policies, and procedures will be established to maintain the design configuration. Management-level policies and procedures are described that will provide the assurance of consistency among design requirements, physical configuration, and facility documentation. These policies and procedures include requirements to perform an analysis and independent safety review of any proposed activity involving IROFS, which ensures that consistency is maintained for new activities or for those changes to existing activities involving licensed material. The applicant's management measures will include the following CM elements:

Design Requirements

The applicant will develop and implement policies, procedures, organizational structure, and personnel responsibilities necessary to effectively implement its CM program.

Design Requirements

The applicant's design requirements and design bases will be documented and be appropriately supported by analyses. All design documentation will be current and maintain consistency with the physical configuration of the facility.

Document Control

Documents, including drawings, will be appropriately controlled, stored, and maintained. Drawings and related documents captured by the document control system will include those documents necessary to adequately describe IROFS.

Change Control

The applicant will achieve and maintain strict consistency among design requirements, documentation, and the physical configuration of the facility. The applicant will develop and implement the policies, procedures, organizational structure, and personnel responsibilities necessary to ensure consistency in design and design bases. These requirements include adequate analysis, review, approval, and implementation of identified changes to IROFS.

Assessments

The applicant will conduct both initial and periodic assessments of its CM program, to assure its effectiveness in meeting its goals and to correct deficiencies.

11.4.2 Maintenance

The applicant ensures the maintenance of IROFS. The applicant's maintenance program implementation strategy adequately address the basic elements required to ensure the availability and reliability of IROFS. These elements include corrective maintenance, PM, functional testing, equipment calibration, and work control for maintenance activities. The applicant's maintenance program will provide assurance that equipment performance will be adequately monitored and assessed, using surveillance testing, functional testing, and maintenance records. The surveillance/monitoring, PM, and functional testing activities described by the applicant provide assurance that the IROFS identified in the ISA Summary will be available and reliable to prevent or mitigate accidents.

The applicant's maintenance program will be based on approved procedures. It will employ work control methods that properly consider personnel safety, awareness of facility operations, QA, and the requirements for corrective maintenance. The ISA Summary will be used to identify

IROFS requiring maintenance, and the determination of initial PM intervals will be based on industry experience, with due consideration to the applicant's equipment reliability goals. The maintenance program will provide for training that emphasizes the importance of IROFS identified in the ISA Summary, as well as regulations, codes, and personnel safety.

Based on the above, the staff concludes that the applicant's maintenance functions meet the requirements of Part 70, and provide assurance of protecting the health and safety of workers, the public, and the environment.

11.4.3 Training and Qualification

Based on review of the applicant's strategy for development and implementation with respect to training and qualification and comparison of these requirements to the review acceptance criteria guidance in NUREG-1520 (NRC, 2002), the NRC staff has concluded that the applicant has adequately described and assessed its personnel training and qualification in a manner that: (1) satisfies regulatory requirements, and (2) is consistent with the guidance in NUREG-1520 (NRC, 2002). The staff's evaluation found that the applicant's description of its training program appropriately covered: (a) training organization and management; (b) analysis and identification of functional and position training requirements; (c) training basis and objectives; (d) organization of instruction; (e) evaluation of trainee learning; (f) conduct of on-the-job training; (g) evaluation of training program effectiveness; (h) personnel qualification; and (i) personnel evaluations. Based on this evaluation, the staff finds the applicant's training program acceptable.

There is assurance based on the NRC's review that implementation of the applicant's training program, as described in the SAR (LES, 2005), will result in personnel who are qualified and competent to design, construct, start up, operate, maintain, modify, and decommission the facility safely. The staff concludes that the applicant's plan for personnel training and qualification meets the requirements of Part 70.

11.4.4 Procedures

Based on the review of the SAR (LES, 2005) and comparison of the applicant's implementation strategy to the review acceptance criteria in NUREG-1520 (NRC, 2002), the NRC staff has determined that the applicant has described a suitably detailed process for the development, approval, and implementation of procedures. The applicant's strategy for developing implementing procedures appropriately cover IROFS, as well as all other items important to protecting the health and safety of workers, the public, and the environment. The staff concludes that the applicant's strategic plan for procedure development will meet the requirements of 10 CFR Part 70.

11.4.5 Audits and Assessments

Based on the review of the SAR (LES, 2005) and comparison of the applicant's detailed plan and processes to the review acceptance criteria in NUREG-1520 (NRC, 2002), NRC staff has concluded that the applicant has adequately described its audit and assessment functions. The staff has reviewed the applicant's implementation process used to describe audits and assessments and the description of its policy directives, plans, and procedural requirements for audits and assessments, taking into consideration such factors as: (1) the general structure of typical audit and assessment activities; (2) facility procedures to be used to control audit and assessment activities; (3) the use of qualified and independent audit and assessment personnel; (4) the planned use of the results of audit and assessment activities; and (5) documentation, planning, and implementation of corrective actions, based on the findings and recommendations of audits and assessments. Based on the above, the staff finds that the applicant's implementation plan describing audits and assessments are acceptable.

The staff concludes that the applicant's plan for audits and assessments meets the requirements of Part 70 and provides assurance of protection of the health and safety of workers, the public, and the environment.

11.4.6 Incident Investigations

The applicant will develop processes to perform incident investigations. The applicant's process which describes the development and implementation methodologies with respect to its incident investigation program include: (1) performing incident investigations of abnormal events that may occur during operation of the facility; (2) determining the root cause(s) and generic implications of the event; (3) recommending corrective actions for ensuring a safe facility and safe facility operations; (4) monitoring and documenting corrective actions to completion; and (5) maintenance of documentation so that "lessons learned" may be applied to future operations at the facility. NRC staff has determined that these developmental and implementational methods adequately follow the review acceptance criteria outlined NUREG-1520 (NRC, 2002). Accordingly, NRC staff concludes that the applicant's description of the incident investigation process complies with applicable NRC regulations and provides assurance of protection of the health and safety of workers, the public, and the environment.

11.4.7 Records Management

The staff has reviewed the applicant's implementation process concerning its records management system against the acceptance criteria in NUREG-1520 (NRC, 2002). Based on this review, NRC staff has concluded that the system will be effective in collecting, verifying, protecting, and storing information concerning the facility, its design, operations, and maintenance and the records management processes will provide for retrieval of the information in readable form for the designated lifetime of the records. The staff verified the adequacy of the applicant's developmental and implementation protocols for using a records storage system that is capable of protecting and preserving health and safety records that are stored at the facility during the mandated periods, as well as the capabilities of the storage system for protecting stored records from loss, theft, tampering, or damage during and after emergencies. The applicant has also provided assurance that any deficiencies in the records management system or its implementation will be detected and corrected in a timely manner. Based on this evaluation, the staff finds that the applicant's description and implementation processes for records management are acceptable.

11.4.8 Other QA Elements

Based on its review of the SAR (LES, 2005) the NRC staff has concluded that the applicant has adequately described the application of other QA elements to IROFS, management measures, and other safety-related items. The NRC staff concludes that the applicant's strategic approach in this area adequately address the implementation of the QA Program before beginning operations at the facility. The staff has also determined that the applicant provides assurance that personnel performing quality-related activities will perform work in accordance with approved procedures and must demonstrate suitable proficiency in their assigned tasks. Additional conclusions the staff reached through its evaluation of the applicant's QA process, procedures, and methods include the following:

1.   The applicant has established, documented, and developed an organizational structure responsible for developing, implementing, and assessing the management measures for providing assurance of safe facility operations, in accordance with the acceptance criteria in Section 11.4 of NUREG-1520 (NRC, 2002).

2.   The applicant has established and documented a program to develop and implement QA elements and administrative measures for staffing, performance, assessing findings, and implementing corrective actions.

3.   The applicant has developed a process for preparation and control of written plant procedures, including a process for evaluating changes to procedures, IROFS, and tests. The process for review, approval, and documentation of procedures will be implemented and maintained.

4.   The applicant will develop and implement a program of surveillance, tests, and inspection that will provide assurance of satisfactory in-service performance of IROFS. Specified standards, acceptance criteria, and testing steps have been described or provided.

5.   Periodic independent audits will be conducted to determine the effectiveness of management measures. Management measures will provide for documentation of audit findings and implementation of corrective actions.

6.   Training requirements have been established and documented for ensuring that employees are provided with the skills necessary to perform their jobs safely. Management measures have been provided for evaluation of the effectiveness of training against predetermined objectives and criteria.

7.   The organizations and personnel responsible for performing QA functions will have the required independence and authority to effectively carry out their QA element functions without undue influence from those directly responsible for process operations.

8.   QA elements cover the IROFS, as identified in the ISA Summary, and measures are established to prevent hazards from escalating into higher-risk events or accidents.

Accordingly, the staff concludes that the applicant's application of other QA elements meets the requirements of 10 CFR 70.62(d), and other applicable regulations (Appendix B of Part 50), and provides the assurance of protection of public health and safety and protection of the environment.

11.5 REFERENCES

(ANSI/ANS, 1996b) American National Standards Institute/American Nuclear Society (ANSI/ANS). ANSI/ANS-8.19, "Administrative Practices for Nuclear Criticality Safety," 1996.

(ANSI/ANS, 1991) American National Standards Institute/American Nuclear Society (ANSI/ANS). ANSI/ANS-8.20, "Nuclear Criticality Safety Training," 1991.

(ASME, 1994) American Society of Mechanical Engineers (ASME). ASME NQA-1, "Quality Assurance Requirements for Nuclear Facility Applications," 1994.

(ASME, 1995) American Society of Mechanical Engineers (ASME). ASME NQA-1a, "Addenda to ASME NQA-1-1994 Edition, Quality Assurance Requirements for Nuclear Facility Applications," 1995.

(LES, 2005) Louisiana Energy Services (LES). "National Enrichment Facility Safety Analysis Report," Revision 6, 2005.

(NRC, 2002) U.S. Nuclear Regulatory Commission (NRC). NUREG-1520, "Standard Review Plan for the Review of a License Application for a Fuel Cycle Facility," 2002.

(NRC, 2004) U.S. Nuclear Regulatory Commission (NRC), letter to Louisiana Energy Services, "Louisiana Energy Services Quality Assurance Program Description for the National Enrichment Facility," April 9, 2004.

12.0 MATERIAL CONTROL AND ACCOUNTING

The Nuclear Regulatory Commission (NRC) staff's review of the applicant's material control and accountability (MC&A) program contains information that has been marked as "Proprietary Information" by the applicant, pursuant to 10 CFR 2.390.

The staff concluded that the applicant provided an acceptable Fundamental Nuclear Materials Control Plan (FNMCP) for the proposed facility that will meet the applicable Part 74 requirements. The FNMCP describes acceptable methods for achieving the performance objectives in 10 CFR 74.33(a) and the system capabilities of 10 CFR 74.33(c). As a result, the staff determined that the applicant meets the requirements in the area of MC&A to operate the proposed facility, under Part 74.

13.0 PHYSICAL PROTECTION

The Nuclear Regulatory Commission (NRC) staff's review of the applicant's Physical Security Plan contains information that has been marked as "Proprietary Information" by the applicant, pursuant to 10 CFR 2.390.

NRC staff reviewed the applicant's Physical Security Plan. The methods and procedures outlined in the Physical Security Plan satisfy the performance objectives, systems capabilities, and reporting requirements specified in 10 CFR 73.67 and 73.71. The Physical Security Plan for the facility is acceptable and meets the NRC requirements for physical protection of SNM of low strategic significance.

In addition to a Physical Security Plan, the applicant submitted Guard Force Training and the Safeguards Contingency Plans. However, Guard Force Training and the Safeguards Contingency Plans are not required for Category III facilities in accordance with 10 CFR 73.67. These plans are only required for Category I licensees in accordance with 10 CFR 73.20 and 73.46. The staff performed an informal review of the LES Guard Force Training and the Safeguards Contingency Plans to ensure there was not information in them that would contradict a requirement in the Physical Security Plan. Since they are not required, the Guard Force Training and the Safeguards Contingency Plans are not considered part of the licensing basis and were not considered as part of the staff's review of the Physical Security Plan.

The protection of classified matter is described in the "National Enrichment Facility Standard Practice Procedures Plan for the Protection of Classified Matter." Evaluation of this plan is discussed in Section 1.2.3.7 of this Safety Evaluation Report.

14.0 PHYSICAL SECURITY OF THE TRANSPORTATION OF SPECIAL NUCLEAR MATERIAL OF LOW STRATEGIC SIGNIFICANCE

The Nuclear Regulatory Commission (NRC) staff's review of the applicant's Transportation Security Plan contains information that has been marked as "Proprietary Information" by the applicant, pursuant to 10 CFR 2.390.

The NRC staff reviewed the applicant's Transportation Security Plan for SNM-LSS shipments originating from, or arriving at, the facility. The approaches and procedures outlined in the Transportation Security Plan satisfy the performance objectives, systems capabilities, and event and advance notification requirements specified in 10 CFR 73.67(c); 73.67(g)(1)-(5); 73.71; 73.73; and 73.74. The NRC staff has concluded that the facility Transportation Security Plan is acceptable and meets the NRC requirements for physical protection of SNM-LSS in transit.

15.0 SAFETY EVALUATION REPORT PREPARERS

The individuals and organizations listed below are the principal contributors to the preparation of this Safety Evaluation Report. U.S. Nuclear Regulatory Commission (NRC) staff directed the effort and contributed to the technical evaluations. Staff also used contractor support from the Center for Nuclear Waste Regulatory Analyses and IEF Consulting in the preparation of this document.

<u>U.S. Nuclear Regulatory Commission Contributors</u>

Timothy C. Johnson, Office of Nuclear Material Safety and Safeguards (NMSS)	NRC Project Manager
Brian W. Smith, NMSS	Section Chief, Gas Centrifuge Facility Licensing Section
Paul Bell, NMSS	Management Measures
Philip Brochman, NSIR	Transportation Security
David Brown, NMSS	Health Physics and Environmental Protection
Oleg Bukharin, Office of Nuclear Security and Incident Response (NSIR)	Transportation Security
Frederick Burrows, NMSS	Electrical Engineering and Instrumentation and Controls
Sherri Cross, NSIR	Material Control and Accounting
Michael Dusaniwskyj, Office of Nuclear Reactor Regulation (NRR)	Financial Qualifications
Stanley Echols, NMSS	Environmental Protection
J. Keith Everly, NSIR	Information Security
Harry Felsher, NMSS	Nuclear Criticality Safety
Alan Frazier, NSIR	Physical Protection
Thomas Fredricks, NMSS	Decommissioning Financial Assurance

Herman Graves, Office of Nuclear Regulatory Research (RES)	Geotechnical Engineering
Edward Johannemann, NSIR	Physical Protection
Joel Klein, NMSS	Health Physics and Emergency Management
Joel Kramer, RES	Human Factors Engineering
Amar Pal, NRR	Electrical Engineering
Julius Persensky, RES	Human Factors Engineering
Thomas Pham, NSIR	Material Control and Accounting
Clayton L. Pittiglio, NRR	Decommissioning Financial Assurance
Roman Shaffer, RES	Instrumentation and Controls
Wilkins Smith, NMSS	Management Measures
William Troskoski, NMSS	Integrated Safety Analysis and Chemical Safety
Rex Wescott, NMSS	Fire Safety

Center for Nuclear Waste Regulatory Analyses Contributors

Asadul H. Chowdhury	Geotechnical Engineering
Sarah H. Gonzalez	Seismic Engineering
Sui-Min (Simon) Hsiung	Geotechnical Engineering
John Stamatakos	Seismic Engineering

IEF Consulting Contributors

Craig Dean	Decommissioning Financial Assurance
Brantley Fry	Decommissioning Financial Assurance

15-2

Elizabeth Gormsen   Decommissioning Financial
                                               Assurance

Jennifer Mayer   Decommissioning Financial
                                               Assurance

APPENDIX A - ACCIDENT ANALYSIS

The operation of the uranium enrichment facility that Louisiana Energy Services (LES) proposes to be located in Lea County, New Mexico, would involve risks to workers, the public, and the environment from potential accidents. The regulations in 10 CFR Part 70, Subpart H, "Additional Requirements for Certain Licensees Authorized to Possess a Critical Mass of Special Nuclear Material," require that each applicant or licensee evaluate, in an "Integrated Safety Analysis" (ISA), its compliance with certain performance requirements. The purpose of this section of the Safety Evaluation Report is to summarize the methods and results the U.S. Nuclear Regulatory Commission (NRC) staff uses to independently evaluate the consequences of potential accidents identified in the applicant's ISA. The accidents the staff has evaluated are a representative selection of the types of accidents that are possible at the proposed facility.

A.1 ACCIDENT ANALYSIS METHODOLOGY

The analytical methods used in this consequence assessment are based on staff guidance for analysis of nuclear fuel cycle facility accidents (NRC, 1990, 1991, 1998, 2001). With the exception of the criticality accident, the hazards the NRC staff evaluated involve the release of uranium hexafluoride (UF_6) vapor from process systems that are designed to confine UF_6 during normal operations. As described below, UF_6 vapor poses a chemical and radiological risk to workers, the public, and the environment.

A.1.1 Selection of Representative Accident Scenarios

The Safety Analysis Report (SAR) (LES, 2005a) and Emergency Plan (EP) (LES, 2005c) describe potential accidents that could occur at the facility. The applicant has provided accident descriptions for two groups, according to the severity of the accident consequences: high-consequence events and intermediate-consequence events. The accident types are summarized in Table A.1-1.

NRC staff selected, for detailed evaluation, a subset of the potential accident scenarios that is intended to encompass the range of possible accidents. The accident sequences the staff selected vary in severity from high- to low-consequence events, and include accidents initiated by natural phenomena, operator error, and equipment failure. The list of accident sequences the staff evaluated are listed in Table A.1-2.

A.1.2 Source Term Methodology

NRC staff evaluated the chemical and radiological hazard to workers, the public, and the environment from accidental releases of UF_6 vapor at the facility (See Section A.2.1 for the source term methodology for the generic inadvertent criticality accident). For most accidents, the UF_6 vapor is assumed to escape its primary confinement system and enter an occupied room at the facility. Staff assumed that UF_6 would mix instantaneously with the air in the room.

Table A.1-1
Accident Types

High-Consequence Events

- Natural Phenomena
    - Earthquake
    - Tornado
    - Flood
- Inadvertent Nuclear Criticality
- Fires Propagating Between Areas
- Fires Involving Excessive Transient Combustibles
- Heater Controller Failure
- Over-filled Cylinder Heated to Ambient
- Product Liquid Sampling Autoclave Heater Failure Followed by Reheat
- Open Sample Manifold Purge Valve and Blind Flange
- Pump Exhaust Plugged - Worker
- UF_6 Sub-sampling Unit Hot Box Heater Controller Failure
- Empty UF_6 Cold Trap (UF_6 Release)
- Cylinder Valve/Connection Failure During Pressure Test
- Chemical Dump Trap Failure
- Worker Evacuation (UF_6 Release due to a seismic event, fire, or other unplanned release)

Intermediate-Consequence Events

- Carbon Trap Failure
- Pump Exhaust Plugged - Public
- Spill of Failed Centrifuge Parts
- Dropped Contaminated Centrifuge
- Fire in Ventilated Room

Table A.1-2
Accident Scenarios Representative of Potential Facility Events

- Generic Inadvertent Nuclear Criticality
- Hydraulic Rupture of a UF_6 Cylinder in the Blending and Liquid-Sampling Area
- Natural Phenomena Hazard - Earthquake
- Fire in a UF_6 Handling Area
- Process Line Rupture in a Product Low-Temperature Takeoff Station

For a constant release rate of UF$_6$, the time-dependent concentration, C(t), of UF$_6$ in a room or workshop at the facility is (NRC, 1990):

$$\frac{dC(t)}{dt} = \frac{R}{V'} - \frac{Q_v f_v C(t)}{V'}$$

Eq. 1

where R = constant UF$_6$ release rate, grams/second
  V' = k x f x V, the effective room volume, m^3
  V = actual room volume, m^3
  k = mixing efficiency (from National Fire Protection Association (NFPA) NFPA 69, Appendix D (NFPA, 2002)), unitless
  f = room free air fraction, unitless
  Q$_v$ = room ventilation rate, m^3/second
  f$_v$ = the fraction of Q$_v$ exhausted to the atmosphere (1-f$_v$ is recycled back into the room), and
  t = time elapsed since start of release, seconds

The values of mixing efficiency, k, and room free air fraction, f, are assumed to be 0.3 and 0.8, respectively. The mixing efficiency is conservatively based on Table D-1 of NFPA 69 (NFPA, 2002), and is for ventilation systems with forced air supplies and single exhaust openings comprised of grills and registers. The staff assumes that the room free air fraction value of 0.8 accounts for the volume of equipment that replaces free air inside the facility. The applicant provided room volumes and ventilation flow rates (LES, 2004). The fraction of air exhaust is 10 percent, which is consistent with the Heating, Ventilation and Air Conditioning (HVAC) descriptions in Chapters 3 and 4 of the SAR (LES, 2005a).

A solution to Equation 1 is:

$$C_1(t) = \frac{R}{Q_v f_v}\left[1 - e^{-\frac{Q_v f_v}{V'}t}\right]$$

Eq. 2

Equation 2 defines the concentration, C$_1$(t), during the period that UF$_6$ is released at a steady-state rate, R, into a room. After T$_i$ = 30 minutes, staff assumes that either the entire material at risk would be released or the release would be stopped when operators intervene. The assumption that operators or affected individuals downwind would respond within 30 minutes is consistent with conservative self-protective criteria NRC uses to evaluate emergency preparedness (NRC, 1988). After T$_i$ = 30 minutes, the room would be ventilated until UF$_6$ is cleared from the room and exhausted to the environment. The room concentration, C$_2$(t), after all the material escapes to the room, or the release is stopped, is:

$$C_2(t) = \frac{R}{Q_v f_v}\left[1 - e^{-\frac{Q_v f_v T_i}{V'}}\right]e^{-\frac{Q_v f_v}{V'}t}$$

Eq. 3

For the seismic event, the applicant has proposed safety-related equipment [i.e., items relied on for safety (IROFS)] that shut down the HVAC systems in certain process areas. With no forced ventilation, the primary means by which UF$_6$, uranyl fluoride (UO$_2$F$_2$) particulate matter, and hydrogen fluoride (HF) vapor enters the environment would be from small cracks and openings in the building.

The volumetric leak rate from small cracks and openings in a building is calculated by evaluating Poiseuille's Law (Baker, 1987):

$$Q_L = -\left(\frac{12\eta d L_s}{C\rho W}\right) + \sqrt{\left(\frac{12\eta d L_s}{C\rho W}\right)^2 + \frac{C_{p,a} v^2 W^2 L_s^2}{C}} \qquad \text{Eq. 4}$$

where Q_L = volumetric leak rate, $m^3 s^{-1}$
L_s = perimeter length of all exterior doors, m
W = width of the opening between door and frame, m
p_1 = pressure of air indoors, N^{-2}
p_2 = pressure of air on building exterior, N^{-2}
η = coefficient of viscosity of air = 1.81×10^{-5} N s $^{-2}$ @ T = 20 C
d = thickness of doors, m
C = 1.5
ρ = density of air = 1.183 kg $^{-3}$ at T=25°C
v = windspeed, m s^{-1}

Further, the values of $C_{p,a}$ depend on the location of the door or opening relative to the direction of the wind (Blevins, 2003):

$C_{p,a}$ = 0.9 windward side of the building
$C_{p,a}$ = -0.3 for leeward side of the building
$C_{p,a}$ = -0.4 for building sides orthogonal to the wind direction

For this assessment, each exterior door in affected process areas of the building is assumed to have a W = 0.16 cm (1/16-in.) opening around both sides and the top, and a W = 0.32cm (1/8-in.) opening at the bottom. The thickness of all doors, d, is estimated to be 5 cm (2-in.). The perimeter length of doors is estimated from drawings in the SAR (LES, 2005a).

The windspeed, v, assumed for the building-leakage calculations, was chosen with consideration of the windspeed and stability class assumed in the derivation of the maximum atmospheric dispersion factor, χ/S. The staff calculated the highest χ/S for the controlled-area boundary (CAB) is 5.4×10^{-5} s $^{-3}$. With corrections for building wake and low wind-speed plume meander, the windspeed for F class stability conditions for which a $\chi/S = 5.4 \times 10^{-5}$ s $^{-3}$ would be derived is 1.8 m s^{-1}. Therefore, a bounding value of v = 2 m s^{-1} is used to estimate building leakage.

Solid UO_2F_2 produced by the reaction of UF_6 with water vapor (i.e., humidity) forms a fine powder that will settle by gravity. Therefore, in addition to removal by exfiltration through door cracks to the environment, solid UO_2F_2 will also be removed from the air by settling on the floor and equipment of the affected process area (NRC, 1998). The concentration in the building is calculated as:

$$C_L(t) = C_{L,O} e^{-\frac{1}{V}(Q_L + v_d A)t} \qquad \text{Eq. 6}$$

where v_d = settling velocity of UO_2F_2 particles in air, m s^{-1}
A = floor area of the affected process area, m^2

From Table 12.4 of U.S. Department of Energy (DOE)/TIC-27601 (DOE, 1984), staff estimated the settling velocity of fine uranium compounds to be approximately 0.0001 cm s^{-1}. The floor areas of the affected process areas are estimated from drawings in the SAR (LES, 2005a).

A.1.3 NRC Performance Requirements

The performance requirements in Part 70, Subpart H, define acceptable levels of risk of accidents at nuclear fuel cycle facilities, such as the proposed facility. The regulations in Subpart H require that the applicant reduce the risks of credible high-consequence and intermediate-consequence events, and assure that under normal and credible abnormal conditions, all nuclear processes are subcritical. Threshold consequence values that define the high- and intermediate-consequence events are described in Table A.1-3 from the facility SAR (LES, 2005a).

Table A.1-3
Definition of High- and Intermediate-Consequence Events

Receptor	Intermediate Consequence	High Consequence
Worker - Radiological	> 25 rem (0.25 Sv)	> 100 rem (1 Sv)
Worker - Chemical (10-minute exposure)*	> 19 mg U / m^3 > 78 mg HF / m^3	> 146 mg U / m^3 > 139 mg HF / m^3
Environment at the Restricted Area Boundary	> 5.4 mg U / m^3, or 24-hour average release greater than 5000 times the values in Table 2 of Appendix B of 10 CFR Part 20	N/A
Individual at the Controlled Area Boundary - Radiological	> 5 rem (0.05 Sv)	> 25 rem (0.25 Sv)
Individual at the Controlled Area Boundary - Chemical (30-minute exposure)	> 2.4 mg U / m^3 > 0.8 mg HF / m^3	> 13 mg U / m^3 > 28 mg HF / m^3

*   Limits on uranium intake are also defined for workers in the immediate proximity of the release. These limits are 10 mg and 40 mg uranium for intermediate- and high-consequence events, respectively.

A.1.4 Consequence-Assessment Approach for Acute Health Effects

The staff evaluated accident consequences for releases of UF$_6$ for the facility worker, the environment outside the restricted area boundary (RAB), an individual at the CAB, and the public beyond the CAB. As stated above, the analytical methods used in this consequence assessment are based on staff guidance for analysis of nuclear fuel cycle facility accidents (NRC, 1990, 1991, 1998, 2001).

Facility Worker Exposure to Uranium and HF

The accident consequences to a facility worker include the risks of toxicological effects of uranium intake, radiation dose from uranium intake, and exposure to HF concentration in air. The amount of uranium a facility worker could inhale (uranium intake) is calculated by assuming the worker is exposed to $C_1(t)$ until $T_1 = 10$ minutes after the start of the release (LES, 2005b). By $T_1 = 10$ minutes, a worker is conservatively assumed to successfully escape the affected room. Staff calculated uranium concentration for comparison with the proposed levels in Table A.1-3. Uranium intake can be determined by multiplying the uranium concentration, time of exposure, and breathing rate. The worker is assumed to inhale at a constant breathing rate of 3.33×10^{-4} m^3/s (20 liters per minute), which is consistent with the breathing rate NRC used in Part 20, Appendix B, for Reference Man performing "light work." Similarly, the HF concentration to which a facility worker could be exposed is calculated by evaluating the time-averaged HF concentration during the first $T_1 = 10$ minutes.

For the uranium and HF exposure calculations, staff assumed that sufficient moisture (i.e., humidity) is present in the room to completely convert released UF_6 gas to UO_2F_2 particulate matter and HF vapor. This assumption results in a conservative estimate of the concentration of HF vapor that would be present in both the affected room of the facility and downwind.

RAB 24-hour Average Uranium Concentration

In accordance with Part 70, Subpart H, the applicant must reduce the environmental risks of accidents. The environmental consequences of accidents are evaluated at the RAB. At the facility, the RAB is a fenced area inside the controlled area that includes the process buildings and the uranium byproduct cylinder pad (LES, 2004). To evaluate whether accidents exceed the environmental performance requirement, staff calculated the 24-hour average uranium concentration at the restricted area boundary. Staff assumed that points of release are the stacks on the roof of the technical services building (TSB).

The total source term for the first phase of the event (before the release is stopped) is S_1. The residual source term from the time that the release is stopped, T_1, until the source is either depleted, or until 24 hours has elapsed, is S_2.

$$S_1 = \int_0^{T_1} S_1(t)\,dt = \int_0^{T_1} C_1(t)\,dt \times Q_v \times f_v = R\left[T_1 - \frac{V'}{Q_v f_v}\left\{1 - e^{-\frac{Q_v f_v T_1}{V'}}\right\}\right]_1, \quad for \ 0 < t \leq T_1$$

Eqs. 7, 8

$$S_2 = \int_{T_1}^{T_2} S_2(t)\,dt = \int_{T_1}^{T_2} C_2(t)\,dt \times Q_v \times f_v = R\left[1 - e^{-\frac{Q_v f_v T_1}{V'}}\right]\left[\frac{V'}{Q_v f_v}\left\{1 - e^{-\frac{Q_v f_v (T_2 - T_1)}{V'}}\right\}\right], \quad for \ T_1 < t \leq T_2$$

To compare downwind concentrations with the applicable performance requirement, staff calculated the uranium concentration downwind as a 24-hour average. For the RAB and the CAB, staff calculated the atmospheric dispersion factor (χ/S) for various distances from the process buildings to the boundary in each downwind sector, using ARCON96 (NRC, 1997). The distance to the RAB and CAB in each compass sector, the persistence of the wind in each direction, and χ/S values calculated using ARCON96, are presented in Table A.1-4. The highest χ/S at the RAB, which would result in the highest downwind concentration, occurs directly east of the TSB. Therefore, the concentration at the RAB is calculated for wind blowing to the east.

Values of Atmospheric Dispersion Factors for the Facility Boundaries

Direction from Facility	Distance from Facility		Frequency of Wind (percent)	RAB χ/S (s/m³)	CAB χ/S (s/m³)
	RAB meters (feet)	CAB meters (feet)			
S	160 (524)	417 (1368)	5.66	2.64E-04	4.84E-05
SSW	168 (552)	417 (1368)	3.98	2.40E-04	4.80E-05
SW	210 (690)	422 (1384)	4.91	1.69E-04	5.37E-05
WSW	261 (856)	503 (1650)	4.87	1.14E-04	4.08E-05
W	261 (856)	769 (2522)	6.29	1.14E-04	2.37E-05
WNW	278 (911)	1071 (3513)	5.52	9.96E-05	1.46E-05
NW	757 (2484)	1072 (3516)	7.52	2.12E-05	1.34E-05
NNW	639 (2098)	995 (3264)	10.80	2.35E-05	1.13E-05
N	589 (1932)	995 (3264)	20.40	2.67E-05	1.18E-05
NNE	530 (1739)	754 (2473)	7.35	3.08E-05	1.77E-05
NE	463 (1518)	581 (1906)	5.46	3.78E-05	2.61E-05
ENE	362 (1187)	540 (1771)	4.68	4.96E-05	2.61E-05
E	109 (359)	540 (1771)	4.45	4.49E-04	2.68E-05
ESE	101 (331)	540 (1771)	2.42	4.26E-04	2.54E-05
SE	143 (469)	487 (1597)	2.69	2.76E-04	3.10E-05
SSE	185 (607)	417 (1368)	3.04	1.70E-04	3.95E-05

The downwind concentration at the RAB is calculated for the downwind sector with the highest $\chi/S|_{RAB}$ using Equation 9.

$$U, \left.\frac{mg}{m^3}\right|_{RAB} = \frac{\left[\int_0^{T_1} S_1(t)dt + \int_{T_1}^{T_2=24hr} S_2(t)dt\right]}{\int_0^{T_2=24hr} dt} \cdot \frac{g}{s} \times \left.\frac{X}{S}\right|_{RAB} \cdot \frac{s}{m^3} \times 10^3 \frac{mg}{g} \times 0.68 \frac{mg\,U}{mg\,UF_6} \quad \text{Eq. 9}$$

A-7

CAB Uranium and HF Exposure

The accident consequences to an individual at the CAB include the risks of toxicological effects of uranium intake, radiation dose from uranium intake, and exposure to HF concentration in air. The uranium concentration at the CAB is calculated for the downwind sector with the highest atmospheric dispersion factor ($\chi/S|_{CAB}$). The highest χ/S at the CAB, which would result in the highest downwind concentration, occurs southwest of the TSB. Therefore, the accident consequences at the CAB are calculated for wind blowing to the southwest.

The 30-minute average uranium concentration at the CAB is calculated using Equation 10.

$$[U], 30\min = \frac{\left[\int_0^{T_1} S_1(t)dt + \int_{T_1}^{T_2=24hr} S_2(t)dt\right], g}{1,800\,s}, \frac{g}{s} \times \left.\frac{X}{S}\right|_{CAB}, \frac{s}{m^3} \times 10^3\,\frac{mg}{g} \times 0.68\,\frac{mg\,U}{mg\,UF_6} \qquad \text{Eq. 10}$$

Similarly, the unmitigated 30-minute average HF concentration is:

$$[HF], 30\min = \frac{\left[\int_0^{T_1} S_1(t)dt + \int_{T_1}^{T_2=24hr} S_2(t)dt\right], g}{1,800\,s}, \frac{g}{s} \times \left.\frac{X}{S}\right|_{CAB}, \frac{s}{m^3} \times 10^3\,\frac{mg}{g} \times 0.23\,\frac{mg\,HF}{mg\,UF_6} \qquad \text{Eq. 11}$$

A.1.5 Consequence-Assessment Methodology for Chronic Health Effects

Staff reviewed the potential, during a release of UF_6, for acute exposures to HF to result in chronic toxicological effects. Earlier studies have indicated that if fatality from suffocation caused by edema (swelling) in the lungs does not occur, the swelling resulting from HF exposure will subside and recovery should be complete. Thus, acute sublethal inhalation of HF is not expected to have long-term effects (NRC, 1991). Therefore, the post-accident chronic health effects the staff evaluated are limited to the toxicological and radiological health effects to members of the public offsite resulting from exposure to uranium compounds.

The risk for chronic health effects is discussed in terms of latent cancer fatalities (LCFs). LCFs are a measure of the expected number of additional cancer deaths in a population (or people dying of cancer) as a result of exposure to radiation. Death from cancer induced by exposure to radiation may occur at any time after the exposure takes place. However, latent cancers would be expected to occur in a population from one year to many years after the exposure takes place (NRC, 2005).

The staff has also previously reviewed human toxicological effects of exposure to soluble uranium compounds (NRC, 1991). The staff concluded that a single acute intake of 10 mg of soluble uranium would produce either minimal or nondetectable effects, either short-term or long-term, in humans. Therefore, if an accident could not result in acute intakes above 10 mg of soluble uranium in any individual at or just beyond the site boundary (CAB), then no long-term health effects would be expected among the exposed population further downwind. At the facility, only one type of event is capable of causing toxicological effects from exposure to soluble uranium among the offsite public -- the rupture of a large UF_6 cylinder from inadvertent overheating or overfilling. The protective measures the applicant proposed to prevent this type of event are described in Section A.2.2.

The staff used GENII v. 1.485 (PNL, 1988) to estimate collective radiation doses (person-rem) to members of the public resulting from post-accident inhalation and ingestion of soluble uranium compounds. The exposure pathways the staff analyzed include inhalation of soluble uranium carried by wind, external radiation from radioactivity deposited on the ground downwind of the facility, and ingestion of contaminated food (produce, meat, and dairy products). The ingestion parameters the staff used to estimate radiological doses to the public from accidents are described in Table A.1-5. Staff used meteorological data for nearby Midland-Odessa (LES, 2005d). Staff obtained demographic information using SECPOP2000, a sector population, land fraction, and economic estimate program prepared for NRC (NRC, 2003). Population data for each of 16 downwind sectors at 10 distance intervals are shown in Table A.1-6. For releases of uranium compounds, staff found that the north sector would have the highest collective doses because Hobbs, New Mexico, is a large population center in the prevailing downwind direction.

Table A.1-5
Ingestion Parameters Used in GENII to Calculate
Collective Radiological Dose to the Public from Accidents

Parameter Values for Consumption of Terrestrial Food

Food Type	General Population			
	Growing Time (days)	Yield (kg/m²)	Holdup Time (days)	Consumption Rate (kg/yr)
Leafy Vegetables	90	1.5	14	15
Root Vegetables	90	4.0	14	140
Fruit	90	2.0	14	64
Grains/Cereals	90	0.8	180	72

Parameter Values for Consumption of Animal Products

Food Type	Consumption Rate (kg/yr)	Holdup Time (days)	Stored Feed				Fresh Forage			
			Diet Fraction	Growing Time (days)	Yield (kg/m²)	Storage Time (days)	Diet Fraction	Growing Time (days)	Yield (kg/m²)	Storage Time (days)
Beef	70	34	0.25	90	0.80	180	0.75	45	2	100
Poultry	8.5	34	1	90	0.80	180	---	---	---	---
Milk	230	3	0.25	45	2	100	0.75	30	1.5	0
Eggs	20	18	1	90	0.80	180	---	---	---	---

A-9

Table A.1-6
Public Population in Sectors Surrounding the Facility

Sector	0-1 mile	1-2 miles	2-3 miles	3-4 miles	4-5 miles
N	0	0	0	0	0
NNE	0	0	0	0	0
NE	0	0	0	0	0
ENE	0	0	0	0	0
E	0	0	0	0	0
ESE	0	0	0	0	0
SE	0	0	0	0	0
SSE	0	0	0	0	0
S	0	0	0	0	0
SSW	0	0	0	4	0
SW	0	0	0	0	0
WSW	0	0	0	0	15
W	0	0	11	53	2099
WNW	0	0	0	0	104
NW	0	0	0	5	2
NNW	0	0	0	0	0
Sector	**5-10 miles**	**10-20 miles**	**20-30 miles**	**30-40 miles**	**40-50 miles**
N	9	14637	12616	273	222
NNE	0	69	217	4760	1120
NE	0	49	995	7464	2809
ENE	0	7	430	972	46
E	0	7	45	351	41
ESE	0	0	105	12351	60
SE	0	23	18	20	848
SSE	0	0	19	8	18
S	0	4	37	3369	3754
SSW	6	4	2033	11	12
SW	17	12	3	1	3
WSW	34	9	13	2	8
W	484	13	2	4	21
WNW	35	20	0	9	8
NW	3	223	33	43	83
NNW	0	5044	4543	10565	1391

For dose calculations to the public, staff assumed that individuals downwind spend 100 percent of the time inside the passing plume (i.e., not sheltered). Staff also evaluated exposure of the offsite population to contaminated soil resuspended by wind using a mass loading factor of 0.1-mg soil per cubic meter of air.

A.2   ACCIDENT ANALYSIS

A.2.1   Generic Inadvertent Criticality

An inadvertent criticality at the facility would result from the unintended accumulation of enriched uranium, leading ultimately to a self-sustaining or runaway nuclear chain reaction. A criticality accident could release large amounts of heat and radiation. An inadvertent criticality could also produce radioactive fission products, such as isotopes of noble gas like xenon and krypton, radioiodines, and radiocesium. As a bounding case at the facility, one process area for which this accident is postulated is the decontamination workshop.

Specifically, the accumulation of uranium in the citric acid tank could cause a criticality accident. For this to occur, the operator would have to fail to control the uranium mass in the tank. A criticality in the solution in the tank could produce an initial burst of 1.0×10^{18} fissions, followed by 47 bursts of 1.92×10^{17} fissions per burst, for a total of 1.0×10^{19} fissions in 8 hours.

The source term (ST) for the inadvertent nuclear criticality was determined using the five-factor formula:

$$ST = MAR \times DR \times ARF \times RF \times LPF \qquad \text{Eq.12}$$

where MAR = material at risk
       DR = damage ratio
       ARF = airborne release fraction
       RF = respirable fraction
       LPF = leak path factor

For the criticality accident, the material at risk (MAR) is the amount of fission product radioactivity that would accumulate during the event (NRC, 1998). The damage ratio (DR) is one, since all of the solution in the tank would be involved in the event. The atmospheric release fraction (ARF) for noble gases is 100 percent. The ARF for radioiodines is 0.25, and the ARF for other fission products is 5×10^{-4} (NRC, 1998). The respirable fraction is assumed to be 100 percent. A leak path factor of 0.001 is used for radioiodines and fission products other than noble gases, since the gaseous effluent vent system (GEVS) in the TSB is equipped with High-Efficiency Particulate Air (HEPA) filters and charcoal filters (LES, 2005b).

Exposure at the CAB was determined using the source terms calculated using Eq. 12, the χ/S value for the SW direction in Table A.1-2, and dose conversion factors from Federal Guidance Report No. 11 (EPA, 1988).

The results of the staff consequence assessment are presented in Table A.2-1. Industry experience with this type of criticality accident indicates that a worker is unlikely to survive the accident if the individual were located in the immediate vicinity of the reaction. However, with increasing distance from the accident, the radiation doses would be lower, and the probability that a worker could survive increases. At the facility, workers have direct access to vessels and other process equipment in which criticality events are possible. Therefore, staff have qualitatively evaluated the accident as a high-consequence event for the worker.

Table A.2-1
Health Effects Resulting from Inadvertent Criticality

Worker (egress after 10 min.)	Environment at RAB (Ratio)	Individual at CAB, SW direction	Collective Dose, West direction	
High	0.66	0.14 rem	person-rem	LCFs
			44	0.03

RAB - Restricted Area Boundary
CAB - Controlled Area Boundary
LCFs - latent cancer fatalities.
To convert rem to sievert, multiply by 0.01.

The environmental consequence is evaluated using the sum of the fractions rule. The concentration at the RAB of each fission product radionuclide generated during a hypothetical uranium solution criticality event in Table 3-14 of NUREG/CR-6410 (NRC, 1998), which was used as a model for the generic inadvertent criticality event, is compared to 5000 times the corresponding values in Appendix B to 10 CFR Part 20. The fractions thus generated (i.e., calculated fission product concentrations divided by their Appendix B limits) are added to yield one value. If that value is less than 1, the accident consequences to the environment are low. Since the sum presented in Table A.2-1 is less than 1, the staff estimates that the postulated criticality event is of low consequence to the environment.

A maximally exposed individual at the CAB in the southwest direction would receive a radiation dose of 1.4 mSv (0.14 rem) total effective dose equivalent. This is a low consequence to this individual. Similarly, the low collective dose to the offsite population, as determined using GENII (PNL, 1988), in the west sector (Eunice, New Mexico) means that the risk of health effects to the offsite public LCFs from this accident is low. The west sector would have the highest radiation doses after a criticality accident because Eunice, New Mexico, lies in closer proximity to the proposed facility than other population centers, and short-lived radionuclides formed during the criticality accident would not have completely decayed before reaching Eunice, New Mexico. Larger population centers in the north sector, such as the city of Hobbs, New Mexico, would receive lower collective doses because the short-lived fission products would decay during the time the plume transits from the proposed facility.

In accordance with the performance requirements of Part 70, Subpart H, the applicant has either identified IROFS to reduce the risk to the facility worker from all criticality accidents or identified safe-by-design components that meet criteria such that they are highly unlikely to fail.

A.2.2 Hydraulic Rupture of a UF$_6$ Cylinder in the Blending and Liquid-Sampling Area

At the product blending system in the blending and liquid-sampling area of the separations building, 30B (2.5-ton) cylinders are filled with product to customer specifications. The transfer of product to 30B cylinders would begin by heating a 14-ton 48Y cylinder containing product UF$_6$, inside a blending donor station, to no more than 61°C (142°F). The heated UF$_6$ gas would be transferred by piping from the heated 48Y cylinder to a blending receiver station containing a 30B cylinder. The blending receiver station is cooled, which allows the UF$_6$ gas to desublime to a solid inside the 30B cylinder, completing the transfer.

An accident is postulated wherein the blending donor station heater controller fails, causing the blending donor heater within the station to remain on. Were this to occur, the 48Y product cylinder could overheat and the cylinder could hydraulically rupture because of the expansion of the liquid UF$_6$. On cylinder rupture, the entire contents of the 48Y product cylinder [12,501 kg (27,560 lbs) of UF$_6$] would be released within the blending donor station. Since the station enclosure is not air-tight, the UF$_6$ would be released to the blending and liquid-sampling area. The UF$_6$, when in contact with air, would produce HF gas and UO$_2$F$_2$. The release into the building would then be released to the environment. The HVAC is conservatively assumed to be operating at the maximum ventilation flow rate. Significant quantities of HF and UO$_2$F$_2$ would be carried by the prevailing wind beyond the CAB.

The results of the staff-consequence assessment are presented in Table A.2-2.

The health and environmental consequences of this accident are high. A facility worker in the vicinity of the blending donor station would, within seconds, be exposed to lethal UF$_6$, UO$_2$F$_2$, and HF concentrations. The environmental consequences are higher than the 5.4 U mg/m^3 threshold for an intermediate consequence. An individual located at the CAB in the southwest sector would suffer high consequences from both uranium and HF exposure. The collective dose to the offsite population in the north sector indicates a risk of several LCFs in the population in the years after the accident.

In accordance with the performance requirements of Part 70, Subpart H, the applicant has identified IROFS to reduce the risk to the facility workers, the public, and the environment from the effects of this accident. To prevent this accident, the applicant will rely on fail-safe hard-wired high-temperature heater trips and redundant independent fail-safe capillary high-temperature heater trips. Each control will be tested annually to ensure its availability and reliability to serve its intended safety function on demand. The purpose of these controls is to ensure that the accident is highly unlikely to occur. In addition, there have been no similar heater control failures at the Urenco facilities in Europe in over 30 years of operation.

In addition to IROFS, the applicant has committed to an EP that includes certain mitigating actions to reduce the consequences of the event. For example, in response to an alarm that indicates the release of UF$_6$, a control room operator can secure the HVAC system for the affected area. The action to secure the HVAC within minutes of the accident would considerably reduce the risk to the public and the environment.

A-13

Table A.2-2
Health Effects Resulting from Hydraulic Rupture of a UF₆ Cylinder

Worker (egress after 10 min.)		Environment at RAB	Individual at CAB, SW direction		Collective Dose, north direction	
U mg/m³ (rem)	HF, mg/m³	U mg/m³	U mg/m³ (rem)	HF, mg/m³	person-rem	LCFs
High		44	250 (0.97)	86	12,000	7

RAB - Restricted Area Boundary
CAB - Controlled Area Boundary
HF - hydrogen fluoride.
LCFs - latent cancer fatalities.

A.2.3 Natural Phenomena Hazard - Earthquake

An earthquake is postulated to breach all UF₆ piping systems and lead to a release of approximately 860 kilograms (kg) (1900 lb) of UF₆ (LES, 2005b). Staff evaluated this accident for the blending and liquid-sampling area; UF₆ handling areas; and the cascade halls. The applicant has committed to ensure that the affected process buildings can withstand the design basis earthquake. Therefore, for this evaluation, staff assumed that the buildings would remain intact. The applicant will also install and maintain an electrical trip system for selected HVAC systems in process areas with large inventories of gaseous UF₆. The trip system will detect earthquakes and secure the HVAC units. Therefore, for this evaluation, staff has assumed that the HVAC in affected process buildings is shut down.

The results of the staff consequence assessment are presented in Table A.2-3.

The consequences to a facility worker shown in Table A.2-3 are for a worker located in one of the cascade halls during the earthquake. Depending on the location of the worker when the event occurs, the large quantity of UF₆ which could be released would result in a high consequence to this individual before the individual could escape the room. However, for seismic events, the worker is assumed to evacuate the area of concern upon detection of a seismic event, which results in a reduced exposure time and an acceptable risk. The consequences to the environment are low. The maximally exposed individual at the CAB in the southwest direction would not be expected to suffer any observable heath effects. Similarly, the low collective dose to the offsite population in the north sector means that the risk of health effects to the offsite public (latent cancer) from this accident is low.

A.2.4 Fire in a UF₆ Handling Area

A fire involving transient combustible material is postulated to breach a UF₆ transfer manifold containing feed vapor from five feed stations in a single UF₆ Handling Area. The release would involve approximately 3.4 kg (7.5 lb) of UF₆ vapor.

Table A.2-3
Health Effects Resulting from an Earthquake

Worker (egress after 10 min.)		Environment at RAB	Individual at CAB, SW direction		Collective Dose, north direction	
U mg/m³ (rem)	HF, mg/m³	U mg/m³	U mg/m³ (rem)	HF, mg/m³	person-rem	LCFs
High		0.11	0.64 (0.0017)	0.22	14	0.008

RAB - Restricted Area Boundary
CAB - Controlled Area Boundary
HF - hydrogen fluoride.
LCFs - latent cancer fatalities.

The results of the staff consequence assessment are presented in Table A.2-4.

Table A.2-4
Health Effects Resulting from Fire in a UF₆ Handling Area

Worker (egress after 10 min.)		Environment at RAB	Individual at CAB, SW direction		Collective Dose, north direction	
U mg/m³ (rem)	HF, mg/m³	U mg/m³	U mg/m³ (rem)	HF, mg/m³	person-rem	LCFs
59 (0.020)	20	0.012	0.070 (0.000072)	0.024	0.92	0.0006

RAB - Restricted Area Boundary
CAB - Controlled Area Boundary
HF - hydrogen fluoride.
LCFs - latent cancer fatalities.

The consequences of this accident are low for the environment, the individual at the CAB, and the public offsite. For the facility worker, the consequences are intermediate for acute chemical exposure to uranium. However, for fires, the worker is assumed to evacuate the area of concern once the fire is detected, which would result in an exposure time much shorter than 10 minutes, thus resulting in acceptable risk.

In accordance with the performance requirements of Part 70, Subpart H, the applicant has identified IROFS to ensure the risk of this type of accident remains low. To reduce the magnitude of fires resulting from the presence of transient combustible material, the applicant will rely on administrative controls. The purpose of these controls is to prevent large fires that could result in the release of large inventories of UF₆ by administratively limiting transient combustible loading in areas containing uranic material to ensure that the integrity of uranic material components/containers is maintained.

A.2.5 Process Line Rupture in a Product Low-Temperature Takeoff Station (LTTS)

Cold traps and chemical traps would be used at the facility to remove residual UF_6 and HF from process lines before discharging exhaust gases from these lines to the GEVS. An accident could occur if a product vent subsystem carbon trap became saturated with UF_6, caused by a small UF_6 leak through a product cold trap valve. Were this to occur, a UF_6 plug could form on the discharge of the vacuum pump, causing high pressure in the vacuum pump and, thus, failing seals, leading to a release of approximately 1.0 kg (2.2 lb) of UF_6 vapor to the UF_6 handling area. The results of the staff-consequence assessment are presented in Table A.2-5.

The consequences of this accident are low for the facility worker, environment, the individual at the CAB, and the public offsite. In accordance with the performance requirements of Part 70, Subpart H, the applicant has identified IROFS to ensure the risk of this type of accident remains low. For this accident, a preventive measure is a fail-safe, hard-wired, high-carbon-trap weight trip of the vacuum pump. This equipment will be tested annually to ensure its availability and reliability to serve its intended safety function.

Table A.2-5
Acute Health Effects Resulting from Process Line Rupture in a Product LTTS

Worker (egress after 10 min.)		Environment at RAB	Individual at CAB, SW direction		Collective Dose, NNW direction	
U mg/m³ (rem)	HF, mg/m³	U mg/m³	U mg/m³ (rem)	HF, mg/m³	person-rem	LCFs
17 (0.022)	5.8	0.0035	0.020 (0.000078)	0.0069	0.97	0.0006

RAB - Restricted Area Boundary
CAB - Controlled Area Boundary
HF - hydrogen fluoride.
LCFs - latent cancer fatalities.

A.3 ACCIDENT ANALYSIS SUMMARY

Results of the accident analysis are summarized in Table A.3-1. Staff selected and evaluated a representative subset of the potential accidents that could occur at the proposed facility. The accident consequences vary in magnitude, and include accidents initiated by natural phenomena, operator error, and equipment failure. Analytical results indicate that accidents at the facility pose acceptably low risks. The most significant accident consequences are those associated with the release of UF_6 caused by rupturing an over filled or over heated cylinder. The facility design reduces the risk (likelihood) of this event by using redundant heater controller trips. In addition, the facility EP addresses this type of event, and all other lower-risk, high- and intermediate-consequence events. The NRC staff concludes that through the combination of plant design, passive and active engineered controls (IROFS), and administrative controls, accidents at the facility pose an acceptably low risk to workers, the environment, and the public.

Table A.3-1
Summary of Health Effects Resulting from Accidents

Accident	Worker[a]		Environment at RAB	Individual at CAB, SW direction		Collective Dose		
	[U], mg/m³ (rem)	[HF], mg/m³	[U], mg/m³	[U], mg/m³ (rem)	[HF], mg/m³	Direction	person-rem	LCFs
Inadvertent Nuclear Criticality	---	High[b]	0.66[c]	(0.14[d])	---	West	44	0.03
Hydraulic Rupture of a UF₆ Cylinder		High[b]	44	250 (0.97)	86	North	12,000	7[e]
Earthquake		High[b]	0.11	0.64 (0.0017)	0.22	North	14	0.008
Fire in a UF₆ Handling Area	59 (0.020)	20	0.012	0.070 (0.000072)	0.024	North	0.92	0.0006
Process Line Rupture	17 (0.022)	5.8	0.0035	0.020 (0.000078)	0.0069	North	0.97	0.0006

[a] Worker exits after 10 minutes.

[b] High consequence could lead to a fatality.

[c] Pursuant to 10 CFR 70.61(c)(3), this value is the sum of the fractions of individual fission product radionuclide concentrations over 5000 times the concentration limits that appear in 10 CFR Part 20, Appendix B, Table 2

[d] The dose to the individual at the CAB is the sum of internal and external doses from fission products released from the Technical Services Building gaseous effluent vent systems stack.

[e] Though the consequences of the rupture of a liquid-filled UF₆ cylinder would be HIGH, redundant heater-controller trips would make this event highly unlikely to occur.

RAB - Restricted Area Boundary

CAB - Controlled Area Boundary

HF - hydrogen fluoride.

LCF - latent cancer fatalities.

mg/m³ - milligrams per cubic meter.

To convert rem to sievert, multiply by 0.01.

A.4   REFERENCES

(Baker, 1987) Baker, P.H., et al. "Air Flow Through Cracks." *Building and Environment.* Vol. 22, No. 4. pp. 293-304. Pergamon. 1987.

(Blevins, 2003) Blevins, Robert D. "Applied Fluid Dynamics Handbook." 2003.

(DOE, 1984) U.S. Department of Energy (DOE). DOE/TIC-27601, "Atmospheric Science and Power Production," 1984.

(EPA, 1988) U.S. Environmental Protection Agency (EPA). EPA-5201/1-88-020 - Federal Guidance Report No. 11, "Limiting Values of Radionuclide Intake and Air Concentration and Dose Conversion Factors for Inhalation, Submersion, and Ingestion," 1988.

(LES, 2004) Louisiana Energy Services (LES), letter to U.S. Nuclear Regulatory Commission, "Response to NRC Request for Additional Information Regarding National Enrichment Facility Safety Analysis Report and Emergency Plan," May 19, 2004.

(LES, 2005a) Louisiana Energy Services (LES). "National Enrichment Facility Safety Analysis Report," Revision 6, 2005.

(LES, 2005b) Louisiana Energy Services (LES). "National Enrichment Facility Integrated Safety Analysis," Revision 4, 2005.

(LES, 2005c) Louisiana Energy Services (LES). "National Enrichment Facility Emergency Plan," Revision 3, 2005.

(LES, 2005d) Louisiana Energy Services (LES). "National Enrichment Facility Environmental Report," Revision 5, 2005.

(NFPA, 2002) National Fire Protection Association (NFPA). NFPA 69, "Standard on Explosion Prevention Systems," 2002.

(NRC, 1988) U.S. Nuclear Regulatory Commission (NRC). NUREG-1140, "A Regulatory Analysis on Emergency Preparedness for Fuel Cycle and Other Radioactive Material Licensees," 1988.

(NRC, 1990) U.S. Nuclear Regulatory Commission (NRC). NUREG/CR-5659, "Control Room Habitability System Review Models," 1990.

(NRC, 1991) U.S. Nuclear Regulatory Commission (NRC). NUREG-1391, "Chemical Toxicity of Uranium Hexafluoride Compared to Acute Effects of Radiation," 1991.

(NRC, 1997) U.S. Nuclear Regulatory Commission (NRC). NUREG/CR-6331, "Atmospheric Relative Concentrations in Building Wakes," 1997.

(NRC, 1998) U.S. Nuclear Regulatory Commission (NRC). NUREG/CR-6410, "Nuclear Fuel Cycle Facility Accident Analysis Handbook," 1998.

(NRC, 2001) U.S. Nuclear Regulatory Commission (NRC). NUREG-1741, "RASCAL 3.0: Description of Models and Methods," 2001.

(NRC, 2002) U.S. Nuclear Regulatory Commission (NRC). NUREG-1520, "Standard Review Plan for the Review of a License Application for a Fuel Cycle Facility," 2002.

(NRC, 2003) U.S. Nuclear Regulatory Commission (NRC). NUREG/CR-6525, "SECPOP2000: Sector Population, Land Fraction, and Economic Estimation Program," 2003.

(NRC, 2005) U.S. Nuclear Regulatory Commission (NRC). NUREG-1767, Vol. 1, "Environmental Impact Statement on the Construction and Operation of a Proposed Mixed Oxide Fuel Fabrication Facility at the Savannah River Site, South Carolina," 2005.

(PNL, 1988) Pacific Northwest Laboratory (PNL). PNL-6584, "GENII - The Hanford Environmental Radiation Dosimetry Software System." Vol. 1. December 1988.

NRC FORM 335 (9-2004) NRCMD 3.7	U.S. NUCLEAR REGULATORY COMMISSION	1. REPORT NUMBER (Assigned by NRC, Add Vol., Supp., Rev., and Addendum Numbers, if any.)
BIBLIOGRAPHIC DATA SHEET *(See Instructions on the reverse)*		NUREG-1827

2. TITLE AND SUBTITLE	3. DATE REPORT PUBLISHED	
Safety Evaluation Report for the National Enrichment Facility in Lea County, New Mexico	MONTH	YEAR
	June	2005
	4. FIN OR GRANT NUMBER	

5. AUTHOR(S)	6. TYPE OF REPORT
	Final
	7. PERIOD COVERED *(Inclusive Dates)*

8. PERFORMING ORGANIZATION - NAME AND ADDRESS *(If NRC, provide Division, Office or Region, U.S. Nuclear Regulatory Commission, and mailing address; if contractor, provide name and mailing address.)*

Division of Fuel Cycle Safety and Safeguards

Office of Nuclear Material Safety and Safeguards

U.S. Nuclear Regulatory Commission

Washington, DC 20555-0001

9. SPONSORING ORGANIZATION - NAME AND ADDRESS *(If NRC, type "Same as above"; if contractor, provide NRC Division, Office or Region, U.S. Nuclear Regulatory Commission, and mailing address.)*

Same as above

10. SUPPLEMENTARY NOTES

11. ABSTRACT *(200 words or less)*

The report documents the U.S. Nuclear Regulatory Commission (NRC) staff review and safety and safeguards evaluation of the Louisiana Energy Services' (LES) (the applicant) application for a license to construct a gas centrifuge uranium enrichment facility and possess and use special nuclear material (SNM), source material, and byproduct material in a gas centrifuge uranium enrichment facility. LES proposes that the gas centrifuge uranium enrichment facility be located in Lea County, New Mexico, near the city of Eunice, New Mexico. The facility will possess natural, depleted uranium, and enriched uranium, and will enrich uranium up to a maximum of 5 weight percent uranium-235.

The objective of this review is to evaluate the potential adverse impacts of operation of the facility on worker and public health and safety under both normal operating and accident conditions. The review also considers physical protection of SNM and classified matter, material control and accounting of SNM, and the management organization, administrative programs, and financial qualifications provided to ensure safe design and operation of the facility.

The NRC staff concludes, in this safety evaluation report, that the applicant's descriptions, specifications, and analyses provide an adequate basis for safety and safeguards of facility operations and that operation of the facility does not pose an undue risk to worker and public health and safety.

12. KEY WORDS/DESCRIPTORS *(List words or phrases that will assist researchers in locating the report.)*	13. AVAILABILITY STATEMENT
Uranium Enrichment Louisiana Energy Services Fuel Cycle Safety Evaluation	unlimited
	14. SECURITY CLASSIFICATION
	(This Page) unclassified
	(This Report) unclassified
	15. NUMBER OF PAGES
	16. PRICE

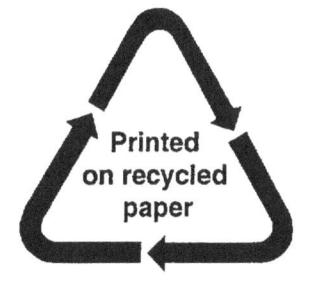

Federal Recycling Program